Drawing Workbook for

AF869013

Engineering Drawing and Design

Sixth Edition

Cecil Jensen
Former Technical Director
R.S. McLaughlin Collegiate and
Vocational Institute
Oshawa, Ontario, Canada

Jay D. Helsel
Professor and Chairman Emeritus
Department of Applied Engineering
and Technology
California University of Pennsylvania
California, Pennsylvania

Glencoe
McGraw-Hill

New York, New York   Columbus, Ohio   Woodland Hills, California   Peoria, Illinois

Glencoe/McGraw-Hill
*A Division of The **McGraw·Hill** Companies*

Drawing Workbook for Engineering Drawing, and Design, Sixth Edition

ISBN: 0-07-824156-1

Send all inquiries to:
Glencoe/McGraw-Hill
8787 Orion Place
Columbus, OH 43240-4027

Printed in the United States of America

1 2 3 4 5 6 7 8 9 045 06 05 04 03

Preface

This *Drawing Workbook* contains worksheets that present a partially worked out solution for assignments in each unit of *Engineering Drawing and Design, Sixth Edition*. Each worksheet has a title block and a reference number that refer to the chapter and unit number of the text. Following the table of contents are instructions that provide an overview for each assignment, and also provide references to the appropriate units. The drawing problems contain both U.S. customary (decimal inch) and metric (millimeter) units of measurement. The worksheets are printed on high-quality white paper to simulate drawings found in actual engineering practice. They are also perforated for easy removal.

These worksheets have a number of benefits. First of all, they save drawing time for both the instructor and the students. For instance, the title blocks and borders are all preprinted, and some of the drawing assignments are partially completed or have preprinted orthographic or pictorial grids. The worksheets also help students correctly position their drawings and focus their attention on the key concepts within the teaching unit. We estimate that these worksheets save instructors time enough to cover 30 percent more teaching material.

Students enrolled in a drafting program who are pursuing other related careers will also benefit from this workbook. For these students, this *Drawing Workbook* provides the opportunity to spend added time studying the material involving the interpretation and understanding of engineering drawing practices.

Students planning a career in drafting will find these sheets especially useful. Supplementing CAD with these sheets will provide the student the opportunity of covering more drafting theory and application as well as developing essential sketching experience.

The positioning points on the worksheets are printed in a light gray that will not be noticeable when drawn over with a pencil. The final drawing will be the student's drawing despite the aids printed on the sheets. Finally, the drawing problem is also provided on the sheet. This means that the student can set up a workspace without the inconvenience of having to keep the textbook on the drawing board.

Another advantage of these worksheets is that they are keyed with the units in the textbook, thereby simplifying the task of selecting the correct problem for the learning material. There are sufficient drawing worksheets to permit the instructor flexibility in making assignments. Sufficient drawing assignments are included to thoroughly cover all basic elements of engineering drawings and the latest drawing practices. In addition, a variety of advanced problems are provided to accommodate individual differences in both students and programs.

The drawing assignments for these worksheets are, in most cases, taken directly from the textbook. The understanding of reading assignments is enhanced by providing worksheet assignments that are closely coordinated with the text units. Because the text drawing problems were mainly designed for use on a B-size drawing sheet, changes to either the scale or some dimensions were often necessary in order to fit the drawing on the worksheet.

Once again, the authors have made a special effort to update all of the items in the *Engineering Drawing and Design* package to conform to the most current drafting standards available. In addition, Chapter 2, *Computer-Aided Drafting (CAD)*; Chapter 19, *Design Concepts*; and Chapter 27, *Electrical and Electronics Drawings* have undergone major revision to remain current with the most recent changes in these fields.

A *Solutions Manual* for the instructor is available for the workbook. Each solution is presented on a separate page for ease in grading students' assignments. For the textbook, *Engineering Drawing and Design, Sixth Edition*, there is a comprehensive *Instructor's Management System* available from Glencoe/McGraw-Hill. It contains all of the textbook drawing solutions. These solutions are also available on the *Instructor's* CD-ROM. Please contact your local Glencoe/McGraw-Hill representative or the publisher for more details.

Although this workbook is keyed to *Engineering Drawing and Design, Sixth Edition*, it can be used with any good text that follows current ANSI and ISO drawing standards. The instructor will need to provide the text references necessary.

We are grateful to Dr. Dennis Short for his contributions to the textbooks and workbooks. We also are grateful to the many instructors who have used *Engineering Drawing and Design* and/or *Fundamentals of Engineering Drawing* over their very successful history. We trust that those instructors and others will find this workbook as useful as the textbooks. As always, we appreciate your comments and suggestions.

Cecil Jensen
Jay D. Helsel

Contents

Chapter 7 Auxiliary Views and Revolutions

Chapter 8 Basic Dimensioning

Chapter 9 Sections

Chapter 10 Threaded Fasteners

Chapter 11 Miscellaneous Types of Fasteners

Chapter 12 Manufacturing Materials

Chapter 13 Forming Processes

Chapter 14 Detail and Assembly Drawings

Overview

Drawing A–1.

Drafting Scales—Measuring.

Reference Unit 1–4.

Use the scales specified at the top of each column and record the length of each of the lettered dimensions on the two drawings. Measure to the nearest calibration on your scale. Draw guidelines .12 in. or 3 mm apart in which to do your lettering. Keep guidelines thin and light, using a hard lead pencil. Use a softer lead for lettering.

Drawing B–1.

Lettering—Vertical.

Reference Unit 4–1.

Use a 4H or 6H lead to draw guidelines for each line of lettering as shown across the top of the sheet. Complete each line of lettering, using an HB or F lead. You may want to draw a few vertical guidelines as an aid in drawing some of the larger letters. Do not erase guidelines.

Drawing B–2.

Lettering—Inclined.

Reference Unit 4–1.

Follow instructions for Drawing B-1, except slant all letters at an angle of 68° to the horizontal. You may use the 68° angle on your Ames lettering guide as an aid in drawing a few of the large letters. Keep all lines on your lettering sharp and black. Do not erase guidelines.

Drawing B–3.

Locating by Coordinates.

Reference Unit 4–1.

Measure directly on each of the drawings, or use the .25 grid to determine the exact location of each of the points. Complete the tables at the right of the drawings for both absolute and relative coordinates. Use guidelines and make all lettering .12 in. high. Letter with an HB or F lead.

Drawing B–4.

Relative Coordinates.

Reference Unit 4–1.

Work directly on the grids shown to establish and label the points given in the tables at the right. Letter neatly, using guidelines if you prefer. Connect the points either by freehand sketching on the grid lines or using a straightedge. Lines should be relatively thick, sharp, and black.

Drawing B–5.

Absolute Coordinates.

Reference Unit 4–1.

Work directly on the grids shown to establish and label the points given in the tables at the right. Connect the points either by freehand sketching on the grid lines or by using a straightedge. Keep lines thick, sharp, and black. Letter neatly.

Drawing B–6.

Polar Coordinates.

Reference Unit 4–1.

Begin at point *A* and work in a clockwise direction to make a full-size drawing of the template. Measure distances and angles as accurately as possible in order to make the figure close back at point *A*. Keep finished lines sharp and black. Corners should be sharp and clean.

Drawing B–7.

Linework—Patterns.

Reference Unit 4–1.

Follow carefully the instructions given for parts 1 through 6. Do all layout work, using light construction lines with a sharp 4H or 6H pencil. Darken all lines using an HB or H lead. Finished lines should be approximately .03 in. (0.7 mm) thick.

Drawing B–8.

Linework—Inlay Designs.

Reference Unit 4–1.

Lay out Figs. A, B, and C, using either inch or millimeter sizes. Millimeter sizes are in brackets. Use a sharp, hard lead pencil for layout. Darken all finished lines, using an HB or H lead. Finished lines should be approximately .03 in. (0.7 mm) thick. Do not dimension.

Drawing B–9.

Linework—Shearing Blank.

Reference Unit 4–1.

Use your metric scale (1:5) and follow the procedure shown at the upper right on your drawing sheet to lay out the shearing blank. Lay out all angles carefully and accurately, using triangles, drafting machine, or protractor. Draw the complete shearing blank. Begin by laying out the drawing, using light construction lines, and finish with thick (.03 in. or 0.7 mm) lines as shown in step 4. Construction lines (layout lines) need not be erased.

Drawing B–10.

Circles and Arcs—Designs.

Reference Unit 4–2.

Lay out Figs. A, B, and C on the centerlines provided. First, lay out each figure, using light construction lines, and finish with thick (.03 in. or 0.7 mm) lines. The crosshatch lines in Fig. C may be drawn as thin (.016 in. or 0.35 mm), sharp, black lines.

Drawing B–11.

Circles and Arcs—Pawl.

Reference Unit 4–2.

Develop the drawing of the pawl around the intersection of the centerlines in the middle of the drawing area. This center point represents the center of the 1.00 in. diameter hole. Measure accurately and lay out the entire figure, using light construction lines. Locate centers of all arcs and mark points of tangency before constructing arcs. Darken all lines (.03 in. or 0.7 mm), using an HB or H lead. Do not erase construction lines.

Drawing B–12.

Circles and Arcs—Anchor Plate.

Reference Unit 4–2.

First, locate and draw all centerlines shown as light solid construction lines. Next, locate the centers of the R65 arcs and mark points of tangency. Complete the drawing by darkening all visible lines with an F or HB lead in your compass for large circles and arcs. Use the same grade of lead in a pencil along with a circle template for small circles. Lay out all circles and arcs with light construction lines before darkening them. This will ensure concentricity, especially with small circles. Darken centerlines (approx. .016 in. or 0.35 mm thick), using long and short dashes (see Figures 4-1-1 through 4-1-3 in text).

Drawing B–13.

Circles and Arcs—Gasket No. 1.

Reference Unit 4–2.

Carefully measure and lay out all centerlines (construction lines). Locate and mark the centers of all areas and points of tangency. Draw all circles and arcs, using light construction lines. Darken all visible lines and centerlines. A circle template may be used for small circles and arcs. Do not erase construction lines or points of tangency.

Drawing B–14.

Circles and Arcs—Gasket No. 2.

Reference Unit 4–2.

Follow the four steps shown in the upper right side of the drawing sheet to draw the gasket around the centerlines shown in the bottom half of the sheet. Begin by locating the centers of circles and arcs with light construction lines. Add the location of all points of tangency. Construct all circles and arcs (thin, light construction lines). Connect arcs with straight lines. Finally, darken all lines, including centerlines. Do not erase construction lines.

Drawing B–15.

Irregular Curves—Patterns.

Reference Unit 4–3.

Establish points on the blank grids through which the irregular curves will pass. Do this by transferring points from the grid at the left to the grid at the right and carefully estimating points where the curved lines intersect the grid lines. Lightly sketch the curved shape through the established points. Use an irregular curve (French curve) to draw the finished lines. The French curve should be matched to as many points as possible in order to achieve a smooth finished curve. Also, on long irregular curved lines, allow the French curve to overlap somewhat from one portion of the curved line to the next.

Drawing B–16.

Sketching—Lines and Circles.

Reference Unit 4–4.

Before darkening any of your lines, carefully sketch each figure using light construction lines.

Drawing B–17.

Sketching—Structural Shapes.

Reference Unit 4–4.

The squares on the grids represent .50 in. Transfer distances from the structural steel shapes to the grid, and sketch each of the figures, using light construction lines. Darken all lines, using thick, black, visible lines. Do not add a dot pattern unless instructed to do so.

Drawing B–18.

Sketching—Shaft Support and Control Arm.

Reference Unit 4–4.

Transfer points from the drawings at the top of the sheet to the grid at the bottom for each part. Carefully sketch circles and arcs followed by straight lines. Darken all visible lines and centerlines.

Drawing C–1

Applied Geometry—Straight Lines.

Reference Unit 5–1.

Follow the instructions carefully in spaces A through F. In all cases, first use light construction lines to completely solve the problem. Do not erase construction lines.

Drawing C–2.

Applied Geometry—Arcs and Circles.

Reference Unit 5–2.

Follow the instructions carefully in spaces A through F.

Space A. Locate the centers of all arcs and mark points of tangency. Use a compass with a hard, sharp lead to lay out arcs. Darken arcs with compass or circle template and pencil with soft lead.

Space B. Bisect any two sides to locate the center of the polygon. Construct the circle touching each corner.

Space C. Locate centers and points of tangency. Draw a light construction line between points *K* and *L* and divide it into three equal parts. The point of tangency between the two arcs is to fall on the mark nearest point *L* on line *LM*. Draw perpendicular bisectors of each segment of construction line *KL* to establish center points for each of the two arcs. A line perpendicular to *JK* upward from point *K* and one perpendicular to *LM* downward from point *L* will also be needed to establish centers for the arcs.

Space D. Locate the arc centers and points of tangency. Use your compass to construct the arcs. You may prefer to use a circle template as a guide when darkening the arcs. Do not erase construction lines.

Space E. Locate centers of arcs and points of tangency. Use your compass to draw the arcs. You may prefer to use a circle template as a guide in darkening the arcs. Do not erase construction lines.

Space F. Draw construction lines (straight lines) that connect *A* with *B* and *B* with C. Draw perpendicular bisectors of *AB* and *BC* to locate the center of the arc. Do not erase construction lines.

Drawing C–3.

Applied Geometry—Polygons.

Reference Unit 5–3.

Follow the instructions carefully in spaces A through F.

Space A. For the large hexagon, begin by constructing a circle whose diameter equals the distance across the corners. Establish points on the circle 60° apart. Connect the points. For the small hexagon, begin by constructing a circle whose diameter equals the distance across the flats. Establish points on the circle 60° apart. Connect the points.

Space B. Begin by constructing 30 mm and 50 mm circles around the center point given. Proceed as in Space A, except in this case divide the circles into eight equal parts rather than six.

Space C. Begin by constructing a 60 mm square and a circle within the square tangent to its sides. Complete the octagon by constructing 45° lines tangent to the circle.

Space D. Construct a 60 mm circle around the center point given. Use construction lines to establish the length of one side and step off the remaining four sides. Connect points to complete the pentagon.

Space E. Using *AB* as the radius, draw a semicircle and divide it into five equal parts. Draw radial lines from point *A* through the three division marks on the right. Using *AB* as the length of each side, locate the intersections of the remaining sides. Connect points and darken the lines representing the pentagon.

Space F. Follow the procedure given for Space E, except divide the semicircle into seven spaces rather than five.

Drawing C–4.

Applied Geometry—Template and Hook.

Reference Unit 5–3.

Template: Locate points on the grid that correspond with points on the drawing. Use appropriate construction methods to develop the geometric shapes. Darken all visible lines and centerlines.

Hook: Use only the dimensions given for part 4 of the table. Locate centers of circles and arcs where possible. Sketch details not given in specific dimensions.

Drawing C–5.

Applied Geometry—Ellipses.

Reference Unit 5–4.

Follow the instructions given in spaces A through C.

Space A. Begin by dividing the circles into any number of equal parts (for example, 30° angles equal 12 parts.) Use radial lines to establish points through which the ellipse will be drawn. Use a French curve to draw the final ellipse.

Space B. Use construction lines to establish the centers of the small and large arcs. Use your compass to construct arcs. Give careful attention to points of tangency.

Space C. Begin by drawing a 60 mm by 120 mm rectangle about the centerlines given. Use construction lines to establish points through which the ellipse will be drawn. Use a French curve to draw a smooth ellipse through the points established.

Drawing C–6.

Applied Geometry—Helix and Parabola.

Reference Unit 5–5.

Space A. Complete the front view of the cylinder on the baseline given. Divide the circle into 12 equal parts. Using the length of one of the 12 dimensions, develop a stretchout of the cylinder beginning at point A. Use construction lines to complete the layout. Darken lines as appropriate. Do not erase construction lines.

Space B. *Parallelogram method.* Divide *OA* and related lines into six equal parts. Use construction lines to connect points to establish a grid. Use a French curve to connect points and complete the parabola.

Offset method. Divide line *OA* into five equal parts. Divisions on the horizontal line beginning at *A* will be established by dividing it into 5^2, or 25, equal parts. Use this method to establish a grid through which the parabolic curve will be drawn.

Drawing D–1.

Orthographic Representation—Absolute Coordinates.

Reference Unit 6–1.

Work directly on the grids shown to establish and label the points given in the tables at the right. Connect the points either by freehand sketching on the grid lines or using a straightedge.

Drawing D–2.

Orthographic Representation—Relative Coordinates.

Reference Unit 6–1.

Work directly on the grids shown to establish and label the points given in the tables at the right. Connect the points either by freehand sketching on the grid lines or using a straightedge.

Drawing D–3.

Orthographic Representation—Two-Dimensional Coordinates.

Reference Unit 6–1.

Part A is developed using the absolute coordinate method. Plot all points and label them directly on the grid. Connect the points to complete the drawing. Part B involves relative coordinates.

Drawing D–4.

Orthographic Representation—Three-Dimensional Coordinates.

Reference Unit 6–1.

Count the squares on the grids for Parts A and B to establish the *X*, *Y*, and *Z* coordinates. Label coordinates on the pictorial drawings to identify points of intersection of the various lines and surfaces.

Drawing D–5.

Orthographic Representation—Corner Bracket.

Reference Unit 6–1.

Study the pictorial drawing carefully before beginning this assignment. Review Unit 6-1 in your text if you are not sure about the difference between third-angle projection and first-angle projection. Either sketch the views or use a straightedge to produce the final drawing.

Drawing D–6.

Orthographic Representation—Location Block.

Reference Unit 6–1.

Review Unit 6-1 in your text. Work from the given view in each case in developing the remaining five views. Sketch the views or use a straightedge to complete the final drawings.

Drawing D–7.

Orthographic Representation—Miter Lines.

Reference Unit 6–2.

Space A. Use light construction lines to draw the front and top views of the angle bracket. Use a 45° miter line to aid in developing the right-side view. Darken all visible lines. Do not erase construction lines.

Space B. In this case, draw the front and right-side views first. Next, add a miter line and develop the top view.

Drawing D–8.

Orthographic Representation—Surfaces Parallel and Visible.

Reference Unit 6–3.

Carefully study each of the pictorial drawings before attempting to sketch or draw the three views of each on the grids.

Drawing D–9.

Orthographic Representation—Hidden Surfaces and Edges.

Reference Unit 6–4.

Carefully study each of the pictorial drawings before attempting to sketch or draw the three views on each grid. Review Unit 6-4 on page 107 of your text before adding hidden lines.

Drawing D–10.

Orthographic Projection—Control Block.

Reference Unit 6–4.

Use construction lines to block in the three views, keeping 20 mm between views. Use a miter line if you prefer. Add hidden lines. Darken all visible and hidden lines. Remember! Visible lines are thick; hidden lines are thin.

Drawing D–11.

Orthographic Projection—Angle Stop.

Reference Unit 6–4.

Block in the front, top, and right-side views of the angle stop. Allow .50 in. between views. Use a miter line if you prefer. Darken lines as appropriate. Are there any hidden lines in the top view, front view, or right-side view? What happens when visible lines and hidden lines coincide?

Drawing D–12.

Orthographic Projection—Guide Bar.

Reference Unit 6–4.

Block in the three views required for third-angle projection. Use a miter line if you prefer. Allow .50 in. between views. Develop each view completely, using construction lines before darkening any lines. Are there hidden lines?

Drawing D–13.

Orthographic Representation—Inclined Surfaces (1).

Reference Unit 6–5.

Draw three normal views of each of the pictorials shown on sheet D-13. Study the details carefully as you transfer information from the pictorial grid to the orthographic grid. Notice that inclined surfaces become foreshortened in certain views.

Drawing D–14.

Orthographic Representation—Inclined Surfaces (2).

Reference Unit 6–5.

Carefully select the front view for each of the six drawings and develop three views of each. Watch for inclined surfaces and hidden lines.

Drawing D–15.

Orthographic Representation—Inclined Surfaces (3).

Reference Unit 6–5.

Block in the three normal views for each of the six pictorials. Notice that the front view is the one that shows the most detail. Watch for inclined surfaces and hidden lines.

Drawing D–16.

Orthographic Representation—Inclined Surfaces (4).

Reference Unit 6–5.

Block in the three normal views for each of the six pictorials. Notice that the front view is the one that shows the most detail. Watch for inclined surfaces and hidden lines.

Drawing D–17.

Orthographic Projection—Base.

Reference Unit 6–5.

Block in the three views with 20 mm between views. A miter line may be useful in this case. Are there hidden lines? Notice that the front view is the one that shows the most detail.

Drawing D–18.

Orthographic Projection—Base Plate.

Reference Unit 6–5.

Lay out the three views of the base plate with 20 mm between views. Use a miter line to develop the top and right-side views. Are there hidden lines in the top view?

Drawing D–19.

Orthographic Representation—Interpreting Drawings (1).

Reference Unit 6–5.

Be sure you first understand which is the front, top, and right-side view of both the bracket and the corner bracket. Letter neatly. Use guidelines if needed.

Drawing D–20.

Orthographic Representation—Interpreting Drawings (2).

Reference Unit 6–5.

Be sure you first understand which is the front, top, and right-side view of both the angled step bracket and the angle stop. Letter neatly. Use guidelines if needed.

Drawing D–21.

Orthographic Representation—Circular Features (1).

Reference Unit 6–5.

These assignments may be freehand sketched or drawn with instruments. First, block in each view, allowing one grid space between views. Use construction lines to complete the detail in each view. Finally, darken all visible lines, centerlines, and hidden lines.

Drawing D–22.

Orthographic Representation—Circular Features (2).

Reference Unit 6–5.

These assignments may be freehand sketched or drawn with instruments. First, block in each view, allowing one grid space between views. Use construction lines to complete the detail in each view. Finally, darken all visible lines, centerlines, and hidden lines.

Drawing D–23.

Orthographic Representation—Interpreting Drawings.

Reference Unit 6–5.

Study the orthographic views carefully before beginning this assignment. Project visually from view to view as you establish the sizes for A through T in each case. Use guidelines for lettering your answers in the space provided.

Drawing D–24.

Orthographic Projection—Rocker Arm.

Reference Unit 6–5.

Draw the top view first and project downward to draw the front view. You may need to refer to Unit 4-2 in your text when locating the centers of the arcs connecting the large and small cylinders. Be sure to mark all points of tangency with light construction lines.

Drawing D–25.

Orthographic Representation—Oblique Surfaces (1).

Reference Unit 6–7.

Study the pictorials carefully. Block in the three views on the grids before adding details. Use a miter line if needed. Are there hidden lines in any of the views?

Drawing D–26.

Orthographic Representation—Oblique Surfaces (2).

Reference Unit 6–7.

Study the pictorials carefully. Block in the three views on the grids before adding details. Use a miter line if needed. Are there hidden lines in any of the views?

Drawing D–27.

Orthographic Representation—One- and Two-View Drawings.

Reference Unit 6–8.

Study each of the pictorials carefully. If needed, sketch the necessary views on a separate sheet of paper before working on your workbook sheet. Your solutions may be freehand sketched or drawn with instruments. Do all of the drawings need hidden lines?

Drawing D–28.

Special Views—Connector.

Reference Unit 6–9.

While a front view, right-side view, and left-side view would generally be sufficient to describe the connector, a top view or partial top view can be used to more clearly show the radii on the bent corners.

Drawing D–29.

Conventional Representation—Locking Plate.

Reference Unit 6–10.

At least three sets of features can be shown simplified on this drawing.

Drawing D–30.

Conventional Breaks—Fixture Base.

Reference Unit 6–11.

With a scale of 1:1, a great deal of foreshortening will need to occur for the views to fit the sheet. Therefore, you will need to remove (break out) some large portions between the center hole and the threaded holes on each end. Use long break lines to show where portions have been removed. Is there another way to accomplish the same objective?

Drawing D–31.

Lines of Intersection—Cylindrical Surfaces.

Reference Unit 6–12.

Grids can be developed to accurately establish the lines of intersection or you can estimate and sketch the finished lines. Your instructor will assign the method.

Drawing D–32.

Foreshortened Projection—Mounting Bracket.

Reference Unit 6–14.

Make front and top views of the mounting bracket. Rotate features into the horizontal plane in order to show all parts in their true size and shape. The front view might best be shown as a half-section. Dimension if instructed to do so. Determine limits for the keyed hole from Appendix Table 43 in your text.

Drawing D–33.

Foreshortened Projection—Clutch.

Reference Unit 6–14.

Make front and top views of the clutch. Rotate features into the horizontal plane as required. Determine limits for the keyed hole from Appendix Table 48.

Drawing D–34.

Intersection of Unfinished Surfaces—Cutoff Stop.

Reference Unit 6–15.

Review Unit 6-15 in your text for a clear understanding of intersections of unfinished surfaces before completing this assignment. Also, see Unit 8-7 for information on surface texture and how it is specified on a working drawing. This drawing should be fully dimensioned.

Drawing D–35.

Intersection of Unfinished Surfaces—Sparker Bracket.

Reference Unit 6–15.

Review Unit 6-15 in your text for a clear understanding of intersections of unfinished surfaces before completing this assignment. Also, see Unit 8-7 for information on surface texture and how it is specified on a working drawing. This drawing should be fully dimensioned.

Drawing E–1.

Primary Auxiliary Views—Angle Bracket.

Reference Unit 7–1.

The auxiliary view need only show the size and shape of the inclined surface. Be sure to allow sufficient space between views for dimensions. Review Unit 8-7 in your text for information on surface texture and how it is specified on a working drawing.

Drawing E–2.

Primary Auxiliary Views—Angle Plate.

Reference Unit 7–1.

Study the drawing setup on drawing sheet E-2 before you begin to draw. Pay attention to the location of views. The auxiliary view need only show the size and shape of the inclined surface. Fully dimension the completed drawing.

Drawing E–3.

Primary Auxiliary Views—Statue Bases.

Reference Unit 7–1.

Project at right angles to the oblique surface to establish the length of the auxiliary views of the two objects. The width dimensions are taken from the top views. Do not dimension.

Drawing E–4.

Circular Features—Link.

Reference Unit 7–2.

Develop the front view and partial top and auxiliary views, using light construction lines. Establish the best location for each of the break lines on the partial views and sketch them in before darkening the remaining lines. Add dimensions and notes.

Drawing E–5.

Multiple Auxiliary Views—Inclined Stop.

Reference Unit 7–3.

Develop the required views, using light construction lines. Sketch the break lines on the three partial views before darkening the remaining lines. Add dimensions and notes. Sizes not given on the pictorial drawing can be measured directly on the front view.

Drawing E–6.

Multiple Auxiliary Views—Connecting Bar.

Reference Unit 7–3.

Develop the required views, using light construction lines. Sketch the break lines on the auxiliary views before darkening the remaining lines. Use an ellipse template to draw the ellipses that appear in the bottom view. Add dimensions and notes.

Drawing E–7.

Multiple Auxiliary Views—Angle Stop.

Reference Unit 7–3.

Draw the top view first. From the top view, project downward to develop the front view. The angular lines on the drawing sheet locate surface A on the auxiliary views. Break the auxiliary views as necessary to fit the sheet. Add dimensions and notes.

Drawing E–8.

Secondary Auxiliary Views—Dovetail Bracket.

Reference Unit 7–4.

The primary auxiliary view is developed from a partially completed top view. The top view is then completed by projecting details back to the top view. The front view is developed by projecting from the top view. Finally, develop the secondary auxiliary view by projecting from the primary auxiliary view. The hexagram hole shown in its true shape on the secondary auxiliary view is then projected to the other view. Add dimensions and notes.

Drawing E–9.

Secondary Auxiliary Views—Pivot Arm.

Reference Unit 7–4.

Complete as much of the front view as possible before projecting to the auxiliary views. The right partial primary auxiliary view should show only the inclined surface (true shape and size) of the right side. The secondary auxiliary view will show only the inclined surface containing the round hole. Add dimensions and notes.

Drawing E–10.

Revolutions.

Reference Unit 7–5.

Use dividers to take sizes from the normal views of parts A, B, and C as you rotate the views as specified. Do not erase construction lines. Do not dimension.

Drawing E–11.

Points and Lines in Space.

Reference Unit 7–6.

Review Unit 7-6 in your text before proceeding. Establish miter lines where appropriate to aid in projecting points and developing lines in the various views required. Label all points and lines. Do not erase construction lines.

Drawing E–12.

Planes in Space.

Reference Unit 7–7.

Review Unit 7-7 in your text before proceeding. Project carefully from view to view to develop an accurate solution to the problem. Do not erase construction lines.

Drawing E–13.

Visibility of Lines in Space.

Reference Unit 7–8.

Review Unit 7-8 in your text before proceeding. Study each of the figures carefully. Try to visualize lines and surfaces in space before attempting to complete the assignments. Do not erase construction lines.

Drawing E–14.

Distance between Lines and Points.

Reference Unit 7–9.

Review Unit 7-9 in your text before proceeding. Study each of the figures carefully. Try to visualize lines and points in space before attempting to complete the assignments. Do not erase construction lines.

Drawing E–15.

Edge and True View of Planes.

Reference Unit 7–10.

Review Unit 7-10 in your text before proceeding. Study each of the figures carefully. Try to visualize lines and points in space before attempting to complete the assignments. Do not erase construction lines.

Drawing E–16.

Angle between Lines and Planes.

Reference Unit 7–11.

Review Unit 7-11 in your text before proceeding. Study each of the figures carefully. Try to visualize lines and points in space before attempting to complete the assignments. Do not erase construction lines.

Drawing F–1

Basic Dimensioning—Notched Block.

Reference Unit 8–1.

Draw the front and right-side views. Allow sufficient space between views for dimensions (approximately 1.50 in.). Do not dimension to hidden lines. Use guidelines for lettering.

Drawing F–2.

Basic Dimensioning—Vertical Guide.

Reference Unit 8–1.

Draw the front, top, and right-side views. Allow sufficient space between views for dimensions (min. 25 mm). Use guidelines for lettering.

Drawing F–3.

Basic Dimensioning—Base.

Reference Unit 8–1.

Draw the front, top, and right-side views. Allow sufficient space between views for dimensions (min. 25 mm). Use guidelines for lettering. Is there more than one way to dimension the angular shapes?

Drawing F–4.

Dimensioning Circular Features—Adjustable Sector.

Reference Unit 8–2.

Review Unit 4-2 in your text for correct procedures in drawing arcs tangent to straight lines and adjacent arcs. Mark all tangent points. Do not erase construction lines. Add dimensions. Use guidelines for lettering.

Drawing F–5.

Dimensioning Circular Features—Bracket.

Reference Unit 8–2.

Draw the front, top, and right-side views. Allow sufficient space between views for dimensions (approx. 1.00 in.). Review Unit 8-2 in your text for information on dimensioning circular features, including holes, slots, circles, and arcs. Use guidelines for all lettering.

Drawing F–6.

Dimensioning Common Features—Handle.

Reference Unit 8–3.

Review Unit 8-3 in your text. You may want to keep your text open to this unit for various examples of acceptable methods for dimensioning common features such as chamfers, knurls, holes, and tapers.

Drawing F–7.

Dimensioning Common Features—Gasket.

Reference Unit 8–3.

Review Unit 4-2 in your text for correct procedures for drawing arcs tangent to straight lines and adjacent arcs. Mark all tangent points. Do not erase construction lines. Add dimensions. Use guidelines for all lettering.

Drawing F–8.

Dimensioning Methods—Transmission Cover.

Reference Unit 8–4.

Review Unit 4-2 in your text for correct procedures for drawing arcs tangent to straight lines and adjacent arcs. Mark all tangent points. Do not erase construction lines. Add dimensions after a careful review of Unit 8-4 in your text. Use rectangular coordinate dimensioning.

Drawing F–9.

Dimensioning Methods—Adapter Plate.

Reference Unit 8–4.

Unit 8-4 in your text describes and illustrates the various methods for dimensioning a drawing. Keep your text open to Unit 8-4 while you complete this assignment. For what purpose is the symmetry symbol used?

Drawing F–10.

Limits and Tolerances.

Reference Unit 6–5.

Study Unit 8-5 in your text. Be sure you understand terms such as basic size, limits of size, and tolerance. Do you need the Appendix Tables on Fits to complete this assignment?

Drawing F–11.

Fits and Allowances—Inch.

Reference Unit 6–5.

Study Unit 8-6, "Fits and Allowances," in your text. Refer to Appendix Tables 43 through 47 for information required to calculate the dimensions needed to complete the assignment. Be sure you determine the correct number of decimal places.

Drawing F–12.

Fits and Allowances—Metric.

Reference Unit 8–6.

Study Unit 8-6, "Fits and Allowances," in your text. Refer to Appendix Tables 51 and 52 for information required to calculate the dimensions needed to complete the assignment. Be sure you determine the correct number of decimal places.

Drawing F–13.

Fits—Inch.

Reference Unit 8–5.

Study Unit 8-6, "Fits and Allowances," in your text. Refer to Appendix Tables 43 through 47 for information required to calculate the dimensions needed to complete the assignment. Be sure you determine the correct number of decimal places.

Drawing F–14.

Fits—Metric.

Reference Unit 8–5.

Study Unit 8-6, "Fits and Allowances," in your text. Refer to Appendix Tables 51 and 52 for information required to calculate the dimensions needed to complete the assignment. Be sure you determine the correct number of decimal places.

Drawing F–15.

Fits and Allowances—Spindle.

Reference Unit 8–5.

Refer to Appendix Tables 43 through 52 for information on standard fits. Show all dimensions either in inches or in millimeters. Do not mix dimensions. Refer to Appendix Table 22 for information on Woodruff keys and Tables 35 through 39 for standard retaining rings.

Drawing F–16.

Surface Texture—Link.

Reference Unit 8–7.

Review Unit 8-7 in your text for information on surface texture. Refer to Appendix Table 46 for information on locational interference fits.

Drawing F–17.

Surface Texture—Column Bracket.

Reference Unit 8–7.

Review Unit 8-7 in your text for information on surface texture. Refer to Appendix Table 43 for information on running and sliding fits, and Table 46 for information on locational interference fits.

Drawing G–1.

Full Sections—Shaft Base.

Reference Unit 9–1.

Review Unit 8-7 on surface textures. Also review Units 8-1 through 8-3 on symbols used in dimensioning before proceeding. In a full section, what portion of the object is theoretically removed? Use general-purpose section lining except in special cases. How many types of cutting-plane lines are there?

Drawing G–2.

Full Sections—Slide Bracket.

Reference Unit 9–1.

Review Unit 8-7 on surface texture. Refer to Appendix Tables 43 through 47 for limit dimensions on the holes showing a precision fit. Can the centerline be used as a cutting-plane line? Use general-purpose section lining except in special cases.

Drawing G–3.

Two or More Sections—Casing.

Reference Unit 9–2.

Refer to Units 8-1 through 8-3 for a review of symbols used in dimensioning and notes. Also, refer to Appendix Table 27 for information on taper pins. What is the purpose of the letters at the ends of the cutting-plane lines on the front view?

Drawing G–4.

Two or More Sections—Housing.

Reference Unit 9–2.

Refer to Units 8-1 through 8-3 for a review of symbols used in dimensioning and notes. Also, refer to Appendix Table 27 for information on taper pins. What is the purpose of the letters at the ends of the cutting-plane lines on the front view?

Drawing G–5.

Half-Sections—Step Pulley.

Reference Unit 9–3.

In a half-sectional view, what portion of the object is theoretically removed? Refer to Appendix Table 21 for information on square keys. Add dimensions.

Drawing G–6.

Half-Sections—Step-V Pulley.

Reference Unit 9–3.

In a half-sectional view, what portion of the object is theoretically removed? Refer to Appendix Table 21 for information on square keys, Add dimensions.

Drawing G–7.

Threads in Section—Pipe Plug.

Reference Unit 9–4.

Refer to Chapter 10, "Threaded Fasteners," for information on thread representation. See Unit 9-4 for specific information on threads in section. Use general-purpose section lining.

Drawing G–8.

Threads in Section—End Plate.

Reference Unit 9–4.

Refer to Chapter 10, "Threaded Fasteners," for information on thread representation. See Unit 9-4 for specific information on threads in section. Use general-purpose section lining.

Drawing G–9.

Assemblies in Section—Flanged Connection.

Reference Unit 9–5.

Refer to Unit 14-5 for information on how to develop an item list. Unit 14-7 will provide you with information on assembly drawings. Refer to Appendix Table 21 for specifications on square keys. Remember that shafts, bolts, pins, and so on, are not generally sectioned even if they are cut by the cutting plane.

Drawing G–10.

Assemblies in Section—Caster Assembly.

Reference Unit 9–5.

Review Chapter 10, "Threaded Fasteners," and refer to Appendix Tables 11 and 14 for information on hex-head bolts and nuts. Refer to Appendix Table 19 for washer sizes. Reverse the direction of section lining on adjacent parts.

Drawing G–11.

Offset Sections—Base Plate.

Reference Unit 9–6.

Review Chapter 10, "Threaded Fasteners," as well as Unit 9-4 for information on drawing threads in section. Do not draw lines on the sectional views to show changes in direction of the offset cutting plane.

Drawing G–12.

Offset Sections—Mounting Plate.

Reference Unit 9–6.

Review Chapter 10, "Threaded Fasteners," as well as Unit 9-4 for information on drawing threads in section. Do not draw lines on the sectional views to show changes in direction of the offset cutting plane.

Drawing G–13.

Ribs, Holes, and Lugs—Two-Post Column Base.

Reference Unit 9–7.

Alternate section lining is often used on ribs to avoid confusion (see Unit 9-7 in your text).

Drawing G–14.

Ribs, Holes, and Lugs—Shaft Support.

Reference Unit 9–7.

Use alternate section lining (see Unit 9-7 in your text) on ribs to avoid confusion. Could the top view be drawn as a half-view with symmetry symbols to save time and space?

Drawing G–15.

Revolved Section—Connector.

Reference Unit 9–6.

Refer to Appendix Table 48 for information on precision fits for the two large holes. Review Unit 8-7 for information on surface texture. Use a template to draw the ellipse.

Drawing G–16.

Revolved Section—Chisel.

Reference Unit 9–6.

Where practical, use revolved sections. Use removed sections only when a revolved section blocks other detail. The octagon shape can be constructed around a circle having a diameter equal to the distance across the flats.

Drawing G–17.

Spokes and Arms—Offset Handwheel.

Reference Unit 9–9.

Spokes and arms are generally not section lined, even if the cutting plane passes through them. Why? Could a revolved section of a spoke be shown just as easily on the side view? Mark points of tangency when drawing the reverse curves on the side view.

Drawing G–18.

Broken-Out Section—Hold-Down Bracket.

Reference Unit 9–10.

A short break line is used to outline a broken-out section. Refer to Appendix Table 36 for information on internal retaining rings. Refer to Appendix Table 48 for information on limits and fits for the hole.

Drawing G–19.

Phantom Section—Bearing Housing.

Reference Unit 9–11.

See Appendix Table 48 for information on limits and fits for the three holes. The dimension (60) represents a reference dimension. What does this mean?

Drawing G–20.

Phantom Section—Housing.

Reference Unit 9–11.

Refer to Appendix Tables 48 and 49 for information on limits and fits for the holes and bushings. Dimension only the precision fits; do not fully dimension the drawing. Remember to reverse the direction of the section lining on adjacent parts.

Drawing G–21.

Section Review—Domed Cover.

Reference Unit 9–12.

Study the three views carefully. Be sure you can visualize all details completely. Refer to Appendix Table 43 for information on fits. Are the six holes on the periphery of the top view to be counterbored or spotfaced?

Drawing G–22.

Section Review—Slide Support and Drill Press Base.

Reference Unit 9–12.

Review Chapter 9 on sections and conventions in your text. These problems are designed to test your general knowledge of this subject.

Drawing H–1.

Simplified Thread Representation—Guide Block.

Reference Unit 10–1.

Study carefully the technique involved in simplified thread representation in Unit 10-1. In particular, study Figs. 10-1-10 and 10-1-11. Fig. 10-1-12 will also be useful as you develop the notes describing the threaded holes.

Drawing H–2. (Two solution pages.)

Simplified Thread Representation—Turnbuckle.

Reference Unit 10–1.

Review Unit 10-1 in your text before proceeding. In particular, study Figs. 10-1-10 through 10-1-12 for specific information on simplified thread representation and notes. You may also find it useful to refer to Chapter 14 for information on detail and assembly drawings as well as item lists.

Drawing H–3. (Two solution pages.)

Simplified Thread Representation—Parallel Clamps.

Reference Unit 10–1.

Review Unit 10-1 in your text before proceeding. In particular, study Figs. 10-1-10 through 10-1-13 for specific information on simplified thread representation and notes. You may also find it useful to refer to Chapter 14 for information on detail and assembly drawings as well as item lists. Chapter 9 contains information on conventional breaks.

Drawing H–4.

Simplified Thread Representation—Terminal Block.

Reference Unit 10–1.

Review Figs. 10-1-10 through 10-1-13 for information on simplified thread representation and notes. See Unit 6-15 for techniques used to show the intersection of unfinished surfaces.

Drawing H–5.

Detailed Thread Representation—Connector and Supports.

Reference Unit 10–2.

Refer to Fig. 10-2-3 in your text for the steps in drawing detailed representations of square and acme threads.

Drawing H–6.

Common Threaded Fasteners.

Reference Unit 10–3.

Unit 10-3 in your text gives a description of the shape, characteristics, and uses for various types of threaded fasteners. See Appendix Tables 13, 14, 17, and 22 for specific information needed to draw the fasteners (including washers) for this assignment. Use simplified thread representation.

Drawing H–7.

Common Threaded Fasteners—Shaft Intermediate Support.

Reference Unit 10–3.

Use simplified thread representation (see Unit 10-1 in your text). Refer to Appendix Table 45 for limits of size for the precision hole. See also Unit 6-15 for information on how to represent the intersection of unfinished surfaces. Review Unit 8-7 in your text for information on surface texture.

Drawing H–8. (Two solution pages.)

Common Threaded Fasteners—Wheel Puller.

Reference Unit 10–3.

Use only simplified thread representation. Review Chapter 14 in your text for information on detail and assembly drawing as well as item lists. See Appendix Table 10 for details on cap screws.

Drawing H–9. (Two solution pages.)

Special Fasteners—Adjustable Shaft Support.

Reference Unit 10–4.

Use simplified thread representation only. Refer to Appendix Tables 48 and 49 for limit dimensions on the precision fits. Information on setscrews can be found in Appendix Table 13 in your text.

Drawing H–10. (Two solution pages.)

Fasteners for Wood—Woodworking Vise.

Reference Unit 10–5.

Use simplified thread representation only. Refer to Appendix Table 13 for information on setscrews. Refer to Appendix Table 10 for information on machine screws. Limit dimensions for the precision fit are found in Appendix Table 43.

Drawing J–1.

Keys, Splines, and Serrations—Key Fasteners.

Reference Unit 11–1.

Use sectional views where appropriate for better clarity. Data on various kinds of keys will be found in Appendix Tables 22 and 23. See Appendix Table 14 for hex nuts. Use schematic representation to show threads.

Drawing J–2.

Pin Fasteners—Draw-Bar Hitch and Cam Follower.

Reference Unit 11–2.

Use sectional views where appropriate for better clarity. See Appendix Table 26 for clevis pins, Table 29 for grooved pins, and Table 25 for cotter pins.

Drawing J–3.

Pin Fasteners—Crane Hook.

Reference Unit 11–2.

See Fig. 10-3 in your text for information on the slotted nut. Appendix Table 28 contains information on spring pins; Table 26, clevis pins; and Table 25, cotter pins. If necessary, refer back to Unit 10-3 in your text for details on how to specify fasteners in your item list.

Drawing J–4. (Two solution pages.)

Retaining Rings—Power Drive Assembly.

Reference Unit 11–3.

Refer to Appendix Table 68 in your text for information and sizes on bearings, Tables 35 through 39 for retaining rings, Table 21 for keys, and Table 13 for setscrews. Take sizes directly from the drawing provided. Add part numbers.

Drawing J–5.

Retaining Rings—Roller Assembly and Viewer Case.

Reference Unit 11–3.

Refer to Appendix Tables 35 through 39 in your text for information on retaining rings and Table 19 for flat washers. Add dimensions and notes to describe the added parts and groove sizes.

Drawing J–6.

Springs—Punch Holder Assembly.

Reference Unit 11–4.

Study Unit 11-4 before proceeding. Information on drawing springs is clearly shown in this unit. See Appendix Tables 35 and 36 for details on retaining rings. Limit dimensions for the precision fit will be found in Appendix Table 43.

Drawing J–7.

Rivets—Structural and Blind Rivets.

Reference Unit 11–5.

Refer to Fig. 11-5-2 in your text for information on drawing rivets. Button head rivets are recommended for the truss assembly. See Fig. 11-5-6 for information on blind rivets.

Drawing J–8.

Rivets—Rivets for Aerospace Equipment.

Reference Unit 11–5.

Refer to Figs. 11-5-4 and 11-5-5 in your text for examples of this type of drawing. Be sure to use the appropriate symbols to represent the aerospace rivets specified.

Drawing J–9.

Welded Fasteners—Pipe and Leakproof Attachments.

Reference Unit 11–9.

Study Unit 11-6 in your text before proceeding. Estimate sizes for the resistance-welded threaded fastener to be used on the pipe attachment. A .25 in. diameter fastener is suggested to hold the bracket to the pipe. A .31 in. diameter stud with spring lockwasher and hex nut are recommended for the leakproof attachment.

Drawing J–10.

Adhesive Fasteners—Joint Design.

Reference Unit 11–7.

Panel 1: See Fig. 11-7-3 in your text. Draw only an edge view, show-ing the tongue and groove feature. Refer to Appendix Table 51 for specification on adhesives.

Panel 2: See Fig. 11-7-3 in your text for joint design. Select adhesive from Appendix Table 51.

Panel 3: See Fig. 11-7-3 in your text. A conventional tongue and groove joint should be considered. Select adhesive from Appendix Table 51.

Panel 4: See Fig. 11-7-3 in your text for the design of a toggle lap joint. Select adhesive from Appendix Table 51.

Drawing J–11. (Two solution pages.)

Fastener Review—Wheel Assembly.

Reference Unit 11–8.

Limit dimensions for the precision fits are found in Appendix Tables 43 and 46. Use sectional views where appropriate. Consider specifying rubber as the material for the wheel.

Drawing J–12.

Fastener Review—Universal Joint.

Reference Unit 11–8.

See Appendix Table 10 for information on machine screws, and Table 28 for information on spring pins. Show a short shaft held in place with a spring pin in each fork. Use simplified thread representation only.

Drawing K–1.

Cast Irons—Door Closing Arm.

Reference Unit 12–1.

Refer to Fig. 12-1-2 in your text for information on a suitable material. Break out a portion of the arm to provide space for the revolved section.

Drawing K–2.

Carbon Steel—Raising Bar.

Reference Unit 12–2.

Refer to Fig. 12-2-4 in your text to help in selecting an appropriate material. Show the acme threads in detailed representation; all others, simplified.

Drawing K–3.

Nonferrous Metals—Outboard Motor Clamp.

Reference Unit 12–3.

Review Unit 8-4 in your text for information on various forms of arrowless dimensioning. Mark all points of tangency for a smooth transition from arc to arc and from arc to straight lines.

Drawing K–4.

Plastics—Connecting Link.

Reference Unit 12–4.

The spacer should be made from a relatively soft material to absorb shock and vibration. The castings need to be made from a relatively inexpensive, moderately strong material.

Drawing K–5.

Rubber—Caster Assembly.

Reference Unit 12–5.

In assembly, show the peened end of the post (PT 2) mushroomed (riveted). This feature can be drawn freehand. A hex nut and lock washer must be shown on the threaded end of PT 3—Shaft Bolt. See Appendix Table 14 for information on drawing the hex nut.

Drawing L–1.

Castings—Connector.

Reference Unit 13–1.

Be sure to allow sufficient material on the top and bottom of the casting for machining. Does the design suggest the need for additional material on the 4.00 in. diameter on the base of the casting for machining?

Drawing L–2.

Castings—Swing Bracket.

Reference Unit 13–1.

Add to the thickness of the material to compensate for the difference in strength between a cast part and a fabricated steel part. Show fillets and rounds as necessary. Review Unit 8-7 in your text for information on surface texture.

Drawing L–3.

Forgings—Bracket.

Reference Unit 13–2.

Holes and similar features are generally machined in a forged part and, therefore, will not show in the forging drawing. Be sure to allow for machining on surfaces with finish marks.

Drawing L–4.

Powder Metallurgy—Bracket.

Reference Unit 13–3.

The machining drawing generally will require only dimensions that specify machining operations.

Drawing L–5.

Plastic Molded Parts—Pivot Arm.

Reference Unit 13–4.

Three threaded inserts are required. Design them to be molded in place with a self-locking feature. Add dimensions and notes.

Drawing M–1.

Functional Drafting—Cable Straps.

Reference Unit 14–2.

Your final design will include flat-end and rounded-end straps in four combinations of sizes and hole spacing. Your drawing should show constant sizes; a table should be included to show variable sizes and hole spacing.

Drawing M–2.

Functional Drafting—Cover Plate.

Reference Unit 14–2.

Refer to Unit 8-4 for a review of arrowless dimensioning. Use a tabulated setup for specifying the size and location of holes.

Drawing M–3.

Detail Drawings—Base Plate.

Reference Unit 14–3.

Allow sufficient space between views for dimensions. Would sectional views be of any value on any of the three views?

Drawing M–4.

Detail Drawings—Swivel Hanger.

Reference Unit 14–3.

Refer to Unit 6-15 in your text for a review of intersections of unfinished surfaces. Mark all tangent points. Allow sufficient space for dimensions.

Drawing M–5.

Detail Drawing—Locating Stand.

Reference Unit 14–3.

Refer to Unit 6-15 in your text for a review of intersections of unfinished surfaces. Refer to Unit 8-7 in your text for a review of surface texture. Use Appendix Table 3 for the conversion of decimal inches to millimeters.

Drawing M–6.

Multiple Detail Drawings—Adjustable Pulley and Pulley Assembly.

Reference Unit 14–4.

Refer to Appendix Tables 43 through 49 for information on limit dimensions. Use single views and partial views of parts where appropriate. Remember that sectional views can often improve clarity. Use simplified thread representation in all cases.

Drawing M–7.

Drawing Revisions—Axle Cap.

Reference Unit 14–5.

Refer to Fig. 14-5-1 as an example of this type of drawing technique. Be sure to change identification numbers near the dimensions on the drawing that require changes.

Drawing M–8.

Assembly Drawings—Tool Post Holder.

Reference Unit 14–6.

Use simplified thread representation. Reverse the direction of section lining on adjacent parts.

Drawing M–9.

Assembly Drawings—V-Block Clamp.

Reference Unit 14–6.

Use simplified thread representation. Review Unit 14-5 for information on item lists.

Drawing M–10.

Exploded Assembly Drawings—Coupling.

Reference Unit 14–7.

Review Unit 14-6 for information on an item list. You may find it more practical to draw the coupling to a somewhat smaller scale. Refer to Appendix Table 23 for appropriate sizes on keyseats.

Drawing M–11.

Detail Assembly Drawings—Sawhorse.

Reference Unit 14–8.

Review Unit 14-6 in you text for information on item lists. Notice that this drawing does not have decimal dimensions. Why? The partial detail of the leg may be added to your drawing if needed for clarity.

Drawing M–12.

Subassembly Drawings—Wheel Assembly.

Reference Unit 14–9.

Use simplified thread representation. Refer to Appendix Tables 11, 14, and 19 for data on bolts, nuts, and washers. Use hidden lines only when needed for clarity.

Drawing N–1.

Isometric Drawings—Flat Surfaces.

Reference Unit 15–1.

Study each of the three-view drawings carefully before attempting the pictorial views. Where inclined lines are involved, simply find the end points and connect them with straight lines.

Drawing N–2.

Isometric Drawings—Step Block and Planter Box.

Reference Unit 15–1.

In each case, begin by blocking in a rectangular solid whose sizes are equal to the overall width, depth, and height of the object.

Drawing N–3.

Isometric Drawings—Curved Surfaces.

Reference Unit 15–2.

Study each of the three-view drawings carefully before attempting the pictorial views. These assignments may be completed by freehand sketching or with instruments (including an ellipse template.)

Drawing N–4.

Curved Surfaces—Base.

Reference Unit 15–2.

First, block in the overall sizes of the object. For inclined lines, locate the end points and connect them with straight lines. Use a template for ellipses. Use unidirectional dimensioning.

Drawing N–5.

Common Features in Isometric—Adapter.

Reference Unit 15–3.

Use an ellipse template if available. Apply the unidirectional dimensioning technique.

Drawing N–6.

Common Features in Isometric—Two-Post Die Set.

Reference Unit 15–3.

Use an ellipse template for isometric circles. Be sure to add part numbers to the pictorial drawing. Review Unit 14-6 for information on item lists. Use only the sizes from the right-hand column in the table.

Drawing N–7.

Oblique Projection—Flat Surfaces.

Reference Unit 16–4.

Study each of the three-view drawings carefully before attempting the pictorial views. You may find it easier to begin by blocking in a rectangular solid whose sizes are equal to the overall width, depth, and height of the object. For inclined lines, locate end points and connect them with straight lines. These assignments may be completed by freehand sketching or with instruments.

Drawing N–8.

Oblique Projection—Spacer Block and Dovetail Guide.

Reference Unit 15–4.

Study the assignment drawings carefully before attempting the pictorial views. Remember that in cabinet oblique, all receding lines are drawn half-size. In cavalier oblique, all lines are drawn full size. Use only unidirectional dimensioning.

Drawing N–9.

Common Features in Oblique—Step Pulley and Connector.

Reference Unit 15–5.

Remember that in cabinet oblique, all receding lines are drawn half-size. Begin at the front of the object and carefully locate each surface (plane) by measuring distances on the receding axis. Use only unidirectional dimensioning.

Drawing N–10.

One-Point Perspective—Vise Base.

Reference Unit 15–6.

Begin by blocking in a perspective rectangular solid (wire frame) whose sides are equal to the width, depth, and height of the vise base. Use unidirectional dimensioning only.

Drawing N–11.

Two-Point Perspective—Support Guide.

Reference Unit 15–7.

Begin by blocking in a perspective rectangular solid (wire flame) whose sides are equal to the width, depth, and height of the vise base. Use unidirectional dimensioning only.

Drawing N–12.

CGS Modeling—Boolean Exercise 1.

Reference Unit 15–8.

Study each of the four figures, A+B, A-B, B-A, and A intersect B. Complete the Boolean operations shown. Darken in the objects and hidden lines for the resulting objects.

Drawing N–13.

CGS Modeling—Boolean Exercise 2.

Reference Unit 15–8.

Study each of the four figures, A+B, results +C, results -D, and results -E. Darken the outline of the object for each step of the Boolean operations. Include hidden lines if needed for clarity.

Drawing N–14.

CGS Modeling—Intersection Profile Exercise.

Reference Unit 15–8.

Study both figures. Draw each of the profiles needed to produce the objects shown by the intersection of the extruded profiles.

Drawing P–1.

Modern Engineering Tolerancing—Deviations and Block Diagram.

Reference Unit 16–1.

Study carefully Unit 16-1 before working on Drawing P-1. An understanding of form variations (deviations) is critical to an understanding of geometric tolerancing (engineering tolerancing.)

Drawing P–2.

Geometric Tolerancing—Straightness Tolerance.

Reference Unit 16–2.

Refer to Fig. 16-2-3 in your text for the size and shape of the feature control frame and the leader that connects it with the view. See also Fig. 16-2-2 for information on geometric characteristic symbols.

Drawing P–3.

Geometric Tolerancing—Flatness Tolerance.

Reference Unit 16–3.

Read Unit 16-3 in your text before proceeding. Refer to Fig. 16-3-l in your text for the size and shape of the flatness symbol. Refer to Fig. 16-3-4 for Part 2, Fig. 16-3-3 for Parts 3 and 4, and Fig. 16-3-2 for Parts 5 and 6.

Drawing P–4.

Straightness of a Feature of Size.

Reference Unit 16–4.

Read Unit 16-4 in your text before proceeding. Refer to Figs. l6-4-l4 and l6-4-15 for examples of this type of problem. What does the diameter symbol in the feature control frame mean? Notice in Fig. 2 that there is a maximum tolerance specified. How does this apply to gage size?

Drawing P–5.

Datums and the Three—Plane Concept.

Reference Unit 16–5.

Add only the geometric tolerances specified; do not fully dimension the views. This assignment combines the concepts covered in previous geometric tolerancing assignments. You may need to review Units 16-1 through 16-4 as you proceed. Also, see Fig. 16-5-5 for information on the placement of the datum feature symbol.

Drawing P–6.

Orientation Tolerancing of Flat Surfaces.

Reference Unit 16–5.

See Fig. 16-2-2 for information on geometric characteristic symbols. Refer to Appendix Table 44 for data on LC3 fits. Add geometric tolerances as specified in the instructions of your assignment sheet. Fully dimension the drawing unless instructed otherwise.

Drawing P–7.

Datum Features Subject to Size Variation.

Reference Unit 16–7.

Read carefully Unit 16-7 before proceeding. What effect does MMC have on gage size? Notice the maximum tolerance given in Examples 5, 6, 11, and 12. How does this apply to gage size?

Drawing P–8.

Orientation Tolerancing for Features of Size.

Reference Unit 16–5.

Review MMC basis. You may want to sketch the solution to this problem on a tissue overlay before adding tolerancing data and dimensions to the final assignment sheet. Refer to Appendix Table 43 for data on the RC7 fit.

Drawing P–9.

Positional Tolerancing.

Reference Unit 16–9.

Positional tolerances have to do with the relative location of features and are most commonly associated with the assembly of parts. Review Fig. 16-9-1 in your text before proceeding.

Drawing P–10.

Projected Tolerancing Zone—Cover Plate.

Reference Unit 16–10.

Read Unit 16-10 in your text before proceeding. You may find it practical to sketch your solution on a tissue overlay before adding tolerancing data and dimensions to the drawing sheet. Remember that standard dimensions enclosed in a rectangle represent theoretically exact dimensions. These are used to establish the theoretically exact location of a feature.

Drawing P–11.

Datum Targets—Bracket Guide.

Reference Unit 16–11.

Read Unit 16-11 in your text before proceeding. You may find it practical to sketch your solution on a tissue overlay before adding tolerancing data and dimensions to the drawing sheet. Refer to Figs. 16-11-1 through 16-11-3 in your text for information on the datum target symbol, the identification of datum targets, and the datum target point symbol. Fully dimension the drawing.

Drawing P–12.

Circularity and Roundness.

Reference Unit 16–12.

Read Unit 16-12 in your text before proceeding. See Fig. 16-2-2 in your text for information on symbols for circularity and cylindricity. Be careful in calculating the geometric tolerances relative to size tolerances.

Drawing P–13.

Profile Tolerancing—Slide.

Reference Unit 16–13.

Refer to Fig. 16-2-2 in your text for information on geometric characteristic symbols. See Fig. 16-13-20 for information on the "all around" symbol. Convert all standard dimensions that relate to geometric tolerances into "theoretically exact" dimensions.

Drawing P–14.

Correlative Tolerances—Adjustable Base.

Reference Unit 16–14.

Read Unit 16-14 in your text before proceeding. Refer to Figs. 16-14-1 and 16-14-2 in your text for examples related to this assignment. Convert all standard dimensions that relate to geometric tolerances into "theoretically exact" dimensions.

Drawing P–15.

Positional Tolerancing of Noncylindrical Features.

Reference Unit 16–15.

Read Unit 16-15 in your text before proceeding. Noncylindrical features in this case involve features such as slots, grooves, and tabs. Refer to Figs. 16-15-1 through 16-15-8 in your text for an example of this concept.

Drawing P–16.

Positional Tolerancing for Multiple Patterns of Features.

Reference Unit 16–16.

Read Unit 16-16 in your text before proceeding. Refer to Figs. 16-16-1 through 16-16-3 for examples of this concept. Convert all standard dimensions that relate to geometric tolerances into "theoretically exact" dimensions.

Drawing P–17.

Formulas for Positional Tolerancing.

Reference Unit 16–17.

Refer to the formulas and calculations in Unit 16-17 of your text as you proceed through this assignment.

Drawing P–18.

Summary of Rules for Geometric Tolerancing—Housing.

Reference Unit 16–18.

Study Unit 16-18 in your text before you proceed. Be sure to allow sufficient space between and around views for dimensions. You may want to draw the two views and then use a tissue overlay to lay out the dimensions and tolerancing data before adding them to the final drawing.

Drawing R–1.

Drawings for Numerical Control—Two-Axis Control Systems.

Reference Unit 17–1.

Read carefully the text material in Unit 17-1. Be sure you understand the difference between coordinate dimensioning and point-to-point dimensioning. Point locations in all cases are determined by counting squares. Each square equals .10 inches.

Drawing R–2.

Drawings for Numerical Control—Three-Axis Control Systems.

Reference Unit 17–2.

Read carefully the text materials in Unit 17-2. Study Fig. 17-2-4 for a clear understanding of how to calculate *Z* distance.

Drawing S–1.

Designing for Welding—Pivot Arm.

Reference Unit 18–1.

When converting from a casting to a weldment, it is often appropriate to reduce material sizes without reducing strength. To minimize cost in production, keep individual parts as simple as possible. Manufacturing irregularly shaped parts is expensive.

Drawing S–2.

Welding Symbols.

Reference Unit 18–2.

Study Fig. 18-2-1 in your text. Be sure you understand the difference between *weld symbol* and *welding symbol*. Also, be sure you understand *arrow side* and *other side*. You may complete this assignment using freehand sketching or instruments.

Drawing S–3.

Fillet Welds—Swing Bracket.

Reference Unit 18–3.

Refer to Fig. 18-3-3 for information on strength of welds. Make sure your welding symbols fully describe the finished welds. Create your own names for each of the parts when you complete the item list.

Drawing S–4.

Groove Welds—Connecting Link.

Reference Unit 16–4.

When converting from a casting to a weldment, keep all individual parts as simple as possible. Complex shapes are costly to manufacture. While shapes and sizes of individual parts can change, hole sizes and center-to-center distances must remain the same. Make up names for individual parts when completing the item list.

Drawing S–5.

Other Basic Welds—Plug, Slot, and Spot Welds.

Reference Unit 18–5.

Read Unit 18-5 in your text. Study Fig. 18-5-1. Sketch the joint preparations and shapes of finished welds in the sectional views. Use instruments to add dimensions and welding symbols.

Drawing T–1.

The Design Process—Toggle Switch Plate.

Reference Unit 19–1.

There are many sizes, styles, and shapes of toggle switch plates. However, simplicity is still the key to good design! Make several freehand sketches of various designs before completing your final drawing.

Drawing T–2.

Assembly Considerations—Conductor Supports.

Reference Unit 19–2.

There are an endless number of solutions to this problem. Your primary objective is to arrive at a solution that is safe, practical, and inexpensive. Make a complete set of working drawings, including an assembly drawing, of your final design.

Drawing U–1.

Belt Drives.

Reference Unit 20–1.

Read Unit 20-1 in your text before proceeding. Refer to the various tables in your text as you complete your calculations.

Drawing U–2.

Chain Drives.

Reference Unit 20–2.

Read Unit 20-2 in your text before proceeding. Follow the steps of the procedure described in your text when developing the solution to the three problems. Refer to the tables as needed.

Drawing U–3.

Gear Drives—Spur Gear.

Reference Unit 20–3.

Read Unit 20-3 in your text before proceeding. Become familiar with spur gear definitions and formulas given in Fig. 20-3-5. A single, full-section view and cutting-data block are all that is needed to fully describe the spur gear. Refer to Appendix Table 21 for data on keyseat sizes.

Drawing U–4.

Power Transmitting Capacity of Spur Gears.

Reference Unit 20–4.

Read Unit 20-4 in your text before proceeding. Follow the examples given in your text for the solutions to similar problems.

Drawing U–5.

Rack and Pinion.

Reference Unit 20–5.

Lay out the two views based on the information given. Show four or five teeth on both the rack and gear. Use phantom lines to represent the remaining gear teeth. Draw the side view as a full section. Refer to Appendix Table 21 in your text for information on keyseats. Complete the cutting-data block.

Drawing U–6.

Bevel Gears.

Reference Unit 20–5.

Draw both gears in full section. Fully dimension. Refer to Appendix Table 21 in your text for information on keyseats. Complete the cutting-data block.

Drawing U–7.

Worm and Gear.

Reference Unit 20–7.

It is not necessary to show the shape (profile) of the teeth in either of the circular views. Use phantom lines to simulate gear teeth. Complete the cutting-data block and fully dimension the working drawing. Refer to Appendix Table 21 in your text for information on keys and keyseats.

Drawing U–8.

Comparison of Chain, Gear, and Belt Drives.

Reference Unit 20–5.

Refer to a standard bearing catalog for specifications on pillow blocks (shaft supports). Refer to Appendix Table 21 in your text for information on keys and keyseats.

Drawing V–1.

Couplings and Flexible Shafts—Fan and Motor Layout.

Reference Unit 21–1.

Estimate sizes not given on the coupling drawing. Your finished drawing should show the motor, gearbox, and housing along with the two couplers and shaft in place. Use a note on the drawing to specify the correct coupling.

Drawing V–2.

Bearings—Gearbox.

Reference Unit 21–2.

Refer to standard gear catalogs for design data on gears. Plain bearings can be found in Appendix Table 66. Appendix Table 64 covers setscrew collars and Table 21 covers keys. Label parts as appropriate.

Drawing V–3.

Antifriction Bearings—Gearbox.

Reference Unit 21–3.

See Appendix Table 21 for information on keys, Table 35 for retaining rings, and Table 68 for radial bearings. You may need to refer to manufacturers' catalogs for some of the standard parts. Label parts, but do not dimension.

Drawing V–4.

Premounted Bearings—Adjustable Shaft Support.

Reference Unit 21–4.

Consider using a full-sectional view or a half-sectional view. However, remember that shafts, bolts, pins, etc., are usually not sectioned even if they fall in the cutting plane. See Appendix Table 13 for setscrews.

Drawing V–5.

Lubricants and Radial Seals.

Reference Unit 21–5.

First, find catalog or appendix references for bearings, seals, etc., and obtain sizes for each. You may prefer to solve the assignment problems through freehand sketching before attempting the final drawing on the workbook sheet. Appendix Table 69 has information on standard tapered roller bearings. You may need to refer to catalogs for other parts.

Drawing V–6.

Static Seals.

Reference Unit 21–6.

Refer to Unit 21-6 in your text for all necessary information on O-rings and gasket-type seals.

Drawing W–1.

Cams—Cam Motions.

Reference Unit 22–1.

Read Unit 22-1 in your text and become familiar with cam terminology. Develop the layout (grid) for the displacement diagram, using thin lines. Only the line representing the path of the follower should be thick.

Drawing W–2.

Cams—Plate Cam.

Reference Unit 22–2.

Develop the displacement diagrams first, followed by the plate cam, and finally, complete the chart showing angular and radial displacement. Follow the line technique shown in Fig. 22-2-2 in your text.

Drawing W–3.

Cams—Positive Motion Cams.

Reference Unit 22–3.

Develop the displacement diagrams first, followed by the plate cam, and finally, complete the chart showing angular and radial displacement. Follow the line technique shown in Fig. 22-2-2 in your text.

Drawing W–4.

Cams—Drum Cam.

Reference Unit 22–4.

Refer to Fig. 22-4-2 in your text for a general understanding of this assignment. Complete the chart on the assignment sheet after you have completed the two views of the drum cam and the displacement diagram.

Drawing W–5.

Indexing—Indexing Drive.

Reference Unit 22–5.

Take sizes directly from the chart at the top of the assignment sheet for indexing drive 6S75. Give only necessary dimensions. Show the direction of rotation on each part. See Appendix Table 23 for information on keys.

Drawing W–6.

Linkages—Simple Crank Mechanisms.

Reference Unit 22–6.

Read Unit 22-6 in your text. Use either a thin solid line, phantom line, or a dashed line to represent the paths of the various points.

Drawing W–7.

Ratchet Wheels—Ratchet and Crank Mechanism.

Reference Unit 22–7.

Read Unit 22-7 in your text. Plot the path of the end of the drive pawl, and mark the positions of the pawl at each 22.5° rotation of the crank. Use a phantom line to connect the points completing the path. Number the points along the path.

Drawing X–1.

Surface Development—Wall Tray.

Reference Unit 23–1.

Use thick lines for visible outlines and thin lines for bend lines. Allow 6 mm for seams and hems. Dimensions are required since the drawing is not full size.

Drawing X–2.

The Packaging Industry—Hexagon Box.

Reference Unit 23–2.

Use dividers to transfer all dimensions from the front and top views to the development. Use thick lines for visible outlines and thin lines for bend lines. Dimensions are not necessary since this is a full-size development. Letter all points of intersection.

Drawing X–3.

Radial Line Development—Truncated Concentric Pyramid.

Reference Unit 23–3.

Use dividers to transfer all dimensions from the front and top views to the development. Use thick lines for visible outlines and thin lines for bend lines. Dimensions are not necessary since this is a full-size development. Letter or number all points of intersection.

Drawing X–4.

Parallel Line Development—Three-Piece Elbow.

Reference Unit 23–4.

Divide the circular view into 30° segments. The length of the 30° segments on the circular view becomes the basis for laying out the development.

Drawing X–5.

Radial Line Development—Conical Surfaces.

Reference Unit 23–5.

True-length lines must be used to lay out the developments. True-length lines are the outside edge lines, not the vertical distances. Add .20 in. seam allowances.

Drawing X–6.

Development of Transition Pieces.

Reference Unit 23–6.

Prepare a true-length diagram to establish the true lengths of the bend lines in the transition. Some lines are true length on the front and top views. Add numbers and letters but do not dimension.

Drawing X–7.

Development of a Sphere—Gore Method.

Reference Unit 23–7.

Read Unit 23-7 in your text carefully. Begin by laying out a front and top view of a 3.00 in. sphere on a separate sheet of paper. Divide the top view into 30° segments and project downward to develop the front view. Follow Fig. 23-7-1 to complete the development. Add appropriate dimensions.

Drawing X–8.

Intersections of Flat Surfaces.

Reference Unit 23–6.

First, complete the lines of intersection on the views given. Use dividers to transfer sizes from the given views to produce the development.

Drawing X–9.

Intersections of Cylindrical Surfaces.

Reference Unit 23–9.

Divide the circle (end view of round pipe) into 30° segments, and project to the other views to establish a series of points (intersections of projection lines) on the front view. Connect the points to complete the front view.

Drawing X–10.

Intersections of Prisms.

Reference Unit 23–10.

Project from the front view to the top view to complete the lines of intersection. You will be using both parallel line development and radial line development to complete the assignment.

Drawing X–11.

Stampings—Swivel Hanger.

Reference Unit 23–11.

Refer to the engineer's sketch of a swivel hanger. Refer also to bending allowances provided on page 882 in the text. Prepare a working drawing and a developed blank of the bracket and connector.

Drawing Y–1.

Orthographic Piping Drawings—Fuel Oil Supply System.

Reference Unit 24–1.

Make a front, top, and side view, single-line drawing representing the pictorial shown on the assignment sheet. Refer to Figs. 24-l-l2 and 24-1-13 as examples of the type of drawing you are expected to do. Add an item list.

Drawing Y–2.

Isometric Piping Drawings—Diesel Engine Starting System.

Reference Unit 24–2.

Convert the three-view orthographic drawing shown on the assignment sheet to a single-line isometric drawing. Refer to Figs. 24-l-2 and 24-1-3 as examples. Add an item list.

Drawing Y–3.

Supplementary Piping Information—Oil-Burning Piping System.

Reference Unit 24–3.

Read Unit 24-3 in your text before you proceed. In particular, study Figs. 24-3-3 and 24-3-6.

Drawing AA–1.

Structural Drafting—Beam Tolerances.

Reference Unit 25–1.

Much of the information needed to complete the tables in this assignment will be found in Figs. 25-1-8 through 25-1-10.

Drawing AA–2.

Structural Drafting—Beams.

Reference Unit 25–2.

The partial design drawings on your assignment sheet contain the information needed to develop the detail drawings. Be sure to include all dimensions and notes on each detail.

Drawing AA–3.

Structural Drafting—Standard Connections.

Reference Unit 25–3.

Locate the north end of beam E3 and the south end of beam K3. Note that the north direction symbol is pointing to the left. Refer to Fig. 25-3-13 as an example. Show all dimensions and welding symbols.

Drawing AA–4.

Structural Drafting—Sectioning.

Reference Unit 25–4.

Carefully study the differences between the two W16 × 40 beams. Use Fig. 25-6-1 in your text as a model for your drawing. The primary difference is that notes will be required to describe the differences between the two beams.

Drawing AA–5.

Structural Drafting—Seated Beam Connections.

Reference Unit 25–5.

Read carefully Unit 25-5 before proceeding. Refer to Fig. 25-5-3 in your text as an example of the type of drawing you will do in this assignment. Show all appropriate dimensions and welding symbols.

Drawing AA–6.

Structural Drafting—Dimensioning.

Reference Unit 25–6.

Refer to Fig. 25-6-2 as an example of the type of drawing you will do in this assignment. Where possible, combine drawings of beams and specify the unique details in each.

Drawing BB–1.

Jigs and Fixtures—Jig Design.

Reference Unit 26–1.

Keep your design as simple and as small as is practical. Complicated designs are more costly to produce. See Fig. 26-3-2 in your text for information on bushings. Is the locking pin required before or after the first hole is drilled? Why?

Drawing BB–2.

Jigs and Fixtures—Drill Jig Components.

Reference Unit 26–2.

Sketch several possible solutions before developing your final drawing. See Fig. 26-3-2 in your text for information on bushings. Keep your design as simple as possible.

Drawing BB–3.

Jigs and Fixtures—Dimensioning Jig Drawings.

Reference Unit 26–3.

Read the unit in your text dealing with dimensioning jig drawings. Study Fig. 26-3-1 before proceeding. Refer to Fig. 26-3-2 for information on bushings. Remember, the accuracy of the final workpiece will only be as accurate as the design and manufacture of the jig.

Drawing BB–4.

Jigs and Fixtures—Milling Fixtures.

Reference Unit 26–4.

Refer to Appendix Table 80. Design your own fixture base similar to that shown in Fig. 26-4-3 in your text. Add components to the fixture base to complete the design. Design the fixture to be used with a horizontal milling machine with dual cutters.

Drawing CC–1.

Schematic Diagrams—Sound-Effects Generator.

Reference Unit 27–2.

Make a freehand sketch of the schematic diagram before doing your final drawing. Keep the diagram compact while allowing sufficient space around components for notes and specifications.

Drawing CC–2.

Wiring Diagrams—Wiring Diagram for Boat.

Reference Unit 27–3.

Keep all components in their same relative positions. Combine pairs or sets of wires into a *highway* (thick line representing more than one wire). Label wire colors.

Drawing CC–3.

Block and Logic Diagrams—Remote Control Housing Wiring.

Reference Unit 27–5.

Refer to Figs. 27-5-4 through 27-5-7 in your text for the shape and size of logic symbols. Label all components.

Drawing CC–4.

Electrical and Electronic Drafting—Component Problem Sheet.

Reference Unit 27–5.

Study the figures shown and the specifications listed in the tables. Answer the seven questions.

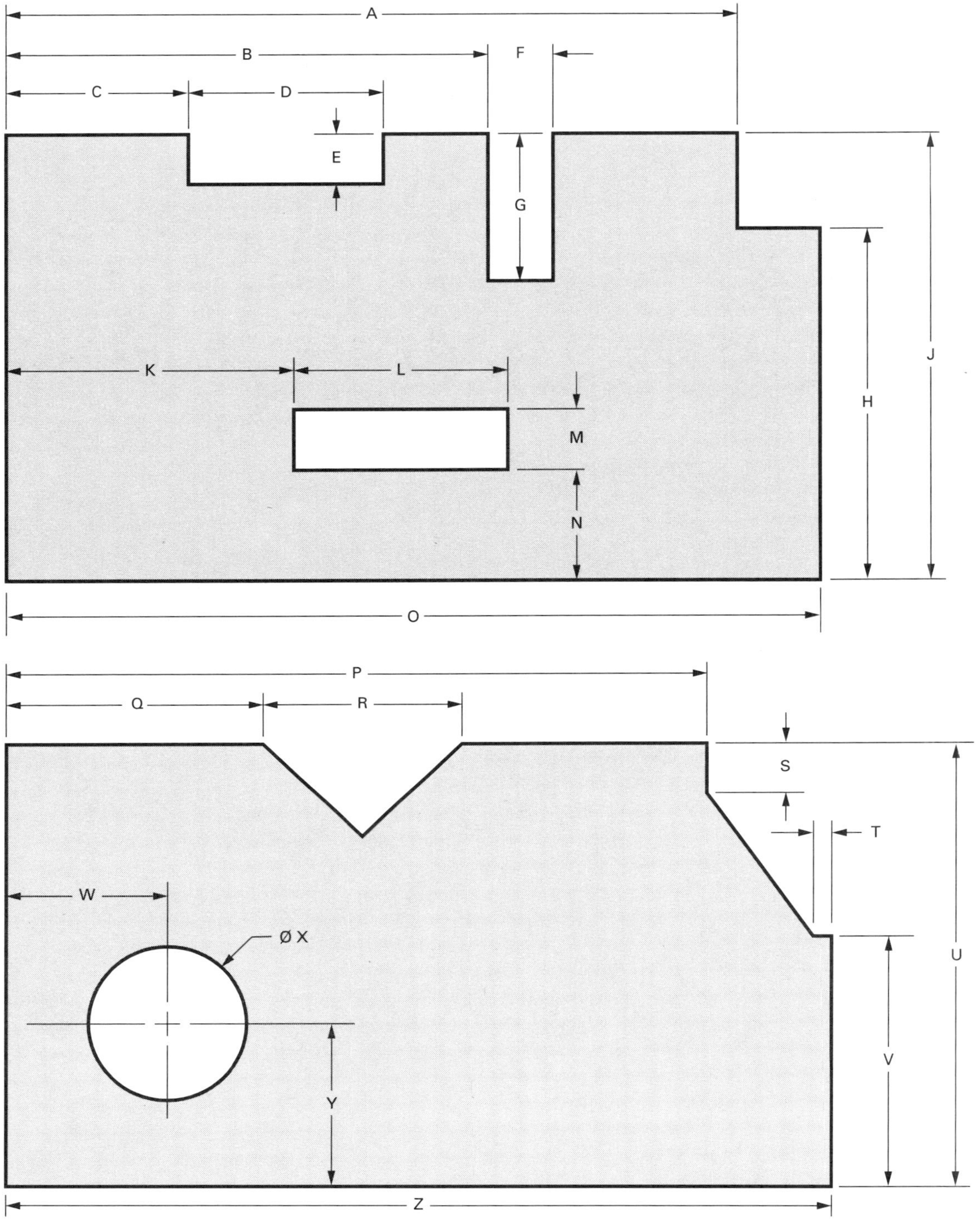

Measure and record the following distances.

Metric Scales (mm)		Decimal Inch Scales		Fractional Inch Scales		Distance
Scale 1:1		Scale 1:1		Scale 1:1		A
						B
						C
						D
						E
						F
						G
						H
						J
Scale 1:2		Scale 1:2				K
						L
						M
						N
				Scale 1:2		O
						P
						Q
						R
Scale 1:5		Scale 1:4				S
						T
						U
						V
						W
						X
						Y
						Z

DRAFTING SCALES— MEASURING	REFERENCE UNIT **1–4**	NAME: COURSE: DATE:	DRAWING **A–1**

A	A	A	N	N	N	I	I	I
B	B	B	O	O	O	2	2	2
C	C	C	P	P	P	3	3	3
D	D	D	Q	Q	Q	4	4	4
E	E	E	R	R	R	5	5	5
F	F	F	S	S	S	6	6	6
G	G	G	T	T	T	7	7	7
H	H	H	U	U	U	8	8	8
I	I	I	V	V	V	9	9	9
J	J	J	W	W	W	10	10	10
K	K	K	X	X	X	ROUNDS AND FILLETS R.10		
L	L	L	Y	Y	Y			
M	M	M	Z	Z	Z			

Complete each line with the size and letter shown.

LETTERING— VERTICAL	REFERENCE UNIT 4–1	NAME COURSE	DATE	DRAWING B–1

A	A	A	N	N	N	1	1	1
B	B	B	O	O	O	2	2	2
C	C	C	P	P	P	3	3	3
D	D	D	Q	Q	Q	4	4	4
E	E	E	R	R	R	5	5	5
F	F	F	S	S	S	6	6	6
G	G	G	T	T	T	7	7	7
H	H	H	U	U	U	8	8	8
I	I	I	V	V	V	9	9	9
J	J	J	W	W	W	10	10	10
K	K	K	X	X	X			
L	L	L	Y	Y	Y			
M	M	M	Z	Z	Z			

ROUNDS AND FILLETS R.10

Complete each line with the size and letter shown.

LETTERING— INCLINED	REFERENCE UNIT 4–1	NAME COURSE DATE	DRAWING B–2

LOCATING BY COORDINATES

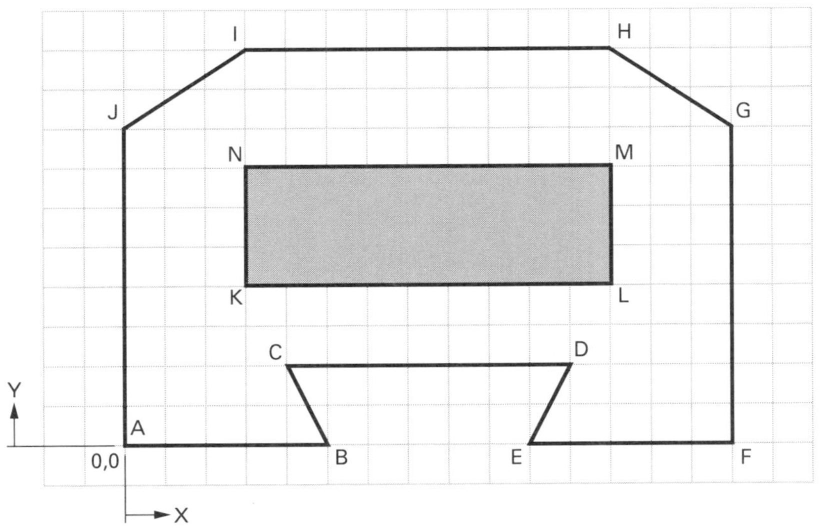

Complete the tables by defining the X and Y coordinates of each point on the part. Note the location of the zero coordinates. Each square on the grid represents one unit of measurement.

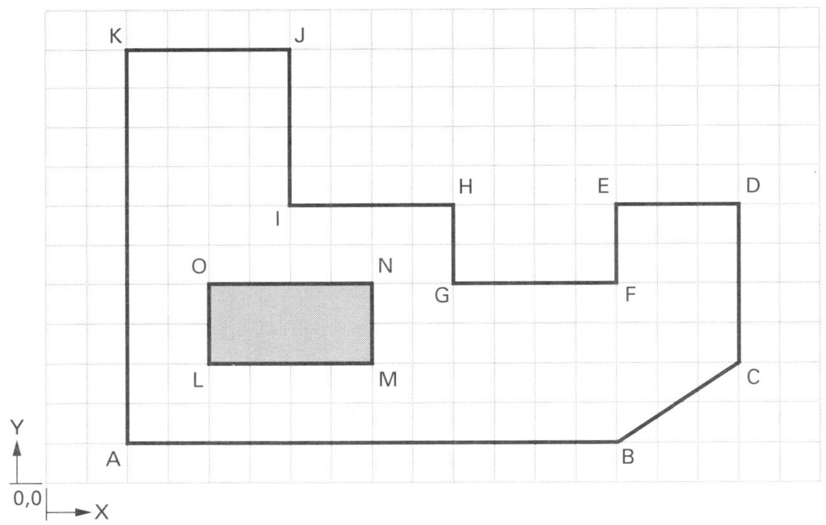

POINT	ABSOLUTE COORDINATES	
	X	Y
A		
B		
C		
D		
E		
F		
G		
H		
I		
J		
K		
L		
M		
N		

POINT	RELATIVE COORDINATES	
	X	Y
A		
B		
C		
D		
E		
F		
G		
H		
I		
J		
K		
L		
M		
N		

POINT	ABSOLUTE COORDINATES	
	X	Y
A		
B		
C		
D		
E		
F		
G		
H		
I		
J		
K		
L		
M		
N		
O		

POINT	RELATIVE COORDINATES	
	X	Y
A		
B		
C		
D		
E		
F		
G		
H		
I		
J		
K		
L		
M		
N		
O		

LOCATING BY COORDINATES	REFERENCE UNIT 4–1	NAME ___ COURSE ___ DATE ___	DRAWING B–3

RELATIVE COORDINATES

REFERENCE UNIT 4–1

NAME

COURSE

DATE

DRAWING B–4

PART A 1.00 IN. GRID (.25 IN. DIVISIONS)

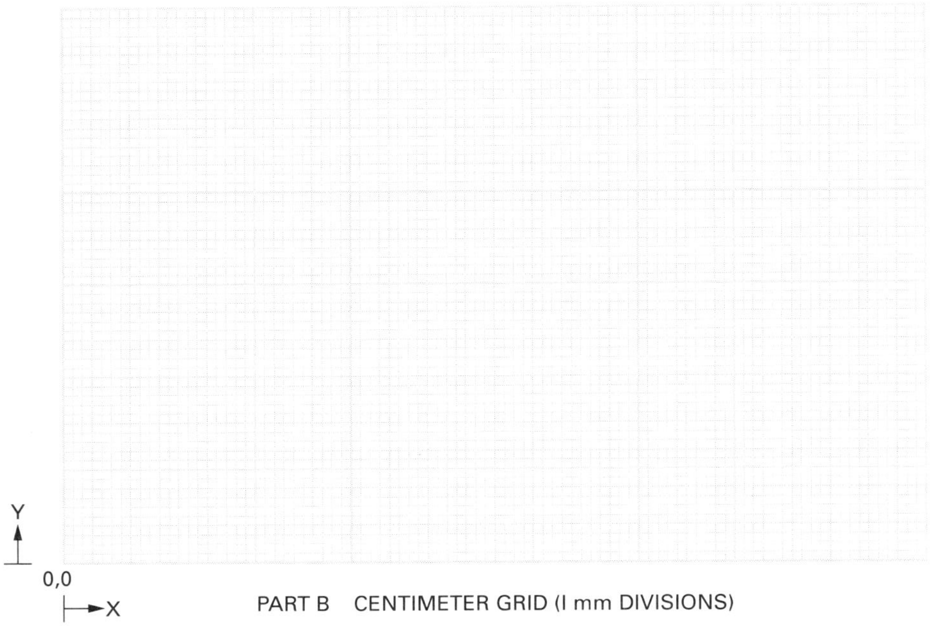

PART B CENTIMETER GRID (I mm DIVISIONS)

Using the coordinates shown, plot parts A and B.
Place the position numbers on the drawings.

RELATIVE COORDINATES (IN.)		
Point	X Axis	Y Axis
1	0	0
2	4.50	0
3	0	.75
4	–.75	0
5	0	.75
6	–.75	0
7	0	.75
8	–3.00	0
9	0	–2.25
New Start—Solid		
10	0	.75
11	3.75	0
New Start		
12	–.75	.75
13	–3.00	0
New Start—Solid		
14	0	1.50
15	4.50	0
16	0	2.25
17	–4.50	0
18	0	–2.25
New Start—Solid		
19	0	.75
20	3.75	0
21	0	1.50
New Start—Solid		
22	–.75	0
23	0	–.75
24	–3.00	0
New Start—Solid		
25	5.25	–4.50
26	2.25	0
27	0	2.25
28	–.75	0
29	0	–.75
30	–.75	0
31	0	–.75
32	–.75	0
33	0	–.75
New Start—Solid		
34	.75	.75
35	1.50	0
New Start—Solid		
36	0	.75
37	–.75	0

Part A: Relative coordinates.

RELATIVE COORDINATES (mm)		
Point	X Axis	Y Axis
1	0	0
2	30	0
3	0	10
4	10	0
5	0	–10
6	30	0
7	0	50
8	–10	0
9	0	–15
10	–50	0
11	0	15
12	–10	0
13	0	–50
New Start—Solid		
14	5	10
15	15	0
16	0	20
17	–15	0
18	0	–20
New Start—Solid		
19	45	0
20	15	0
21	0	20
22	–15	0
23	0	–20

Part B: Relative coordinates.

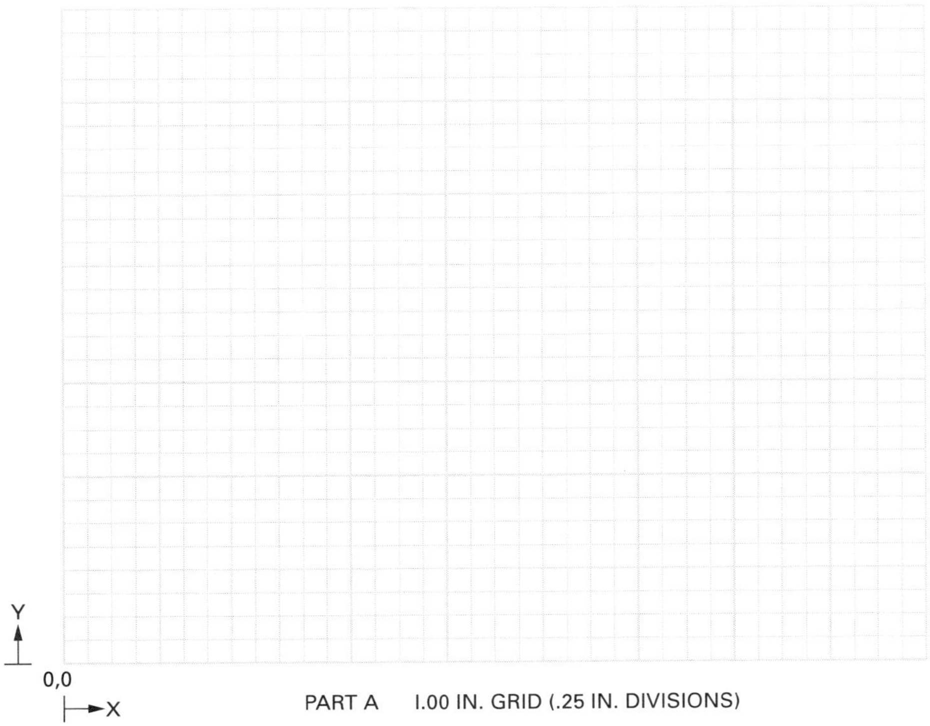

PART A   1.00 IN. GRID (.25 IN. DIVISIONS)

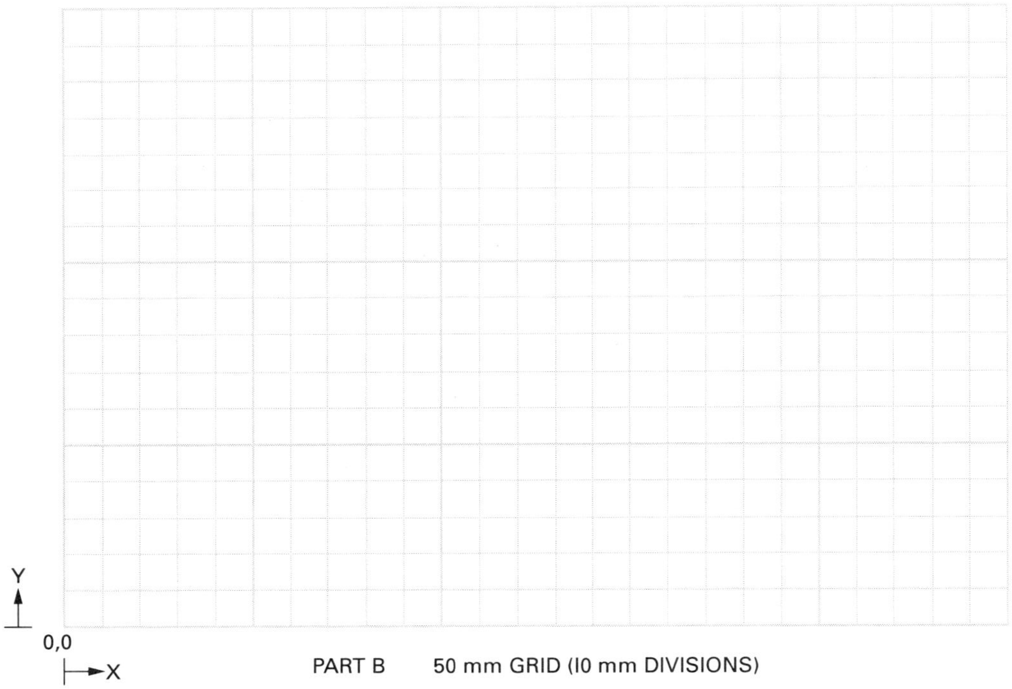

PART B   50 mm GRID (10 mm DIVISIONS)

ABSOLUTE COORDINATES (IN.)

Point	X Axis	Y Axis
1	.25	.25
2	7.00	.25
3	8.50	1.00
4	8.50	2.25
5	7.00	3.00
6	8.50	3.75
7	8.50	5.25
8	7.00	6.25
9	5.50	6.25
10	5.50	4.50
11	4.75	4.50
12	4.75	6.25
13	3.50	6.25
14	3.50	5.50
15	1.50	5.50
16	1.50	6.25
17	.75	6.25
18	.25	5.50
19	.25	.25
	NEW START	
20	.75	1.50
21	2.25	3.50
22	.75	3.50
23	.75	1.50
	NEW START	
24	3.00	2.25
25	5.75	2.25
26	5.75	3.75
27	3.00	3.75
28	3.00	2.25
	NEW START	
29	2.75	.75
30	6.00	.75
31	5.25	1.50
32	3.50	1.50
33	2.75	.75

PART A: Absolute Coordinates.

ABSOLUTE COORDINATES (mm)

Point	X Axis	Y Axis
1	10	10
2	50	10
3	50	20
4	120	20
5	120	10
6	150	10
7	180	30
8	220	30
9	220	100
10	160	100
11	160	130
12	140	130
13	140	160
14	110	160
15	120	140
16	90	140
17	70	100
18	40	120
19	60	160
20	40	160
21	10	140
22	10	80
23	20	40
24	10	40
25	10	10
	NEW START	
26	40	50
27	160	50
28	120	90
29	50	70
30	40	50

PART B: Absolute Coordinates.

Using the coordinates shown, plot Parts A and B. Place the point numbers on the drawings.

ABSOLUTE COORDINATES

REFERENCE UNIT **4–1**

NAME

COURSE

DATE

DRAWING **B–5**

Draw the template shown to the scale 1:1. Do not dimension, except for line M–A. Show length and angle with horizontal start line.

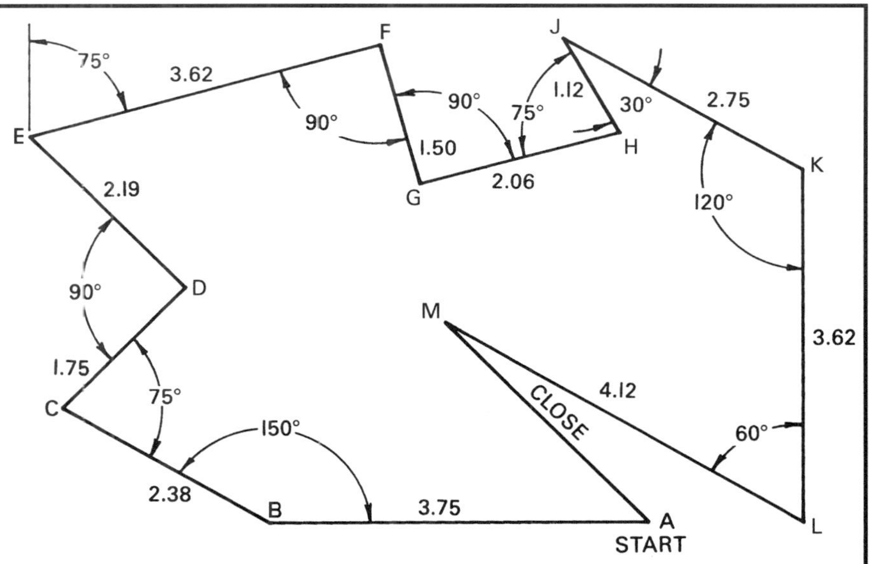

A

POLAR COORDINATES

REFERENCE UNIT **4–1**

NAME

COURSE

DATE

DRAWING **B–6**

1. Fill the space below with lines drawn at 45° to the horizon. Using the 45% triangle, draw parallel lines .10 in. apart.

2. Fill in the space below with lines drawn at 60° with the horizontal. Use the 30-60 triangle and draw parallel lines .10 in. apart.

3. Fill in the space below with .20 in. squares at 15° and 75° to the horizon.

4. Fill the space below with the basket-weave pattern shown.

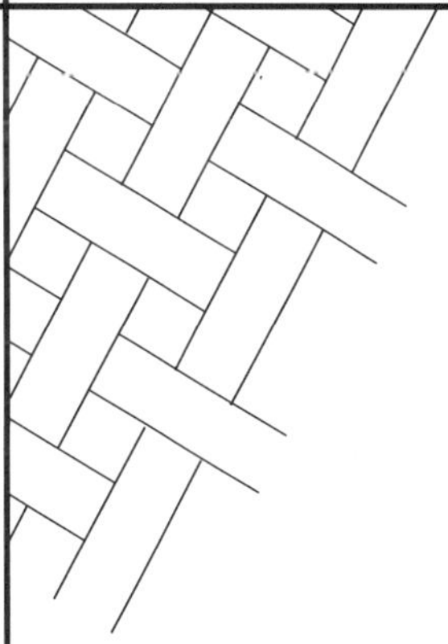

5. Draw rectangles around the rectangle shown below. Each successive rectangle should be 10 mm wider and taller than the preceding one.

6. Draw increasingly larger squares around the square shown below. Each successive square should be rotated 45° and its four sides should touch the corners of the preceding square.

LINEWORK— PATTERNS	REFERENCE UNIT **4–1**	NAME COURSE DATE	DRAWING **B–7**

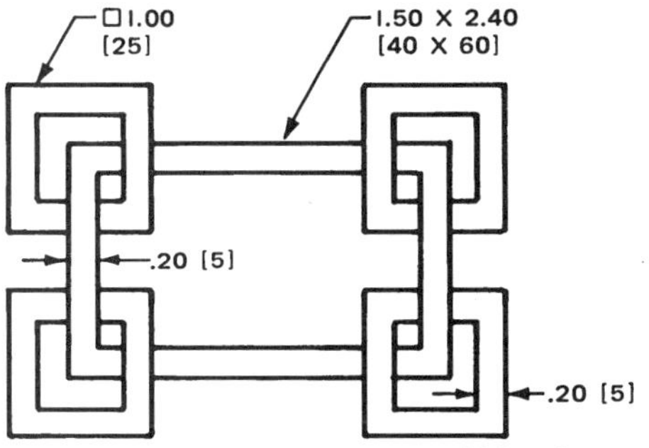

(A)

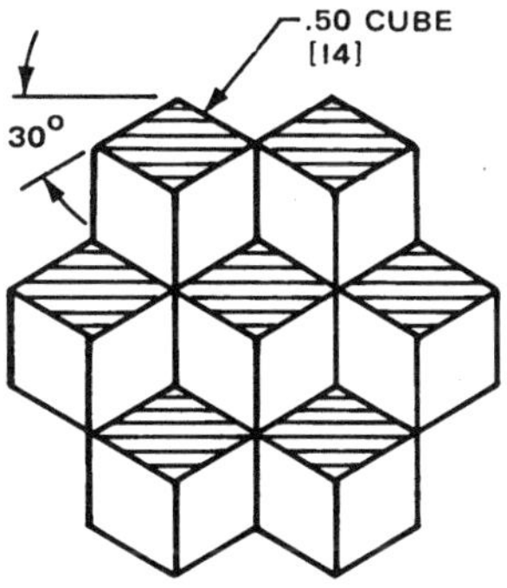

(B)

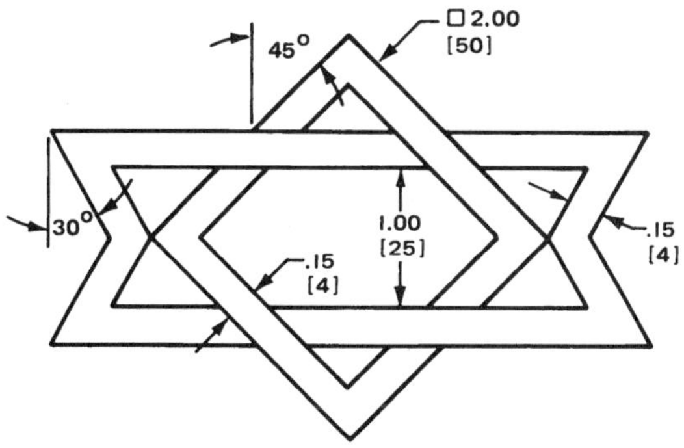

(C)

Draw the inlay designs shown to the scale 1:1. Do not dimension.

LINEWORK— INLAY DESIGNS	REFERENCE UNIT 4–1	NAME COURSE DATE	DRAWING B–8

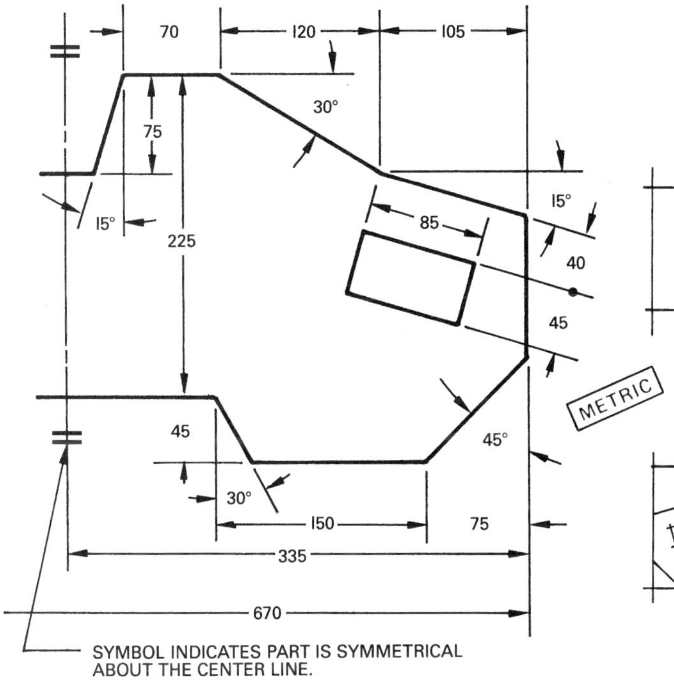

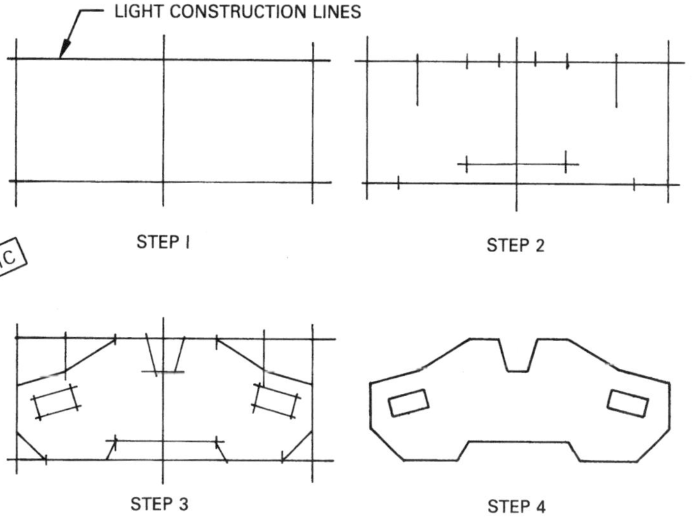

SEQUENCE OF STEPS FOR DRAWING THE SHEARING BLANK

Draw the shearing blank shown above to the scale of 1:5. Do not dimension.

LINEWORK— SHEARING BLANK	REFERENCE UNIT **4–1**	NAME COURSE	DATE	DRAWING **B–9**

IN. [mm]

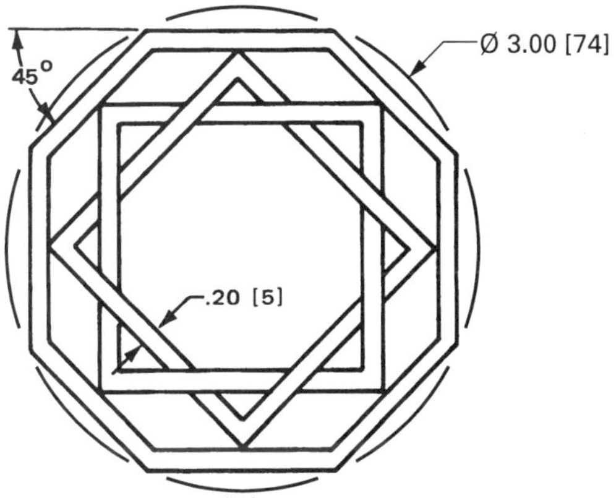

(A)

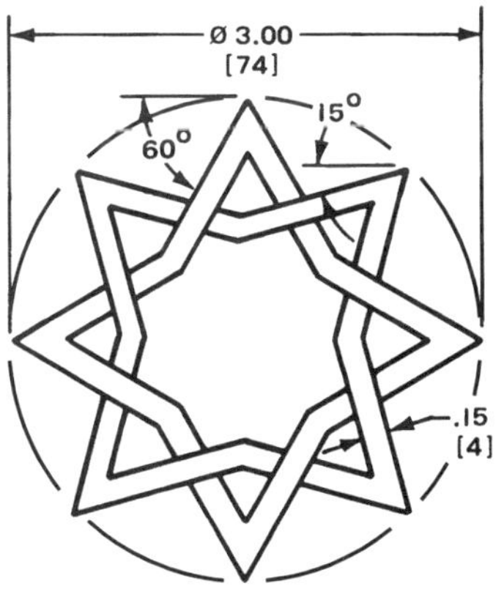

(B)

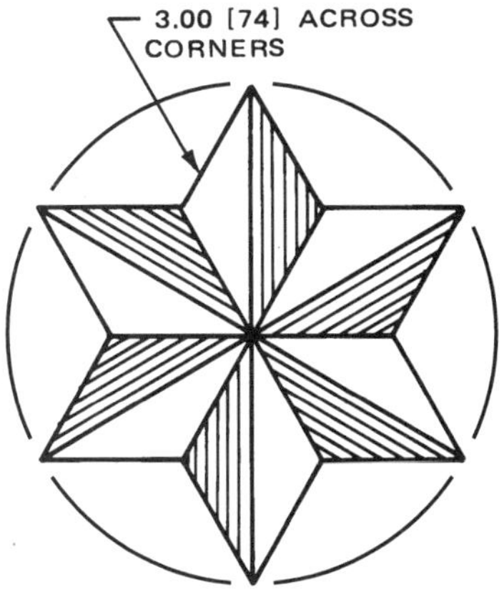

(C)

Draw the designs shown. Do not dimension. The scale 1:1.

CIRCLES AND ARCS— DESIGNS	REFERENCE UNIT **4–2**	NAME COURSE DATE	DRAWING **B–10**

Draw the pawl shown to the scale 1:1. Do not dimension.

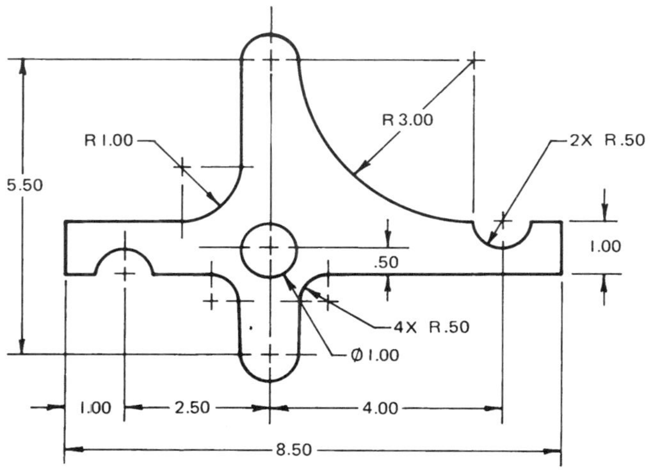

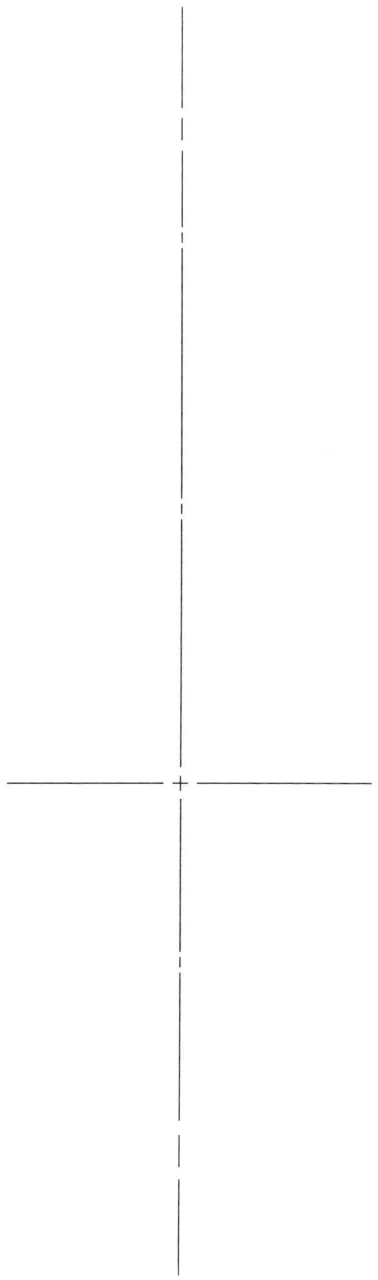

CIRCLES AND ARCS— PAWL	REFERENCE UNIT 4–2	NAME COURSE DATE	DRAWING B–11

Draw the anchor plate shown to the scale 1:1. Do not dimension.

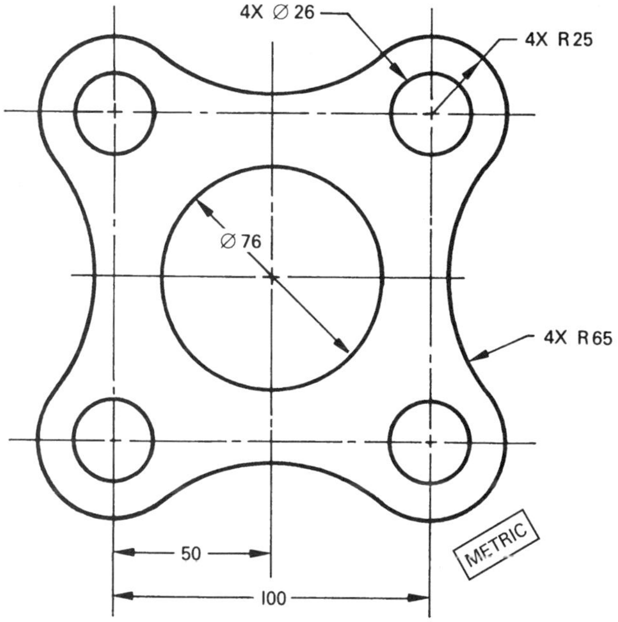

CIRCLES AND ARCS— ANCHOR PLATE	REFERENCE UNIT 4–2	NAME COURSE DATE	DRAWING B–12

Draw the carburetor gasket shown to the scale of 1:1. Do not dimension.

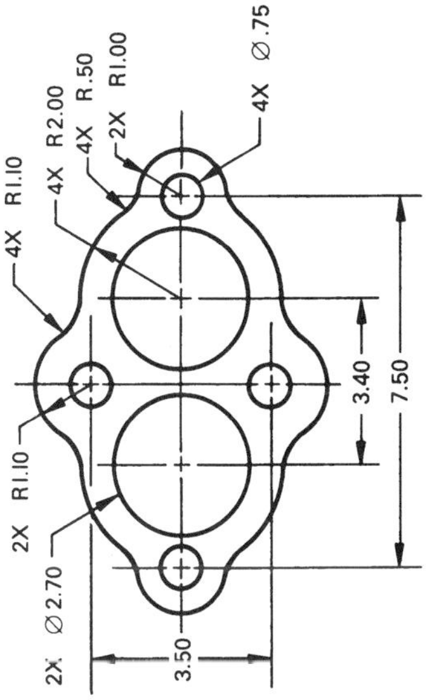

CIRCLES AND ARCS— GASKET NO. 1	REFERENCE UNIT 4–2	NAME COURSE	DATE	DRAWING B–13

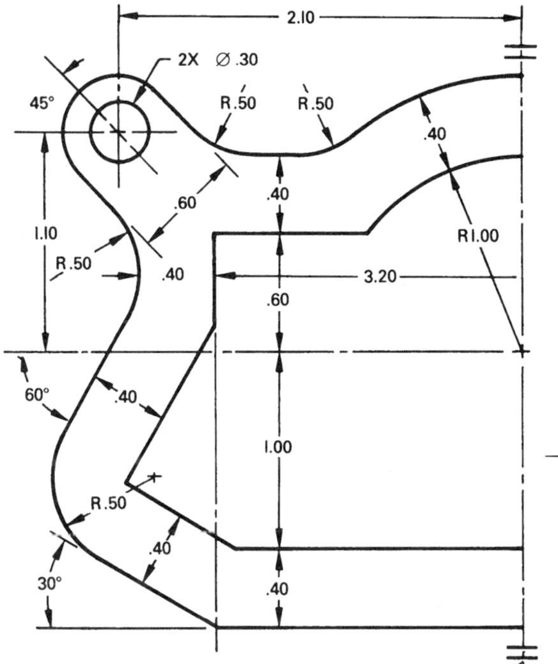

SYMBOL INDICATES SYMMETRICAL
ABOUT THIS CENTER LINE

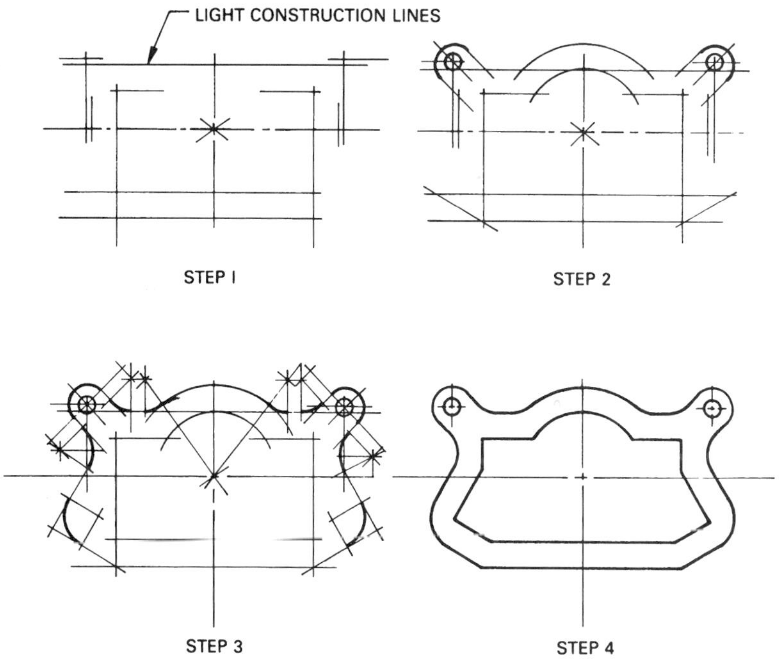

SEQUENCE OF STEPS FOR MAKING DRAWING OF GASKET

Draw the gasket shown to the scale of 1:1. Do not dimension.

CIRCLES AND ARCS— GASKET NO. 2	REFERENCE UNIT **4–2**	NAME COURSE DATE	DRAWING **B–14**

On the grids provided, duplicate the parts shown.
The letters shown on the grids locate the parts.

(A)

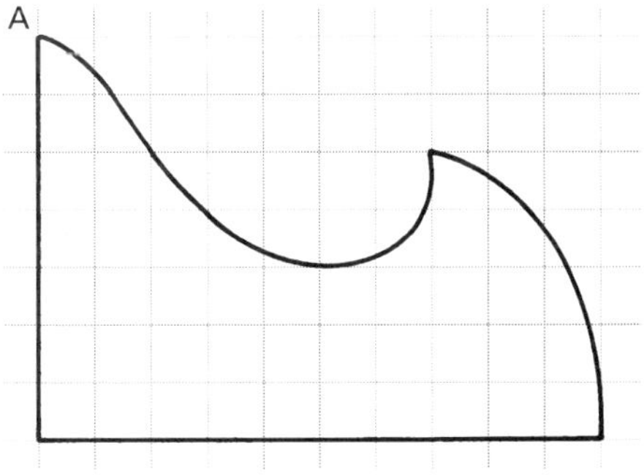

(B)

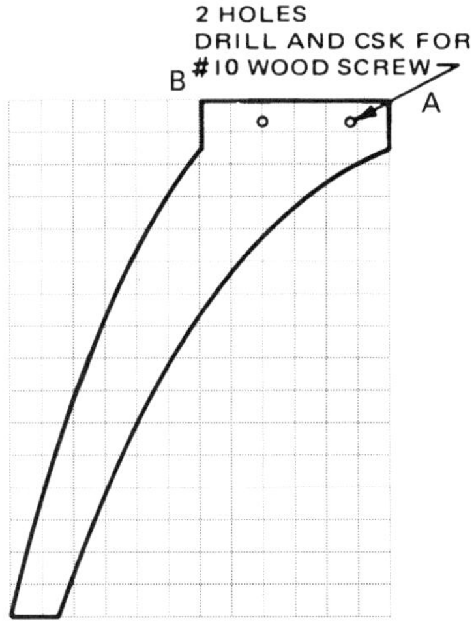

(C)

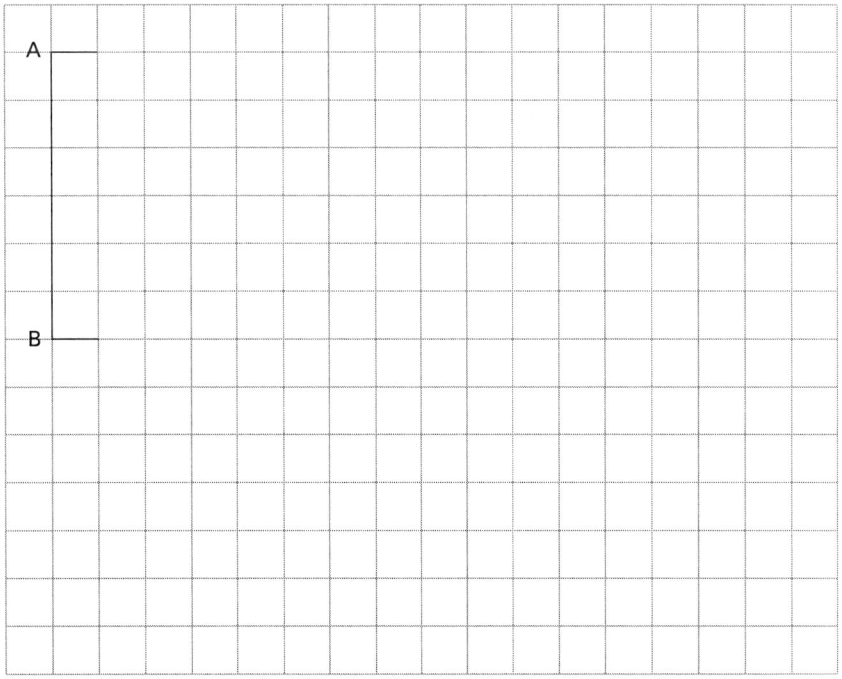

IRREGULAR CURVES— PATTERNS	REFERENCE UNIT **4–3**	NAME ___ COURSE ___ DATE ___	DRAWING **B–15**

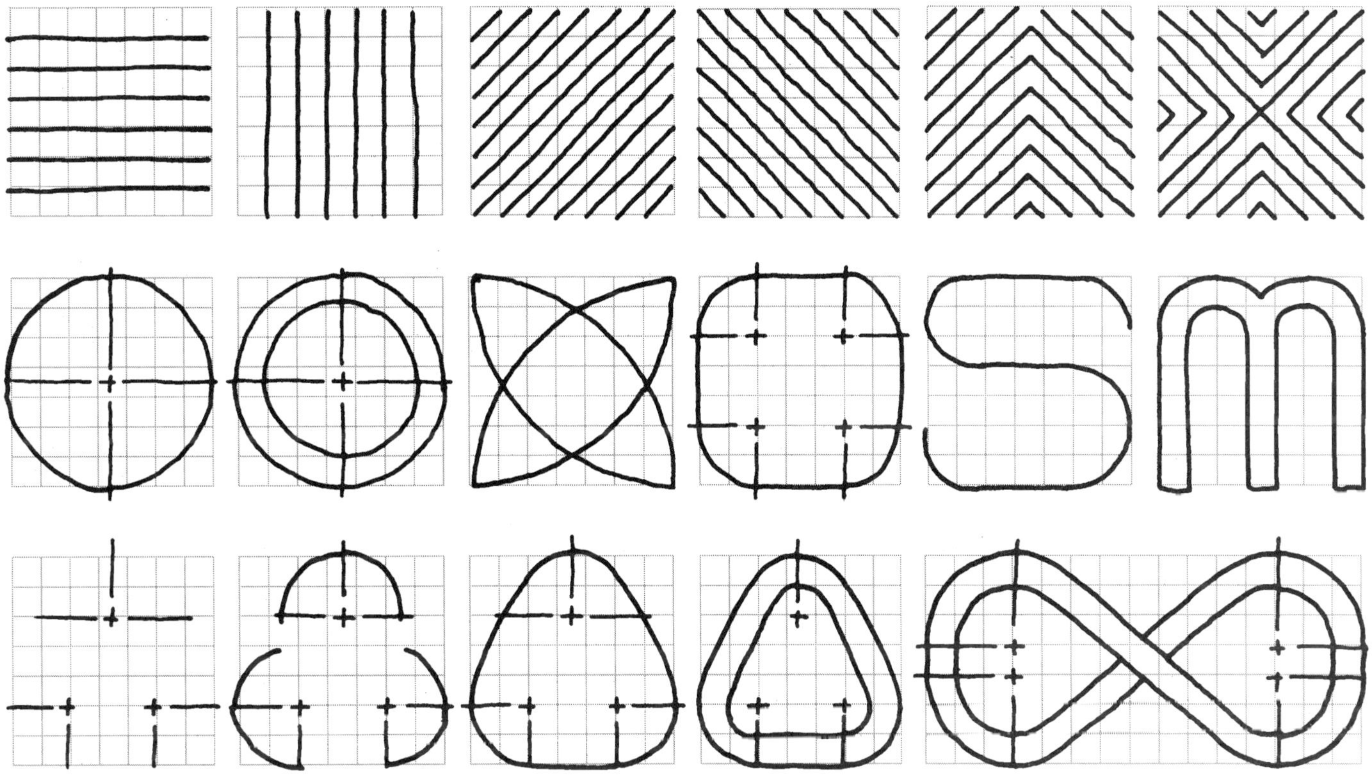

Sketch the patterns shown.

SKETCHING— LINES AND CIRCLES	REFERENCE UNIT 4–4	NAME COURSE DATE	DRAWING B–16

Sketch the structural steel shapes shown to the scale 1:2. Do not dimension. The grids are .50 in.

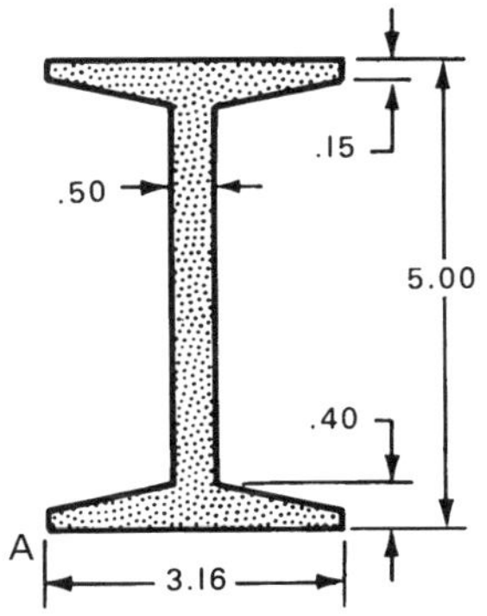

I BEAM

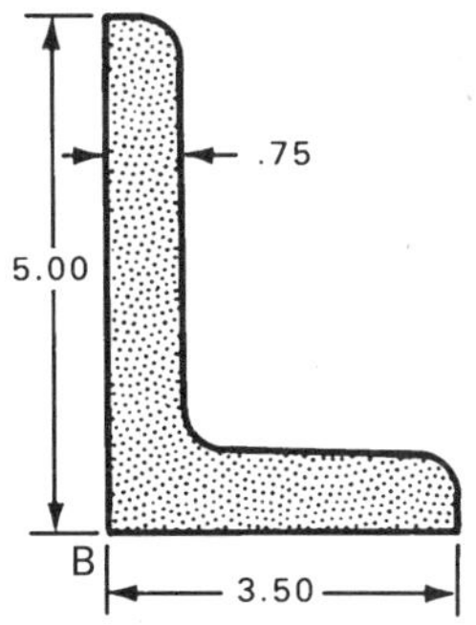

ANGLE

A I BEAM

B ANGLE

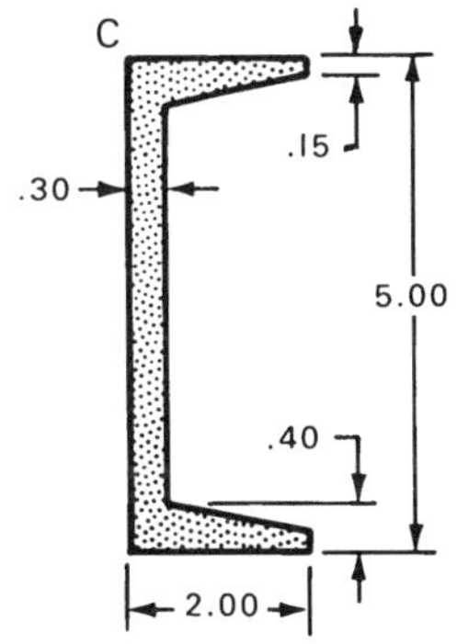

CHANNEL

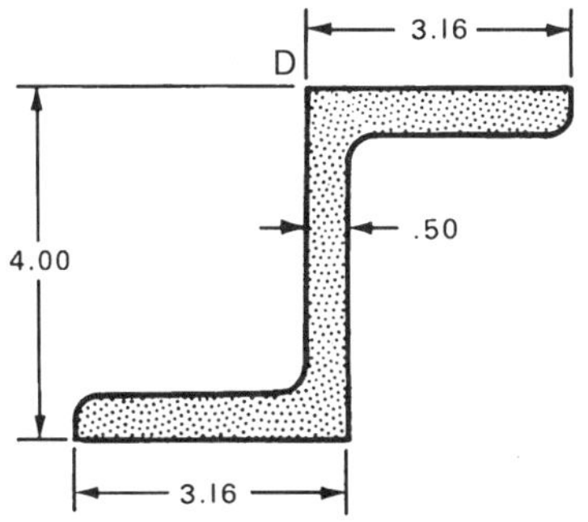

Z-BAR

C

D

CHANNEL

Z-BAR

SKETCHING— STRUCTURAL SHAPES	REFERENCE UNIT **4–4**	NAME_ COURSE_ DATE_	DRAWING **B–17**

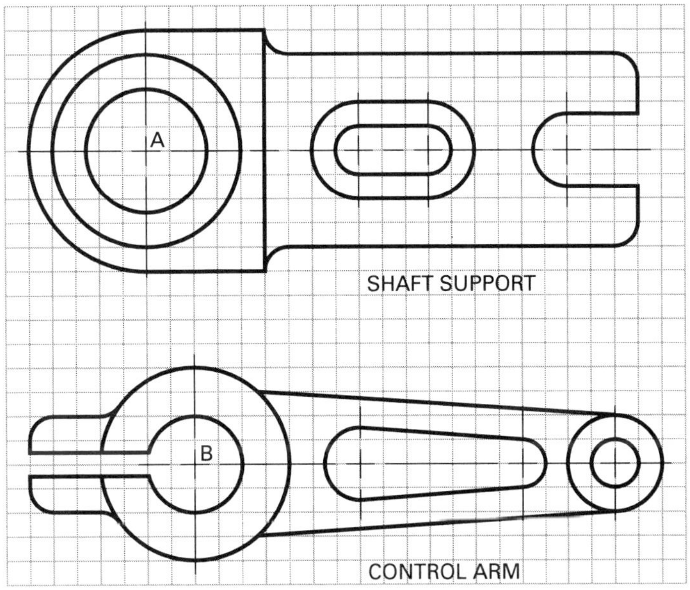

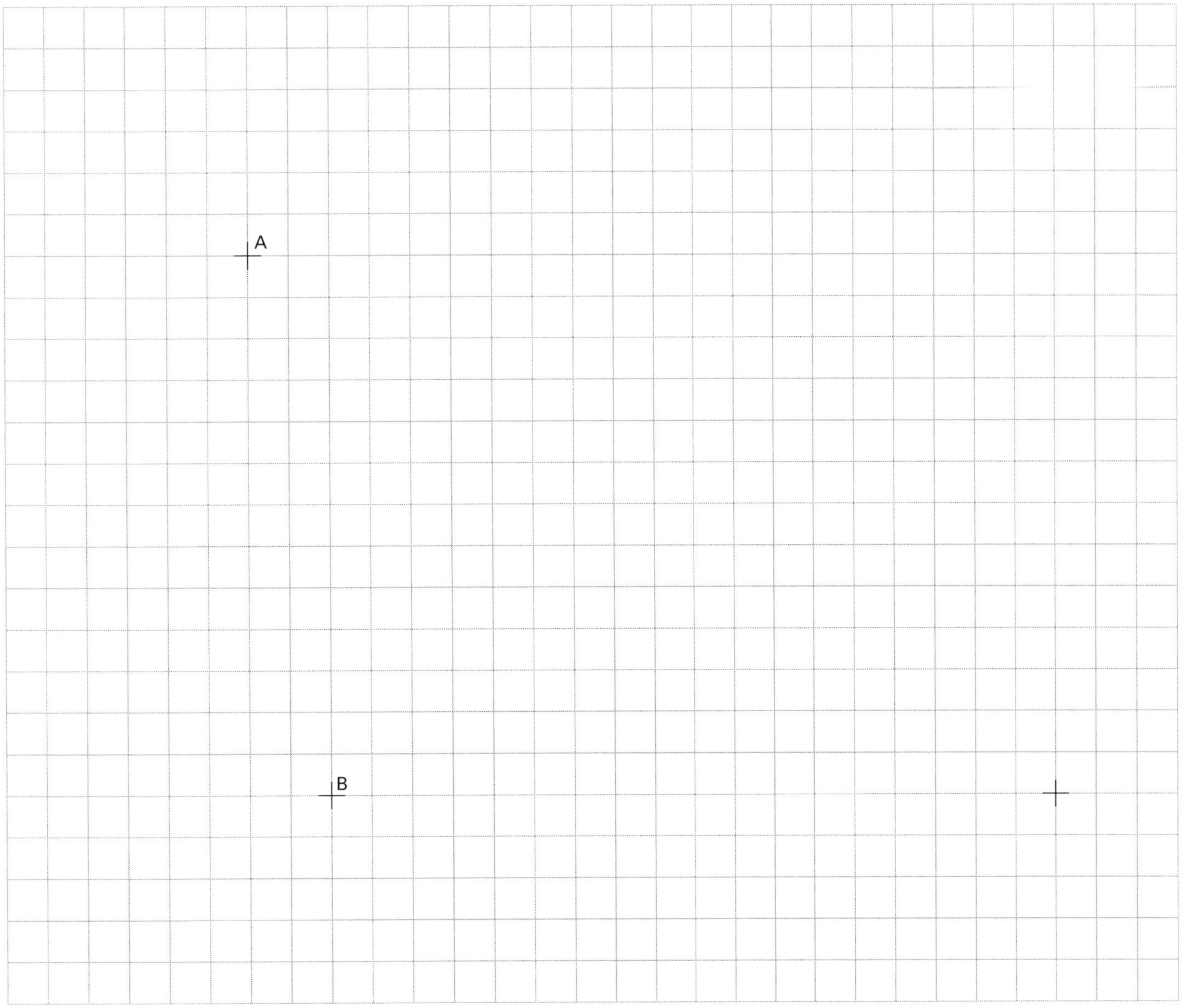

Make freehand sketches of the parts shown. Points A and B are located on the drawing grid. Refer to unit 4–2 for center line application.

SKETCHING— SHAFT SUPPORT AND CONTROL ARM	REFERENCE UNIT **4–4**	NAME COURSE DATE	DRAWING **B–18**

A. Below line A–B, draw five lines equally spaced .10 in. apart, and parallel and equal in length to line A–B. Above line A–B and starting at A, draw .85 in. long lines equally spaced .20 in. apart, and perpendicular to and touching line A–B.

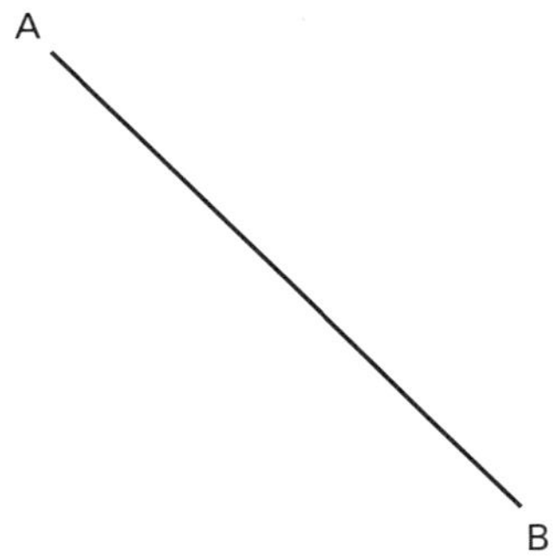

B. Draw straight tangent lines (both sides) to circles C and D and circles D and E.
Circle C = Ø.60
Circle D = Ø.90
Circle E = Ø1.10

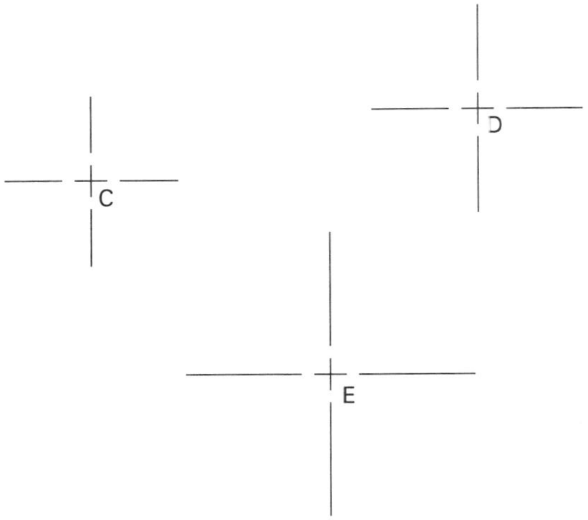

C. Bisect lines F–G, G-H, and H–F.

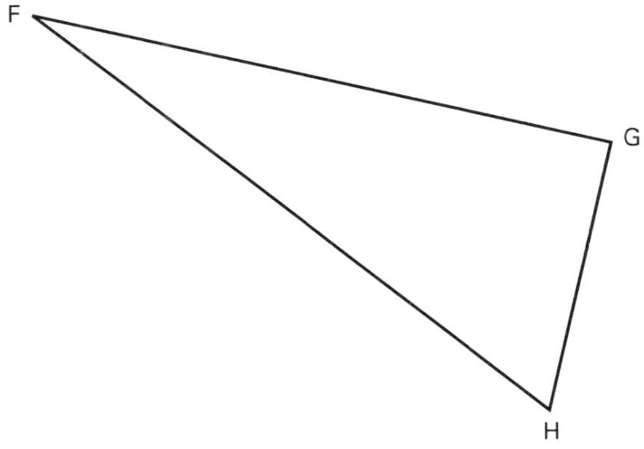

D. Bisect arcs J–K and L–M.

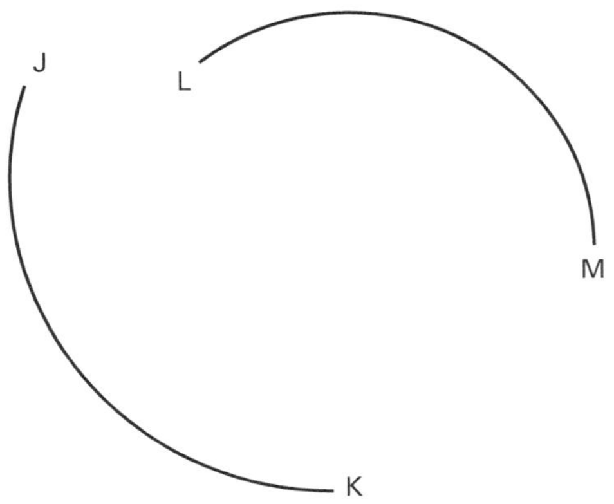

E. Bisect acute angle A–B–C, and obtuse angle D–E–F.

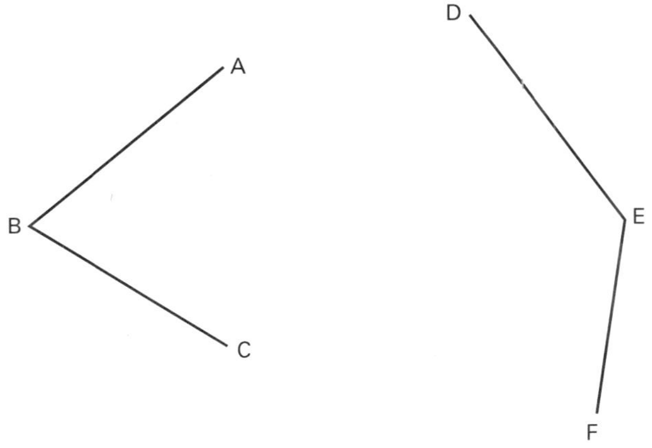

F. Divide line R–S into seven equal parts.
Divide line T–U into five equal parts.

APPLIED GEOMETRY— STRAIGHT LINES	REFERENCE UNIT **5–1**	NAME COURSE DATE	DRAWING **C–1**

A. Draw the following arcs to the lines shown.
R.40 between lines A–B and C–D.
R.70 between lines C–D and E–F.
R.30 between lines E–F and G–H.
R.60 between lines G–H and A–B.

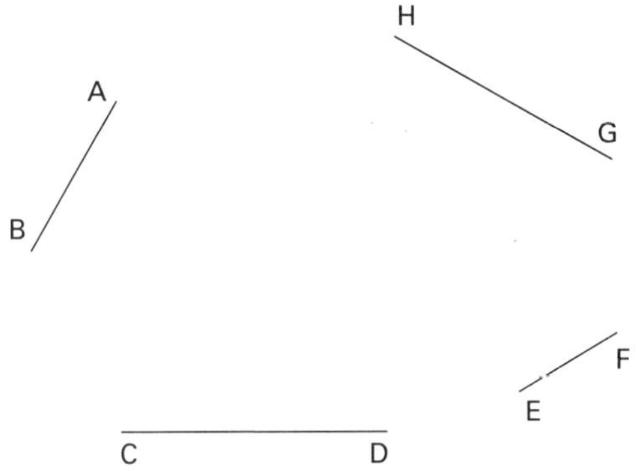

B. Construct a circle about the polygon shown.

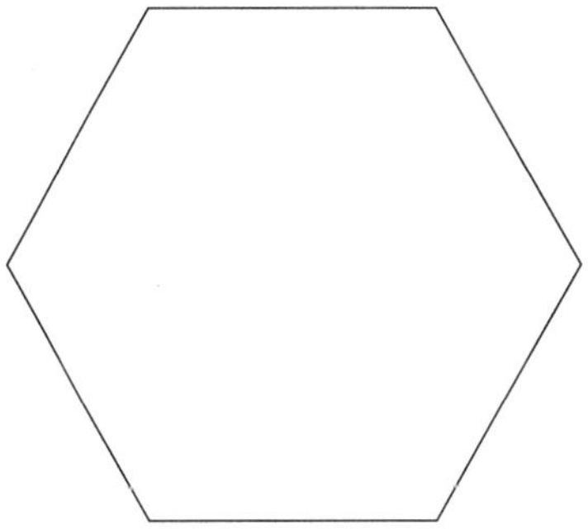

C. Join lines J–K and L–M with an ogee curve. Draw a light construction line between points K and L. The arcs should intersect at a point one-third the length of this line from point L.

D. Draw a Ø1.00 circle using center N. Join line R–S and the circle with a R.60. Join line T–U and the circle with a R.80.

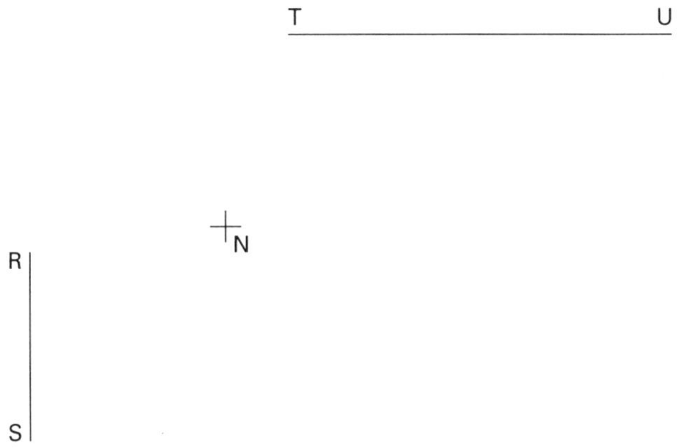

E. Draw a Ø1.50 using center X. Draw a Ø1.20 using center Y. On the top side, join the circles with a R.50. On the bottom side, join the circles with a R.70.

F. Construct an arc through points A, B, and C.

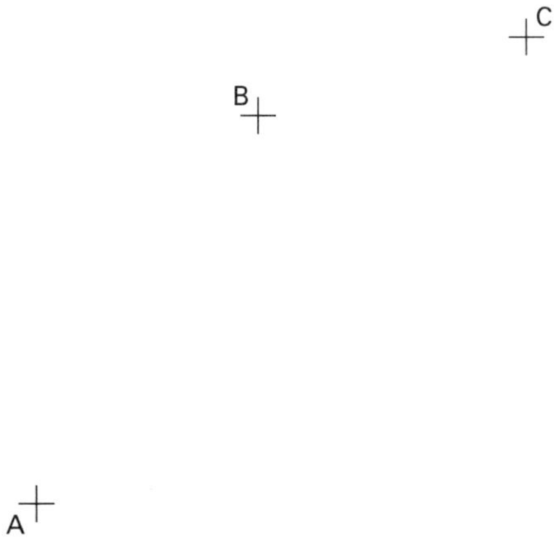

APPLIED GEOMETRY— ARCS AND CIRCLES	REFERENCE UNIT **5–2**	NAME _ COURSE _ DATE _	DRAWING **C–2**

A. Given the center below, construct a hexagon 30 mm across corners. On the same center, construct a hexagon 50 mm across flats.

B. Given the center below, construct an octagon 30 mm across flats. On the same center, construct an octagon 50 mm across corners.

C. Construct an octagon in a 60 mm square.

D. Inscribe a regular pentagon in a 60 mm circle.

E. Given the length of one side of a pentagon (distance A–B), draw the pentagon.

A B

F. Given the length of one side of a seven-sided polygon (distance C–D), draw the polygon.

C D

APPLIED GEOMETRY— POLYGONS	REFERENCE UNIT **5–3**	NAME COURSE DATE	DRAWING **C–3**

TEMPLATE: Draw the template on the grid provided. Do not dimension. The scale is 1:1.

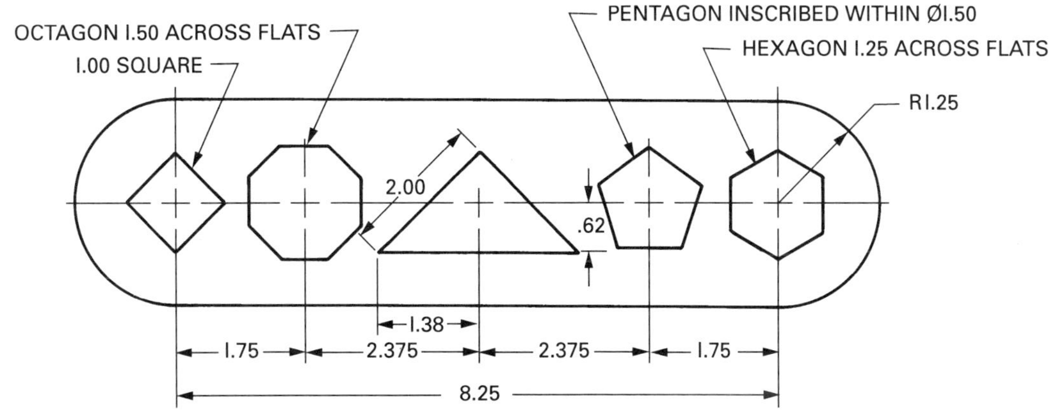

GRID – I.00 IN. (.25 IN. DIVISIONS)

HOOK: Draw the hook (PT4) on the grid provided. Use your judgment for dimensions not given. Do not dimension. The scale is 1:1.

GRID I.00 IN. (.I0 IN. DIVISIONS)

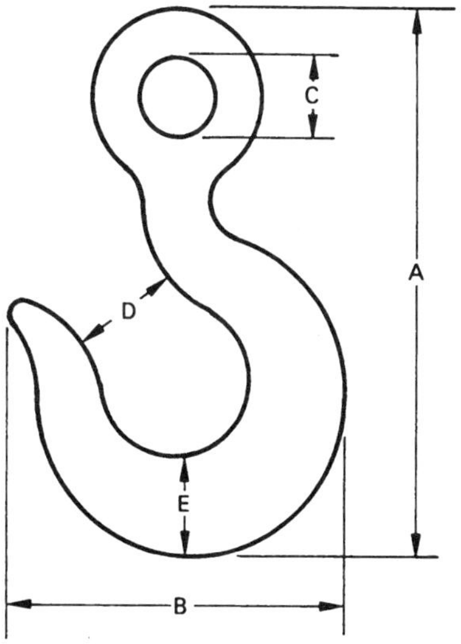

PT	A	B	C	D	E
I	4.90	3.20	.90	I.06	.84
2	5.40	3.50	I.00	I.I0	.90
3	6.25	4.I0	I.I0	I.24	I.I0
4	6.90	4.54	I.24	I.40	I.30

APPLIED GEOMETRY— TEMPLATE AND HOOK	REFERENCE UNIT **5–3**	NAME COURSE DATE	DRAWING **C–4**

A. ELLIPSE—TWO CIRCLE METHOD
Given two circles of Ø40 and Ø60 mm, draw an ellipse.

B. ELLIPSE—FOUR CENTER METHOD
Given a major diameter of Ø60 mm and a minor diameter of Ø30 mm, draw an ellipse. Place the major axis on the vertical center line.

C. ELLIPSE—PARALLELOGRAM METHOD
Given a major diameter of Ø120 mm and a minor diameter of Ø60 mm, draw an ellipse.

APPLIED GEOMETRY— ELLIPSES	REFERENCE UNIT **5–4**	NAME COURSE DATE	DRAWING **C–5**

A. HELIX: Given a cylinder of Ø1.50 × 2.50 in. long, construct a helix with a lead of 2.50 in. starting at point 1. Plot points every 30°.

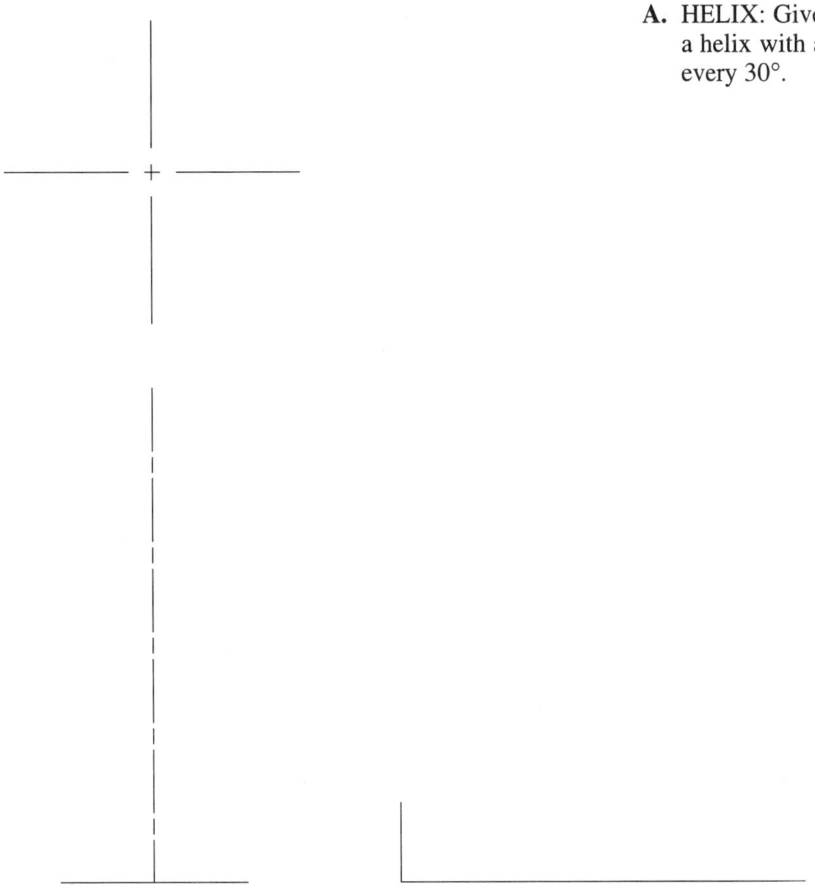

DEVELOPMENT OF CYLINDER

B. PARABOLA: Given the enclosed rectangles, construct parabolas. For the parallelogram method, divide distance O–A into six equal spaces. For the offset method, divide distance O–A into five equal spaces.

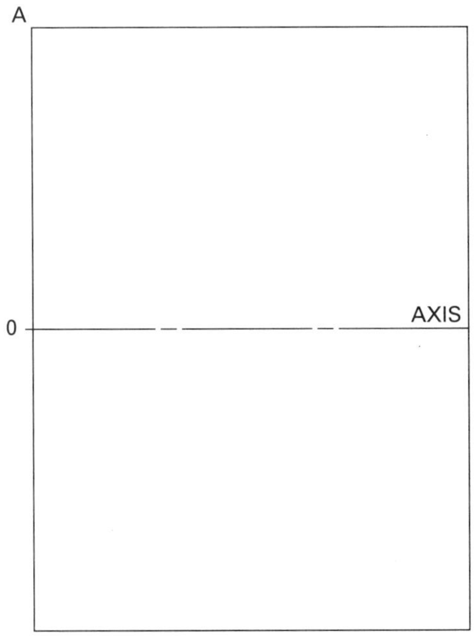

PARALLELOGRAM METHOD

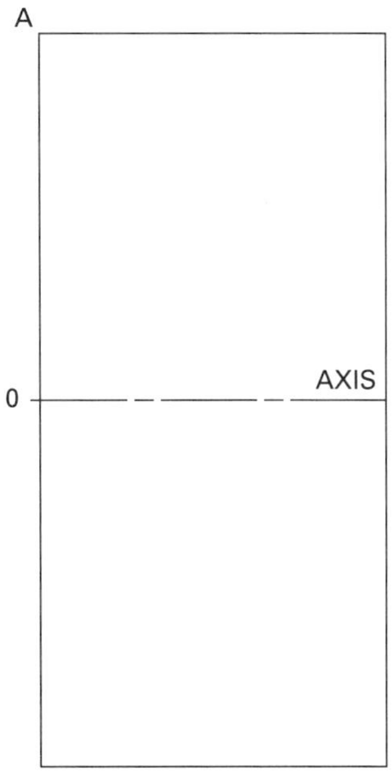

OFFSET METHOD

APPLIED GEOMETRY— HELIX AND PARABOLA	REFERENCE UNIT **5–5**	NAME ______ COURSE ______ DATE ______	DRAWING **C–6**

ORTHOGRAPHIC REPRESENTATION—
ABSOLUTE COORDINATES

REFERENCE UNIT 6–1

NAME

COURSE

DATE

DRAWING D–1

PART A I.00 IN. GRID (.I0 IN. DIVISIONS)

PART B I.00 IN. GRID (.I0 IN. DIVISIONS)

ABSOLUTE COORDINATES (IN.)		
Point	**X Axis**	**Y Axis**
1	0	0
2	3.50	0
3	3.50	1.00
4	2.00	1.00
5	2.00	2.00
6	0	2.00
7	0	0
	NEW START	
8	0	.50
9	3.50	.50
	NEW START	
10	1.50	0
11	1.50	.50
	NEW START	
12	0	2.50
13	1.50	2.50
14	1.50	3.00
15	3.50	3.00
16	3.50	4.50
17	0	4.50
18	0	2.50
	NEW START	
19	0	3.50
20	3.50	3.50
	NEW START	
21	2.00	3.50
22	2.00	4.50
	NEW START	
23	4.00	0
24	6.00	0
25	6.00	2.00
26	5.00	2.00
27	5.00	.50
28	4.00	.50
29	4.00	0
	NEW START	
30	5.00	1.00
31	6.00	1.00
	NEW START	
32	4.50	0
33	4.50	.50

Part A: Absolute coordinate (inches)

ABSOLUTE COORDINATES (IN.)		
Point	**X Axis**	**Y Axis**
1	0	0
2	2.50	0
3	2.50	.50
4	0	.50
	NEW START	
5	2.00	.50
6	2.00	1.00
7	0	1.00
	NEW START	
8	1.50	1.00
9	1.50	1.50
10	0	1.50
11	0	0
	NEW START	
12	0	2.00
13	2.50	2.00
14	2.50	3.50
15	0	3.50
16	0	2.00
	NEW START	
17	0	2.50
18	2.00	2.50
19	2.00	3.50
	NEW START	
20	0	3.00
21	1.50	3.00
22	1.50	3.50
	NEW START	
23	4.50	.50
24	3.00	.50
25	3.00	0
26	4.50	0
27	4.50	1.50
28	4.00	1.50
29	4.00	1.00
	NEW START	
30	4.50	1.00
31	3.50	1.00
32	3.50	.50

Part B: Absolute coordinate (inches)

Using the coordinates shown, plot Parts A and B. Place the point numbers on the drawing.

PART A I.00 IN. GRID (.I0 IN. DIVISIONS)

PART B I.00 IN. GRID (.I0 IN. DIVISIONS)

Using the coordinates shown, plot parts A and B. Place the point numbers on the drawings.

RELATIVE COORDINATES (IN.)		
Point	**X Axis**	**Y Axis**
1	0	0
2	3.00	0
3	0	.50
4	–2.00	0
5	0	1.50
6	–1.00	0
7	0	–2.00
	NEW START	
8	0	1.50
9	1.00	0
	NEW START	
10	1.50	0
11	0	.50
	NEW START	
12	0	2.50
13	1.50	0
14	0	.50
15	1.50	0
16	0	1.50
17	–3.00	0
18	0	–2.00
	NEW START	
19	1.00	2.50
20	0	2.00
	NEW START	
21	0	3.50
22	1.00	0
	NEW START	
23	3.50	0
24	2.00	0
25	0	2.00
26	–1.00	0
27	0	–.50
28	–1.00	0
29	0	–1.50
	NEW START	
30	3.50	.50
31	2.00	0
	NEW START	
32	4.00	0
33	0	.50

Part A: Relative coordinate (inches)

RELATIVE COORDINATES (IN.)		
Point	**X Axis**	**Y Axis**
1	0	0
2	3.00	0
3	0	1.00
4	–1.00	0
5	0	.50
6	–2.00	0
7	0	–1.50
	NEW START	
8	1.50	0
9	0	.50
10	–1.00	0
11	0	1.00
	NEW START	
12	0	2.00
13	0	1.50
14	3.00	0
15	0	–.50
16	–2.50	0
17	0	–1.00
	NEW START	
18	0	2.00
19	1.50	0
20	0	1.00
	NEW START	
21	2.00	3.00
22	0	.50
	NEW START	
23	3.50	0
24	1.50	0
25	0	1.50
26	–1.50	0
27	0	–1.50
	NEW START	
28	4.50	0
29	0	1.50
	NEW START	
30	3.50	.50
31	1.00	0
	NEW START	
32	4.50	1.00
33	.50	0

Part B: Relative coordinate (inches)

ORTHOGRAPHIC REPRESENTATION—
RELATIVE COORDINATES

REFERENCE UNIT **6–1**

NAME

COURSE

DATE

DRAWING **D–2**

ORTHOGRAPHIC REPRESENTATION—

TWO-DIMENSIONAL COORDINATES

REFERENCE UNIT **6–1**

NAME

COURSE

DATE

DRAWING **D–3**

PART A 10 mm GRID (2 mm DIVISIONS)

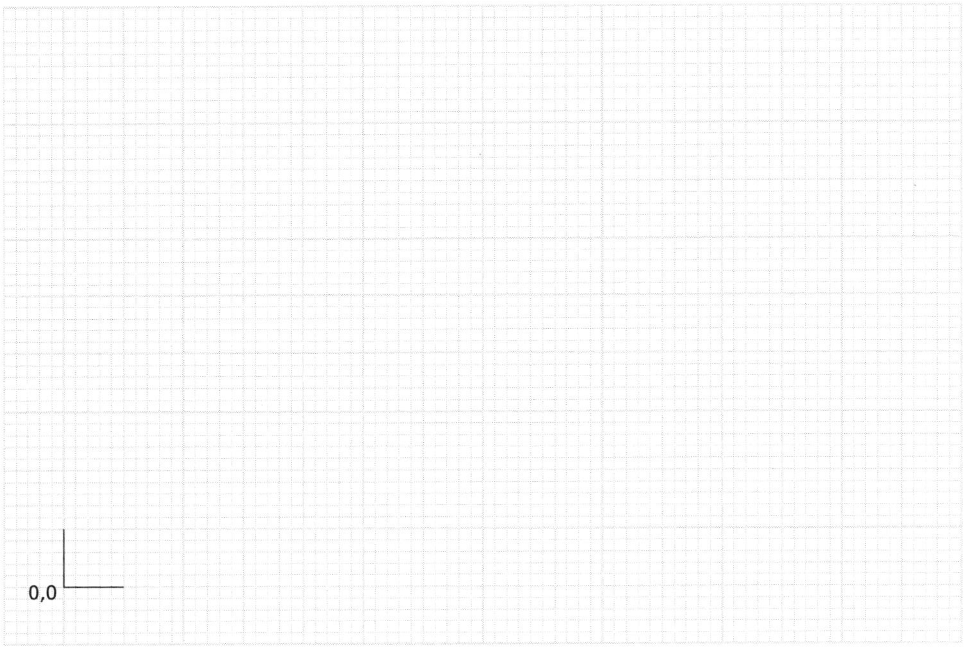

PART B 10 mm GRID (2 mm DIVISIONS)

Using the coordinates shown, plot parts A and B. Place the point numbers on the drawings.

RELATIVE COORDINATES (mm)		
Point	X Axis	Y Axis
1	0	0
2	70	0
3	0	20
4	–70	0
NEW START		
5	0	2
6	0	30
7	30	0
8	0	–10
NEW START		
9	20	30
10	0	–10
NEW START		
11	40	20
12	0	–20
NEW START		
13	0	40
14	40	0
15	0	20
16	30	0
17	0	20
18	–70	0
19	0	–40
NEW START		
20	0	60
21	20	0
22	0	10
23	10	0
24	0	10
NEW START		
25	120	20
26	–40	0
27	0	–20
28	40	0
29	0	30
30	–20	0
31	0	–30
NEW START		
32	110	20
33	0	10

Part A: Relative coordinate (metric)

ABSOLUTE COORDINATES (mm)		
Point	X Axis	Y Axis
1	0	0
2	90	0
3	90	10
4	0	10
NEW START		
5	0	0
6	0	40
7	50	40
8	50	10
NEW START		
9	70	0
10	70	10
NEW START		
11	0	50
12	70	50
13	70	70
14	90	70
15	90	90
16	0	90
17	0	50
NEW START		
18	0	80
19	50	80
20	50	90
NEW START		
21	140	10
22	100	10
23	100	0
24	140	0
25	140	40
26	130	40
27	130	10
NEW START		
28	120	0
29	120	10

Part B: Absolute coordinate (metric)

ORTHOGRAPHIC REPRESENTATION—

THREE-DIMENSIONAL COORDINATES

REFERENCE UNIT	NAME	DRAWING
6–1	COURSE / DATE	D–4

A.

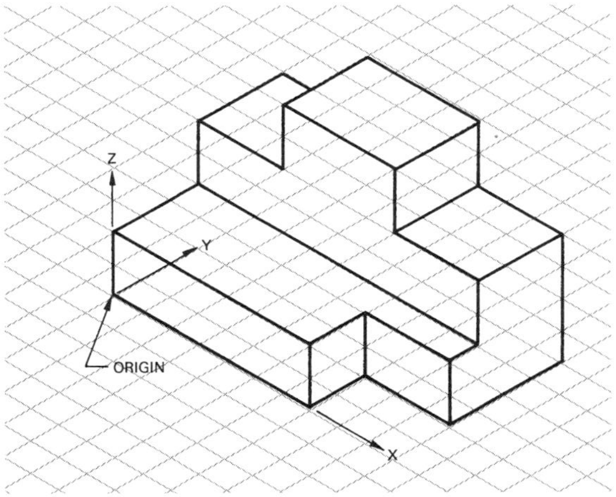

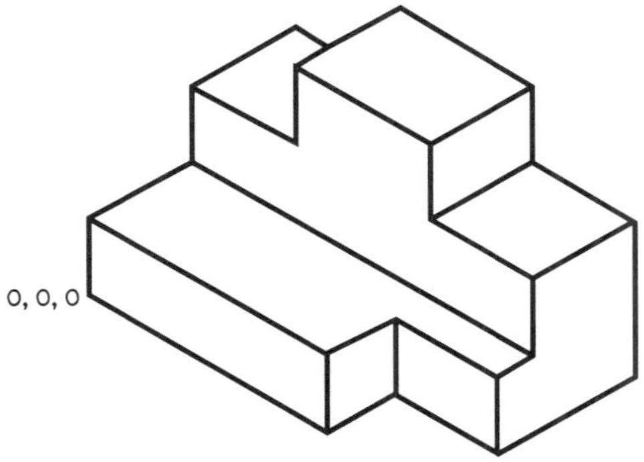

Decimal inch coordinates: .50 in. grid.

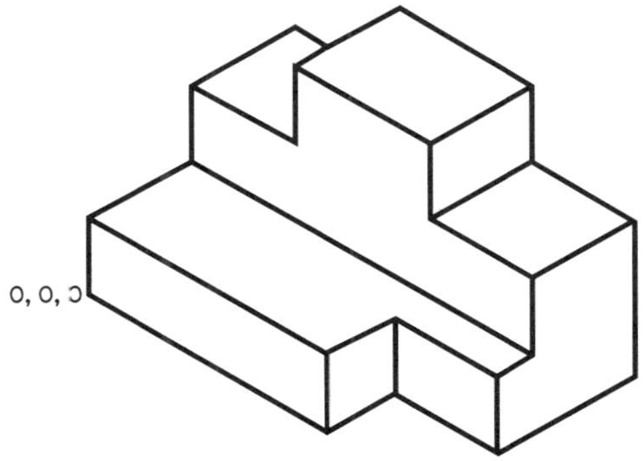

Millimeter coordinates: 10 mm grid.

B.

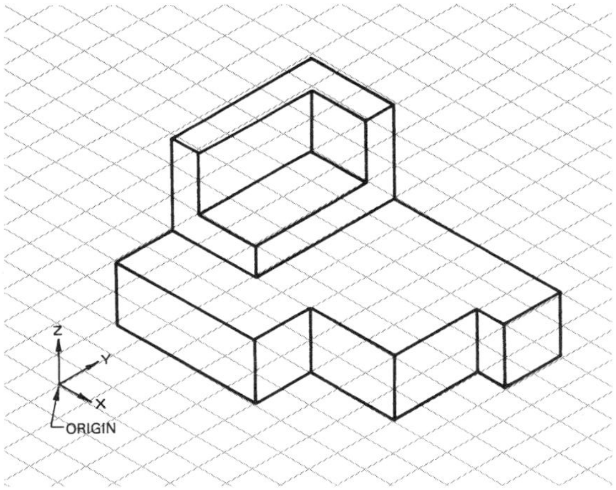

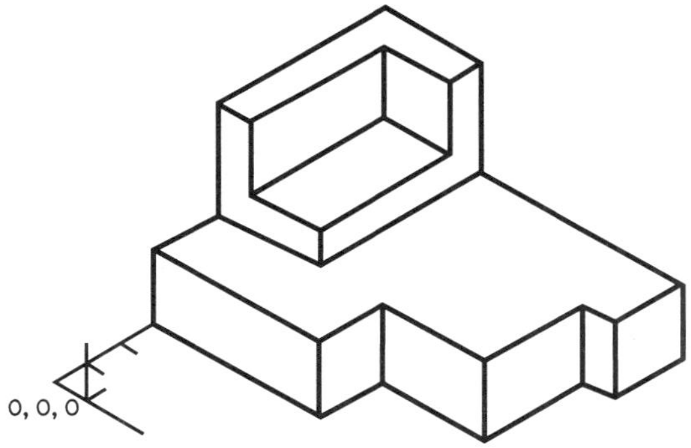

Decimal inch coordinates: .50 in. grid.

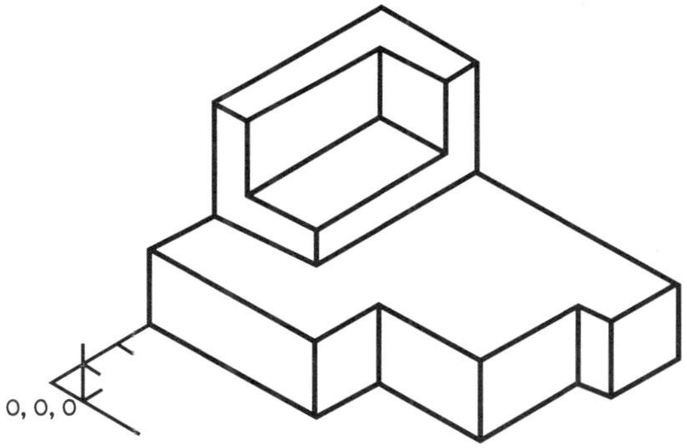

Millimeter coordinates: 10 mm grid.

Add the *X*, *Y*, and *Z* coordinates where the lines intersect one another. Identify only those intersections that can be seen. Note the location of the origin for each part.

Draw the six views of the corner bracket using the following methods of representation: (A) third-angle projection, (B) first-angle projection, (C) reference arrow layout. The principal view is shown. Do not attempt to show any hidden features. The grid is 10 mm. Identify the views as shown in Figs. 6–1–3 through 6–1–5.

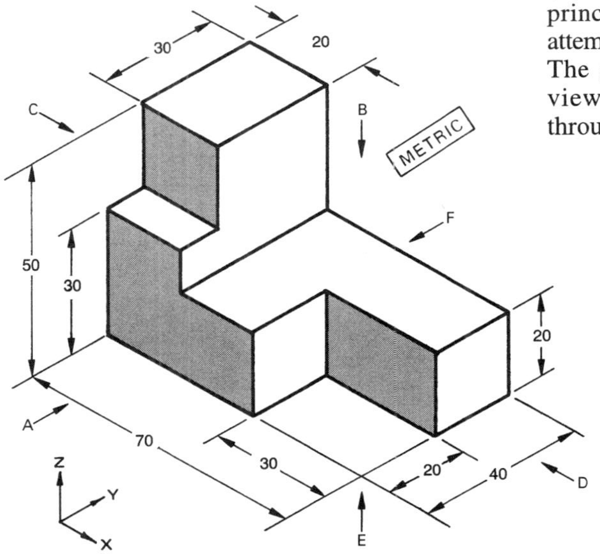

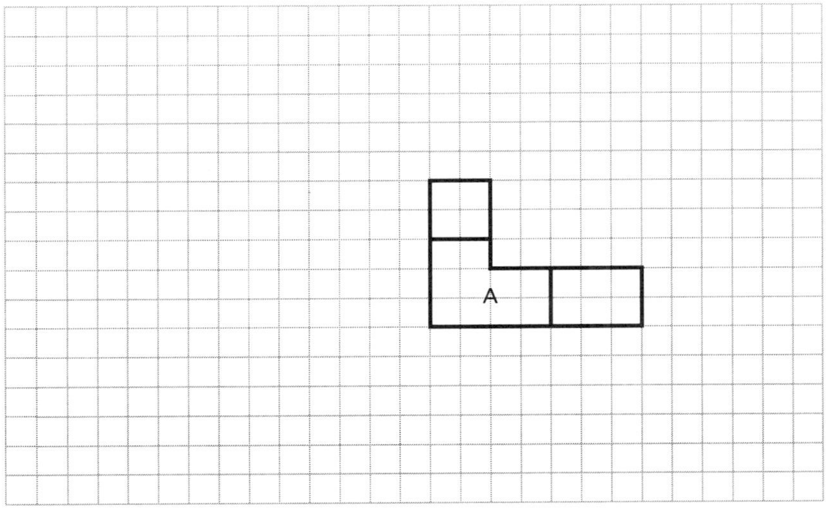

A. THIRD-ANGLE PROJECTION:
Allow 1 grid space between views.

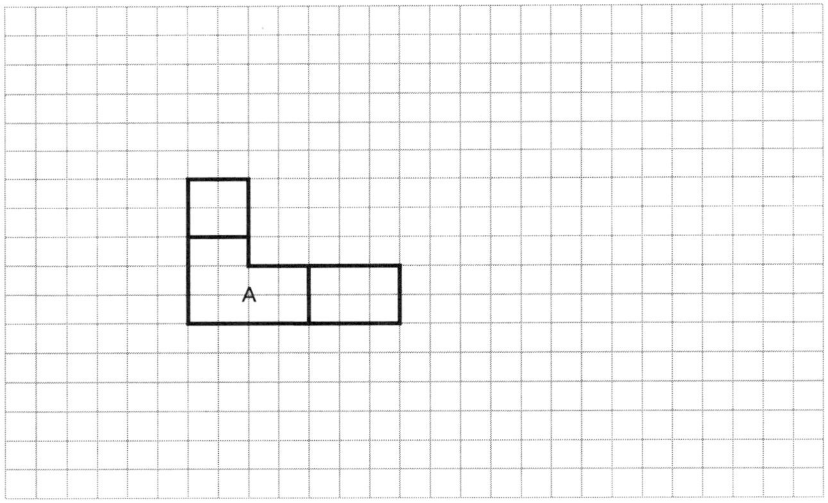

B. FIRST-ANGLE PROJECTION:
Allow 1 grid space between views.

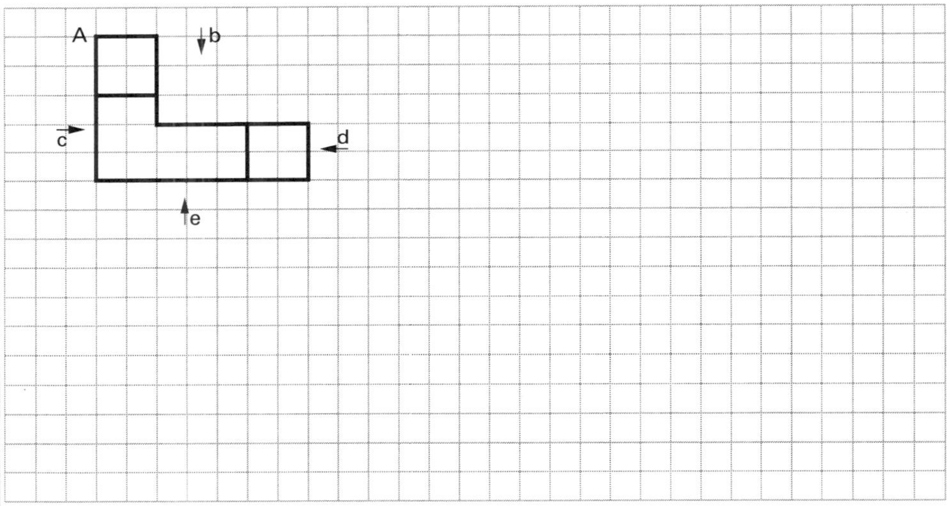

C. REFERENCE-ARROW LAYOUT:
Allow 3 grid spaces between views.

ORTHOGRAPHIC REPRESENTATION—
CORNER BRACKET

REFERENCE UNIT **6–1**

NAME

COURSE

DATE

DRAWING **D–5**

Draw the six views of the location block using the following methods of representation: (A) third-angle projection, (B) first-angle projection, (C) reference arrow layout. The principal view is shown. Do not attempt to show any hidden features. The grid is .25 in. Identify the views as shown in Figs. 6–1–3 through 6–1–5.

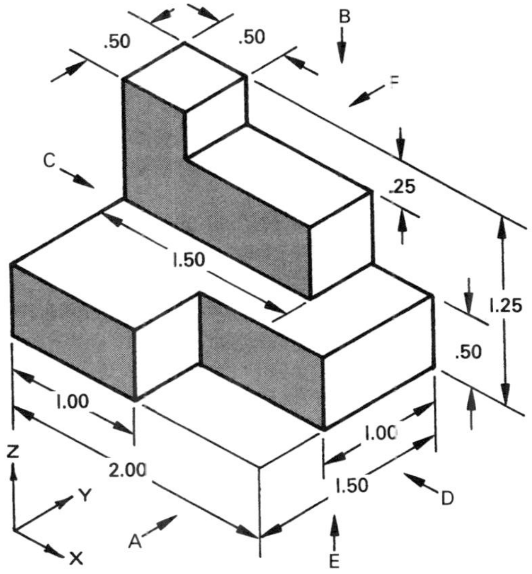

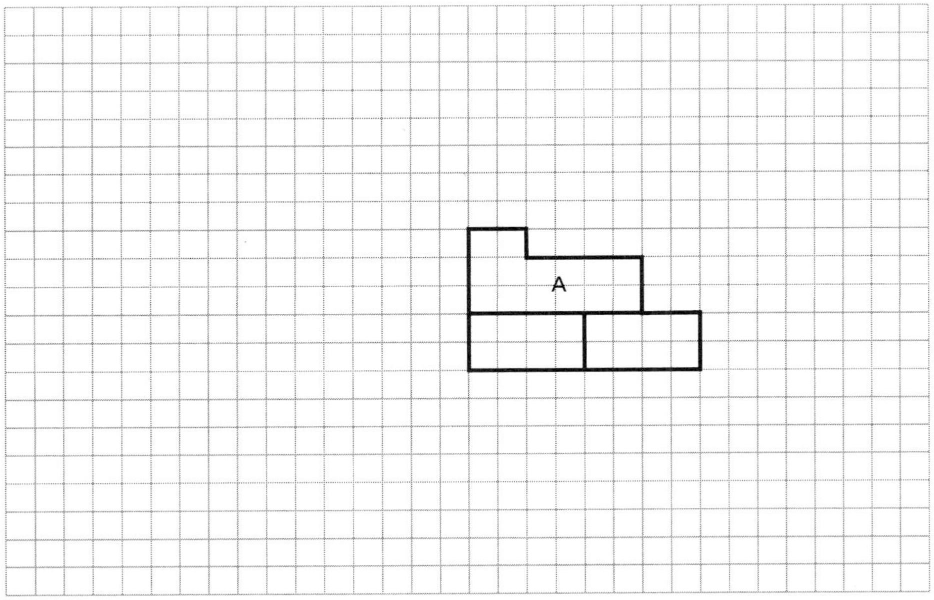

A. THIRD-ANGLE PROJECTION: Allow 1 grid space between views.

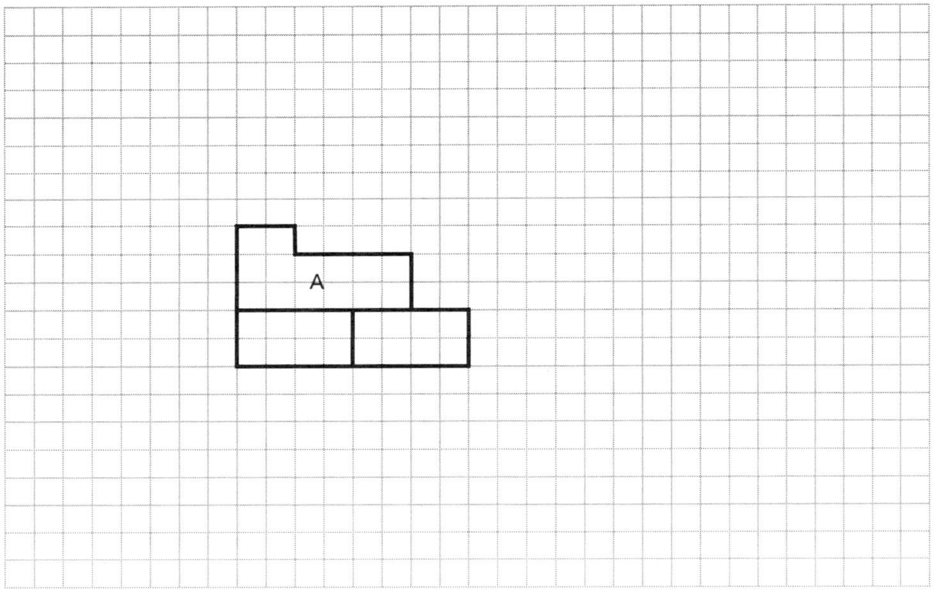

B. FIRST-ANGLE PROJECTION: Allow 1 grid space between views.

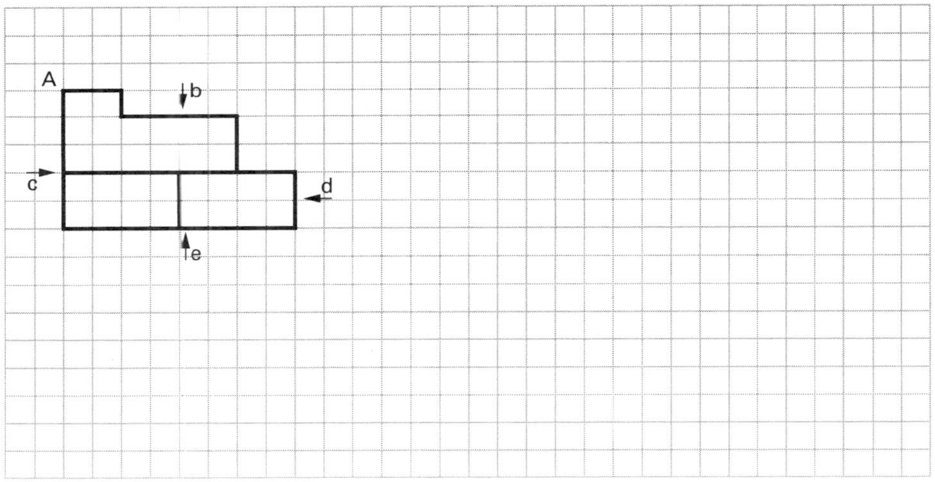

C. REFERENCE-ARROW LAYOUT: Allow 3 grid spaces between views.

ORTHOGRAPHIC REPRESENTATION—

LOCATION BLOCK

REFERENCE UNIT **6–1**

NAME

COURSE

DATE

DRAWING **D–6**

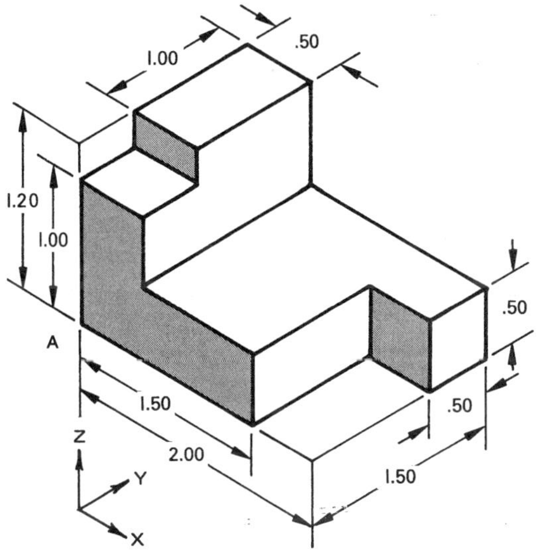

ANGLE BRACKET

A

From the information shown in the figure, draw the front and top views using third-angle projection. Point A locates the position of the front view. Allow .50 in. between views. Using a miter line, construct the right side view. The scale is 1:1.

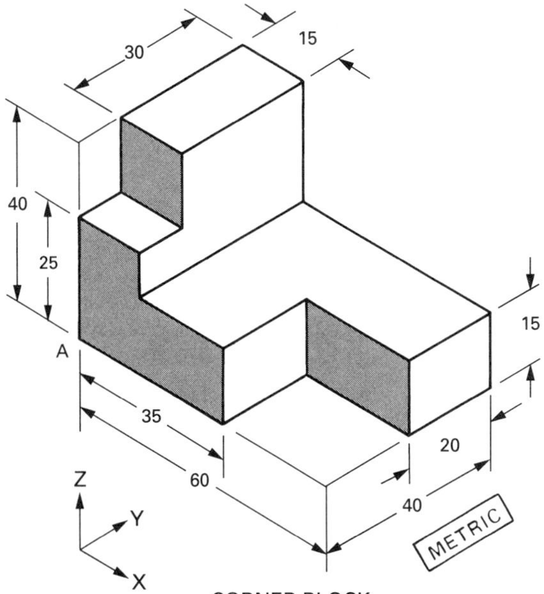

CORNER BLOCK

A

From the information shown in the figure, draw the front and right-side views using third-angle projection. Point A locates the position of the front view. Allow 10 mm between views. Using a miter line, construct the top view. The scale is 1:1.

ORTHOGRAPHIC REPRESENTATION— MITER LINES	REFERENCE UNIT **6–2**	NAME COURSE DATE	DRAWING **D–7**

1

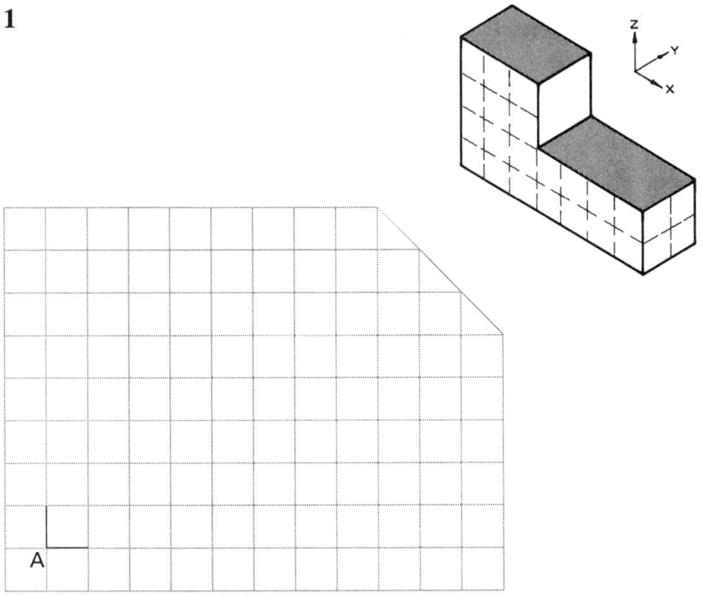

4

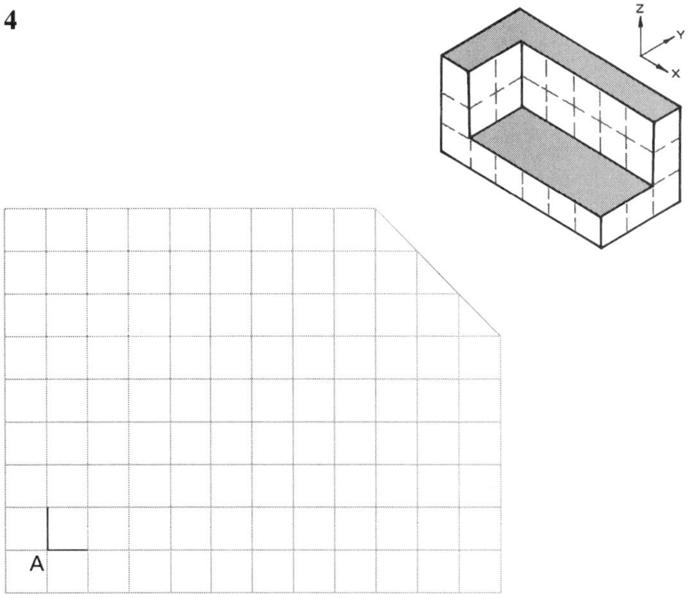

2

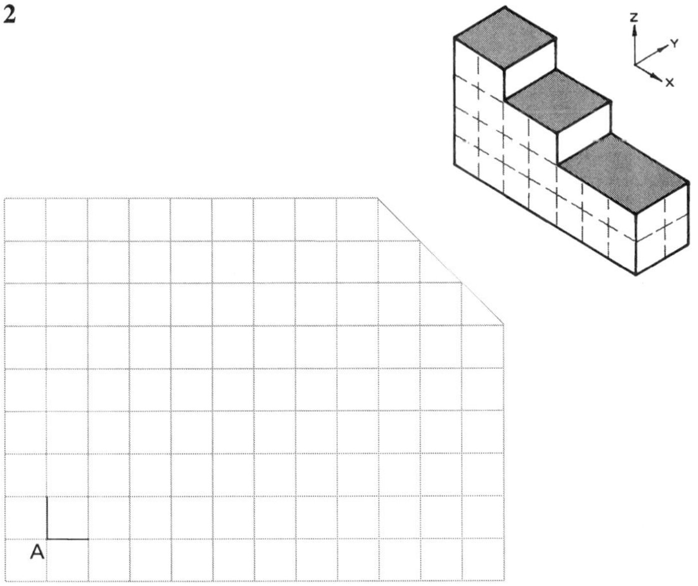

5

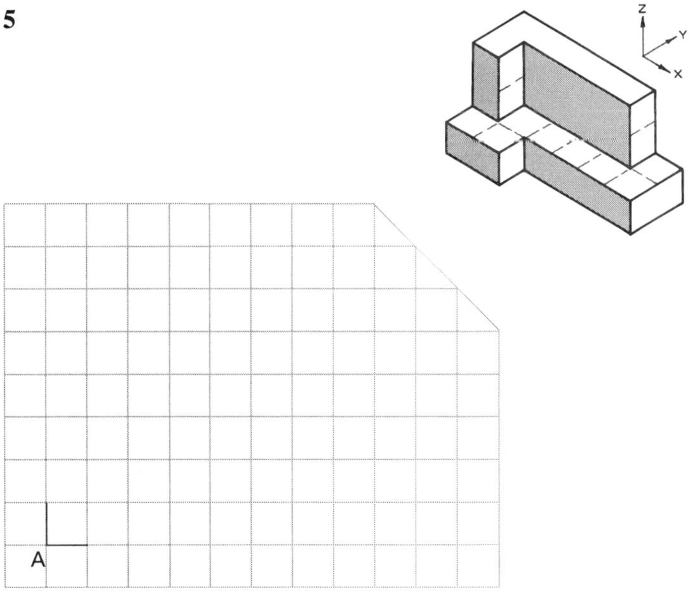

3

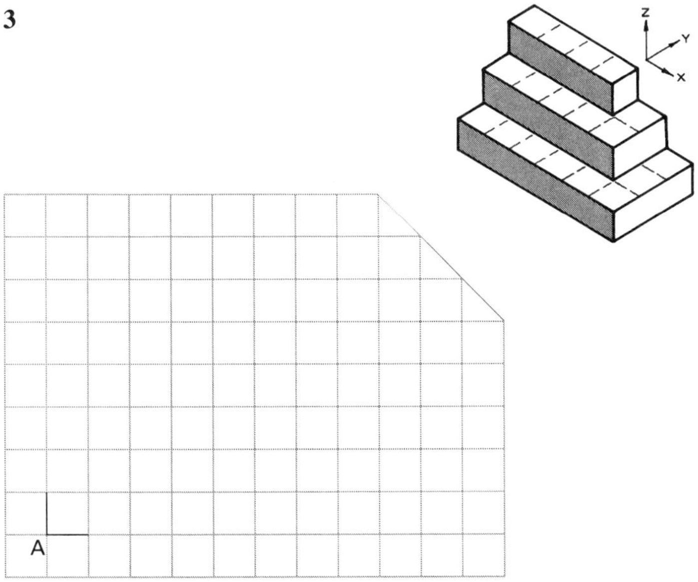

6

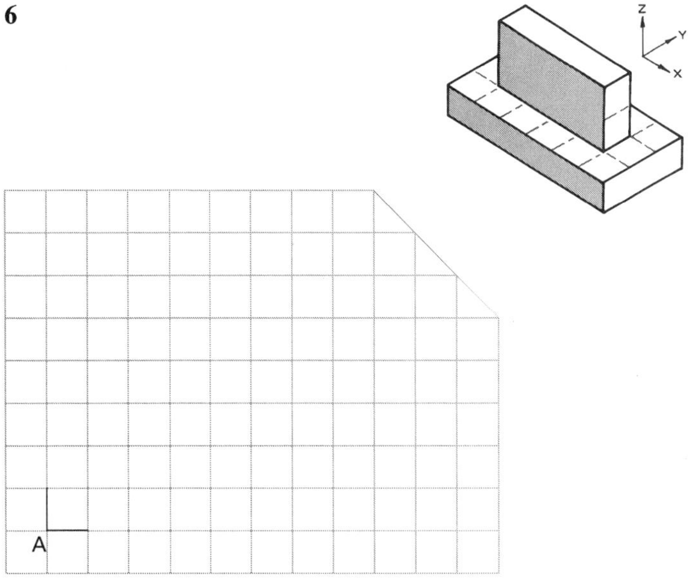

From the information shown in the figures, draw the top, front, and right-side views using third-angle projection. Point A locates the position of the bottom left corner of the front view. Allow 1 grid space between views.

ORTHOGRAPHIC REPRESENTATION— SURFACES PARALLEL AND VISIBLE	REFERENCE UNIT 6–3	NAME / COURSE / DATE	DRAWING D–8

1

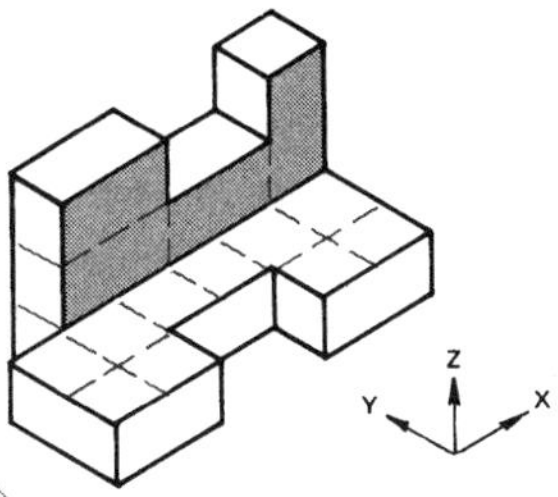

A

4

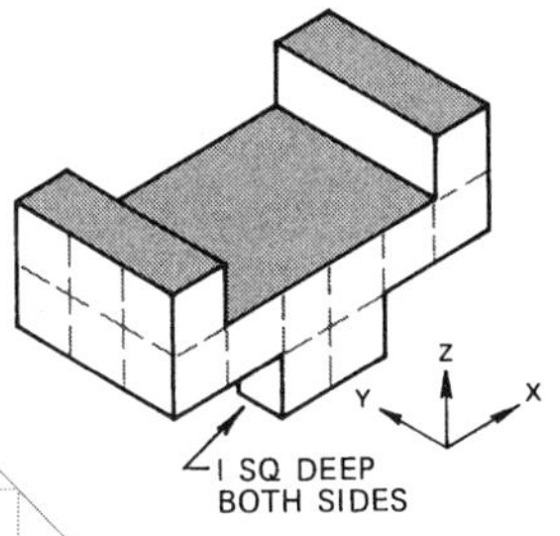

A

2

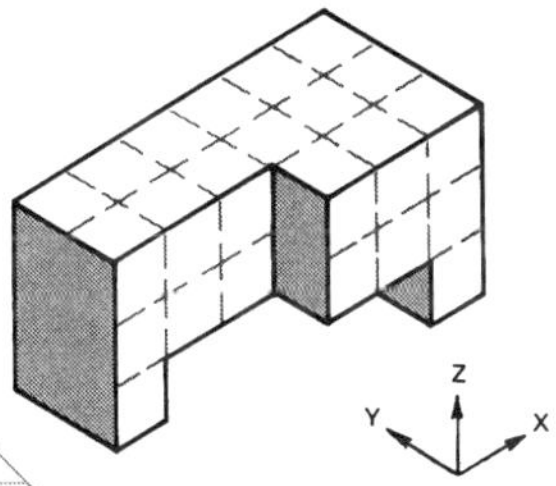

A

5

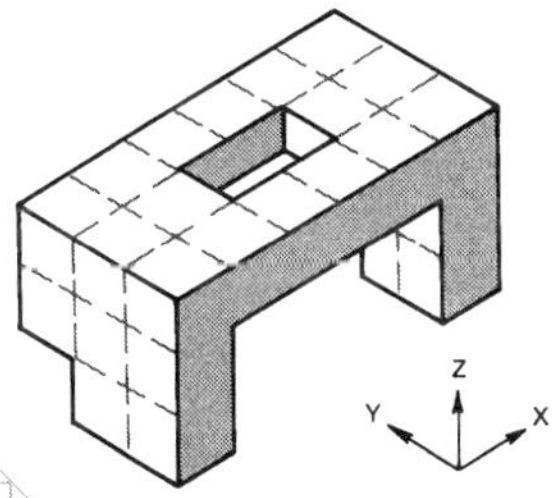

A

3

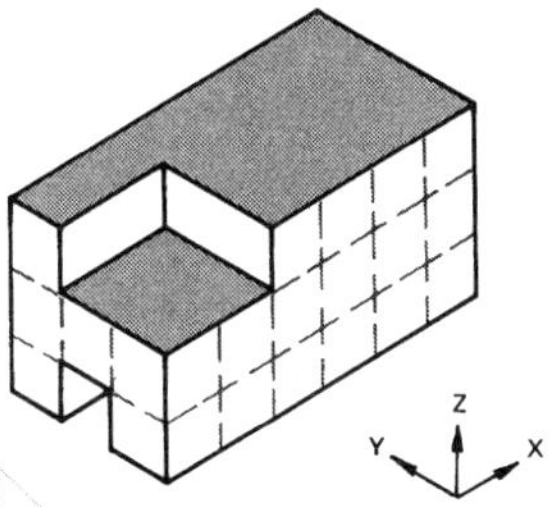

A

6

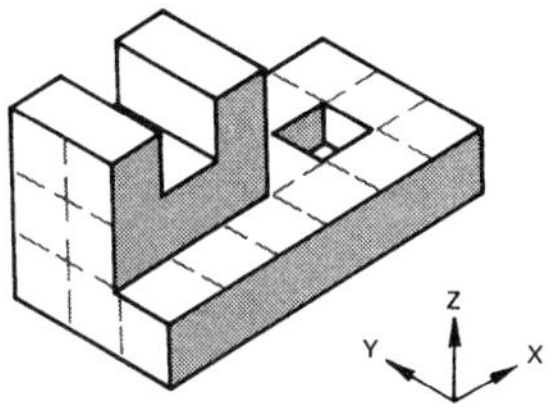

A

From the information shown in the figures, draw the top, front, and right-side views using third-angle orthographic projection. Point A locates the position of the bottom left corner of the front view. Show the hidden lines. Allow 1 grid space between views.

ORTHOGRAPHIC REPRESENTATION— HIDDEN SURFACES AND EDGES	REFERENCE UNIT 6–4	NAME ___ COURSE ___ DATE ___	DRAWING D–9

In the space below, draw the top, front, and right-side views of the control block. Use third-angle orthographic projection. Allow 20 mm between views. Do not dimension. The scale is 1:2.

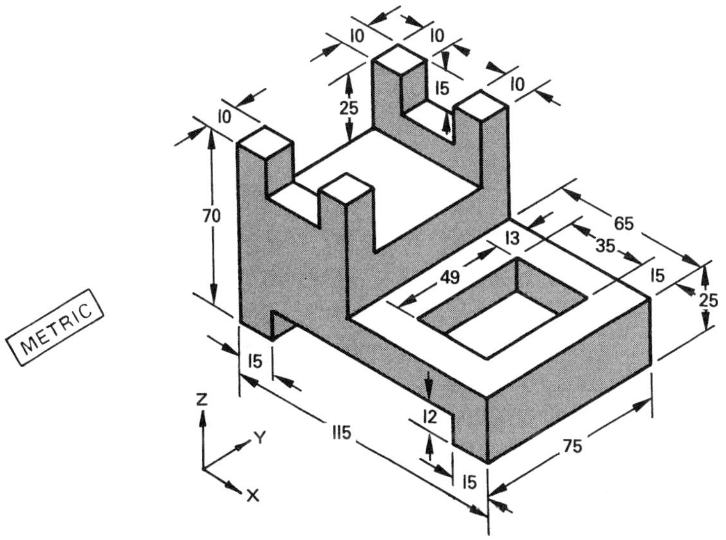

ORTHOGRAPHIC PROJECTION— CONTROL BLOCK	REFERENCE UNIT 6–4	NAME COURSE	DATE	DRAWING D–10

In the space below, draw the top, front, and right-side views of the angle stop. Use third-angle orthographic projection. Allow .50 in. between views. Do not dimension. The scale is 1:1.

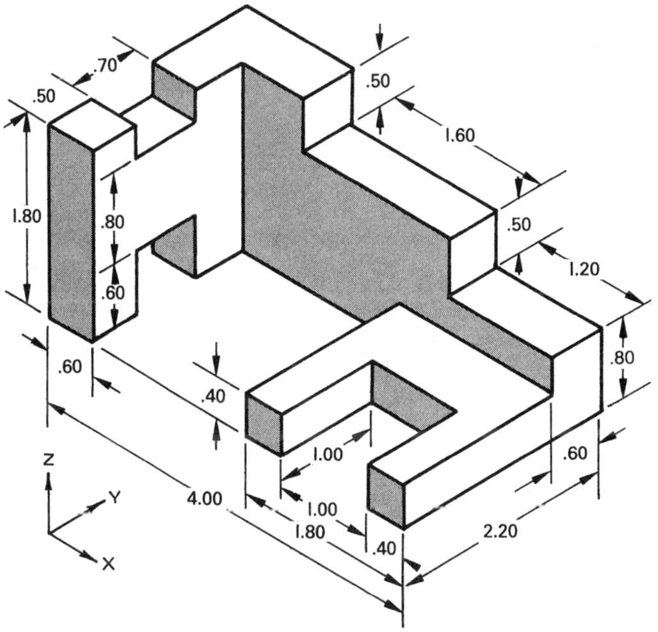

ORTHOGRAPHIC PROJECTION— ANGLE STOP	REFERENCE UNIT 6-4	NAME COURSE DATE	DRAWING D-11

In the space below, draw the top, front, and right-side views of the guide bar. Used third-angle orthographic projection. Allow .50 in. between views. Do not dimension. The scale is 1:1.

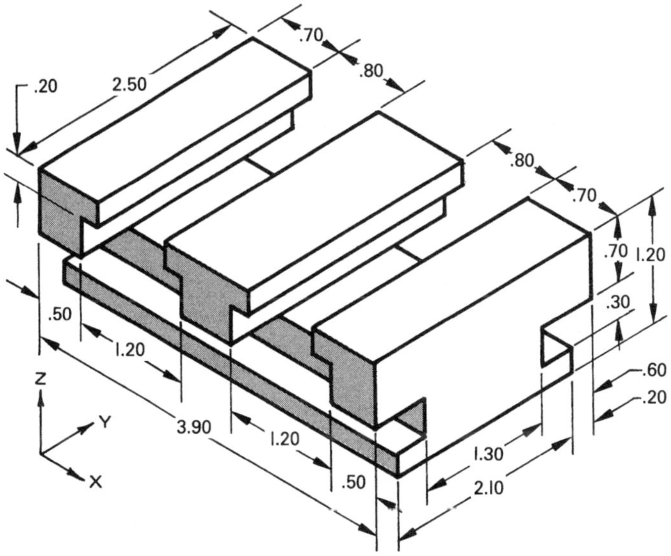

ORTHOGRAPHIC PROJECTION— GUIDE BAR	REFERENCE UNIT 6–4	NAME COURSE DATE	DRAWING D–12

1

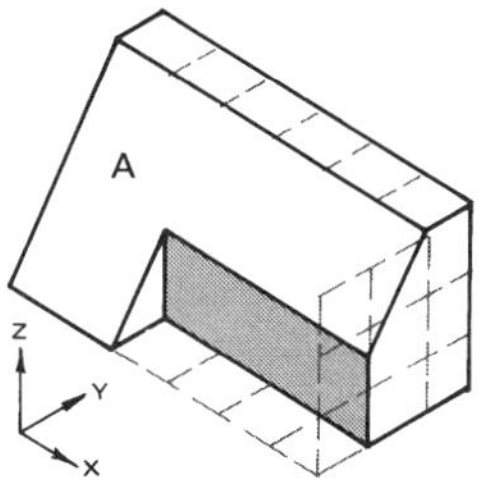

4

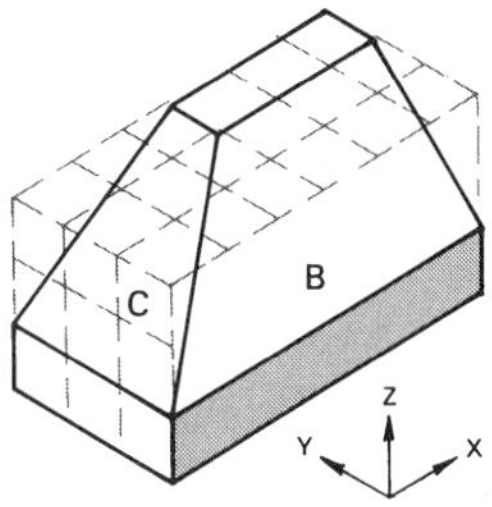

2

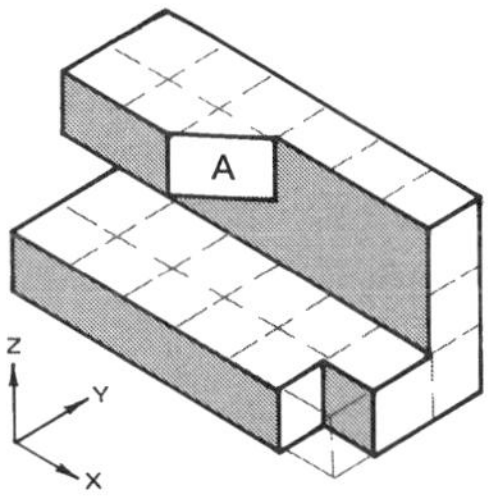

5

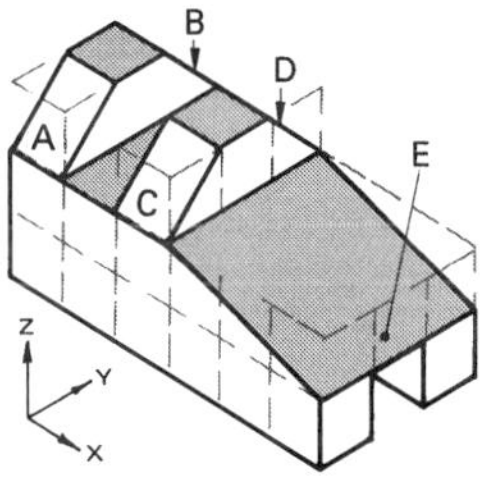

3

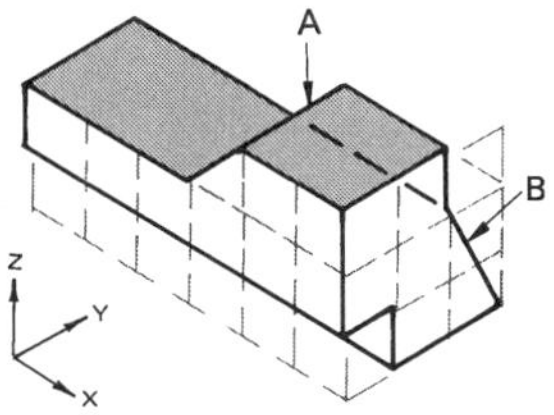

6

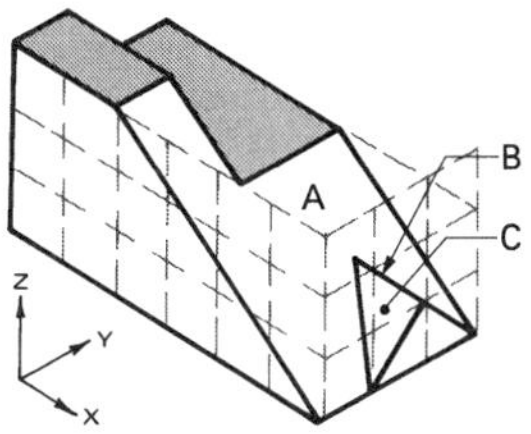

From the information shown in the figures, draw the top, front, and right-side views using third-angle projection. The bottom left corner of the front view is indicated on the drawing area. Allow 1 grid space between views. Label the inclined surfaces on all three views using the letters shown in the figures.

ORTHOGRAPHIC REPRESENTATION— INCLINED SURFACES (1)	REFERENCE UNIT **6–5**	NAME COURSE	DATE	DRAWING **D–13**

1

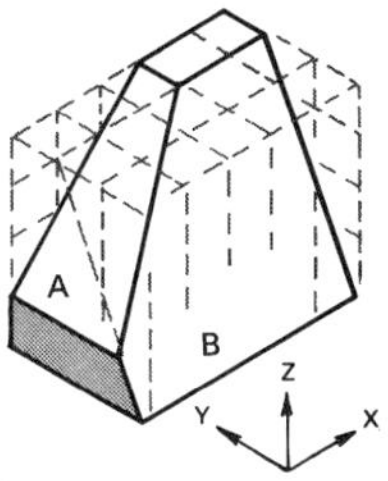

4

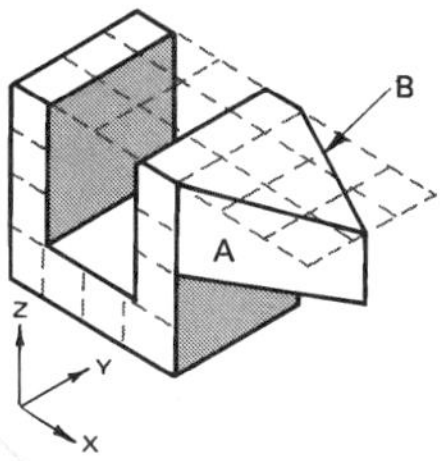

2

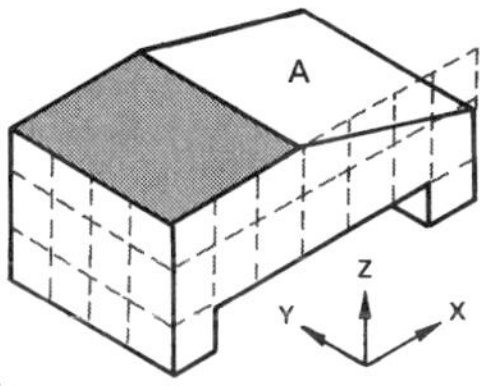

5

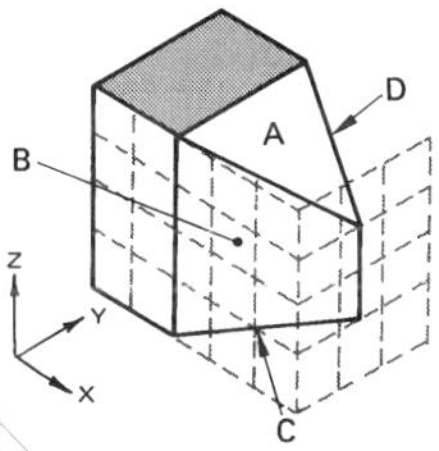

3

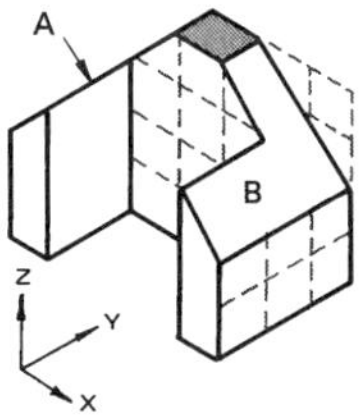

6

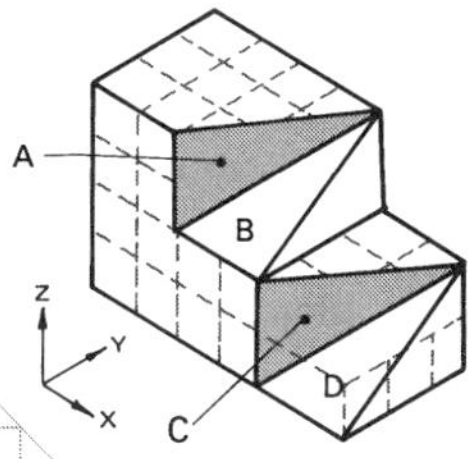

From the information shown in the figures, draw the top, front, and right-side views using third-angle projection. The bottom left corner of the front view is indicated on the drawing area. Allow 1 grid space between views. Label the inclined surfaces with the letters shown in the figures.

ORTHOGRAPHIC REPRESENTATION— INCLINED SURFACES (2)	REFERENCE UNIT **6–5**	NAME COURSE DATE	DRAWING **D–14**

1

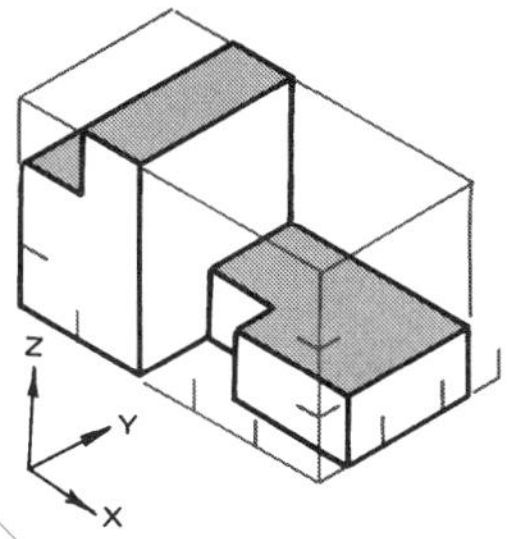

4

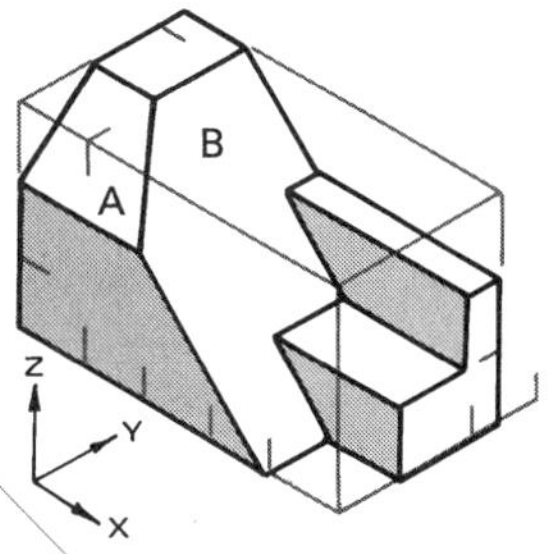

2

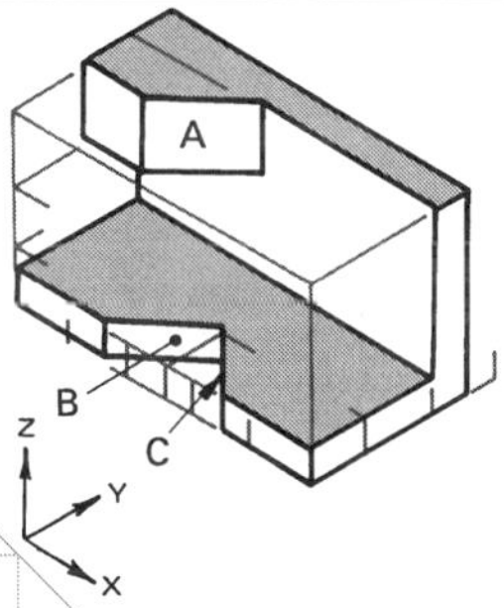

5

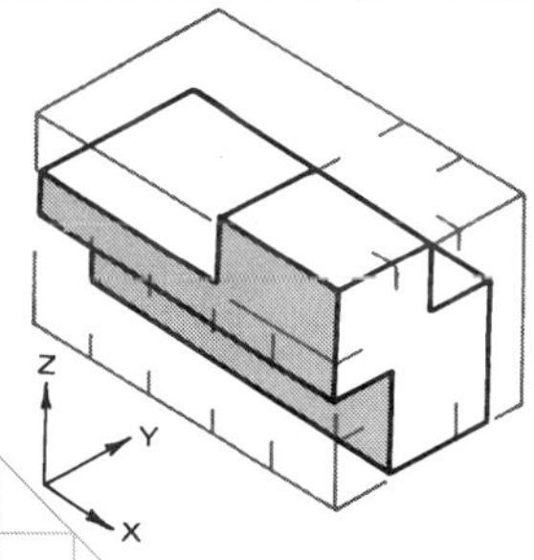

3

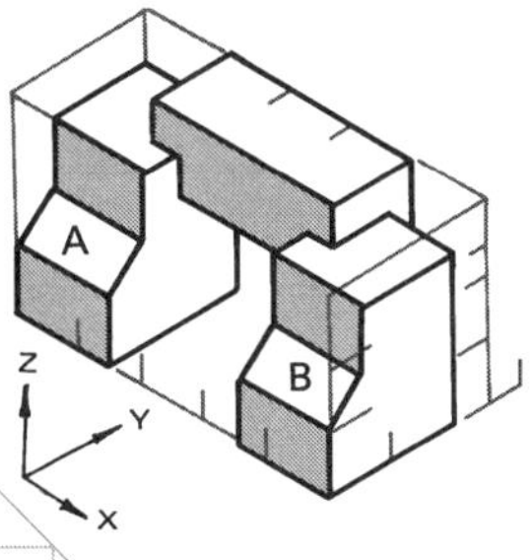

6

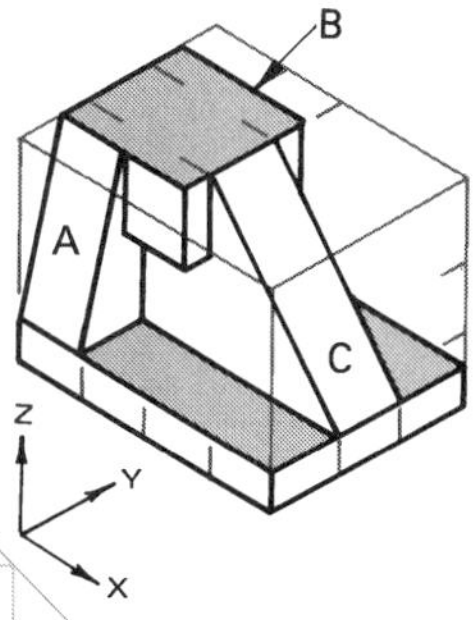

From the information shown in the figures, draw the top, front, and right-side views using third-angle projection. The bottom left corner of the front view is indicated on the drawing area. Allow 1 grid space between views. Label the inclined surfaces on all three views using the letters shown in the figures.

ORTHOGRAPHIC REPRESENTATION— INCLINED SURFACES (3)	REFERENCE UNIT **6–5**	NAME COURSE DATE	DRAWING **D–15**

1

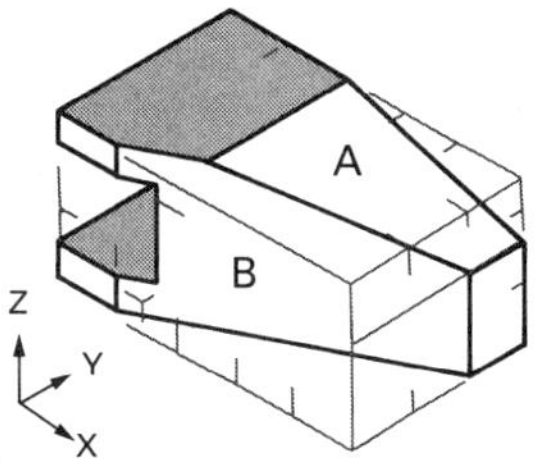

4

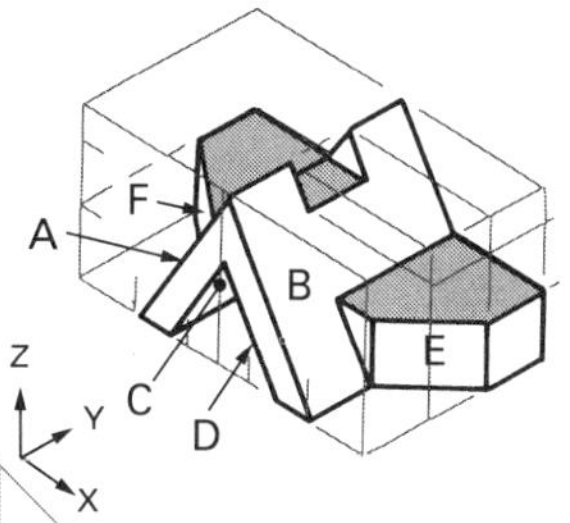

2

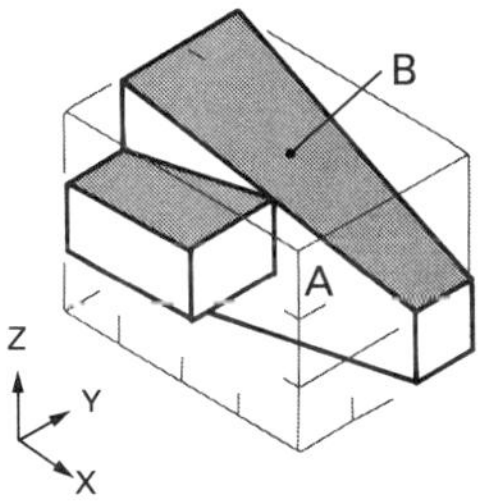

5

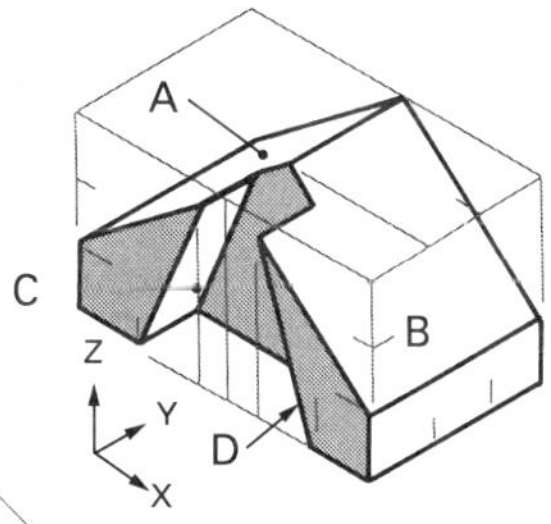

3

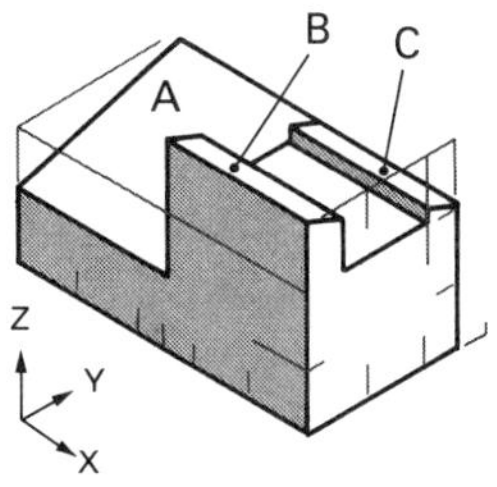

6

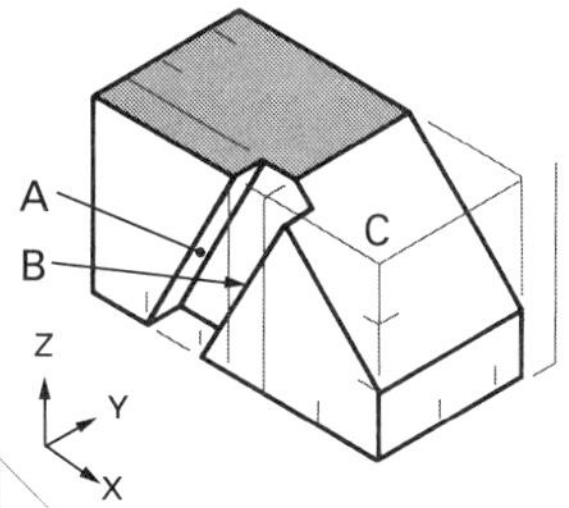

From the information shown in the figures, draw the top, front, and right-side views using third-angle projection. The bottom left corner of the front view is indicated on the drawing area. Allow 1 grid space between views. Label the inclined surfaces on all three views using the letters shown in the figures.

ORTHOGRAPHIC REPRESENTATION— INCLINED SURFACES (4)	REFERENCE UNIT **6–5**	NAME _ COURSE _ DATE _	DRAWING **D–16**

In the space below, draw the top, front, and right-side views of the base. Use third-angle orthographic projection. Allow 20 mm between views. Do not dimension. The scale is 1:2.

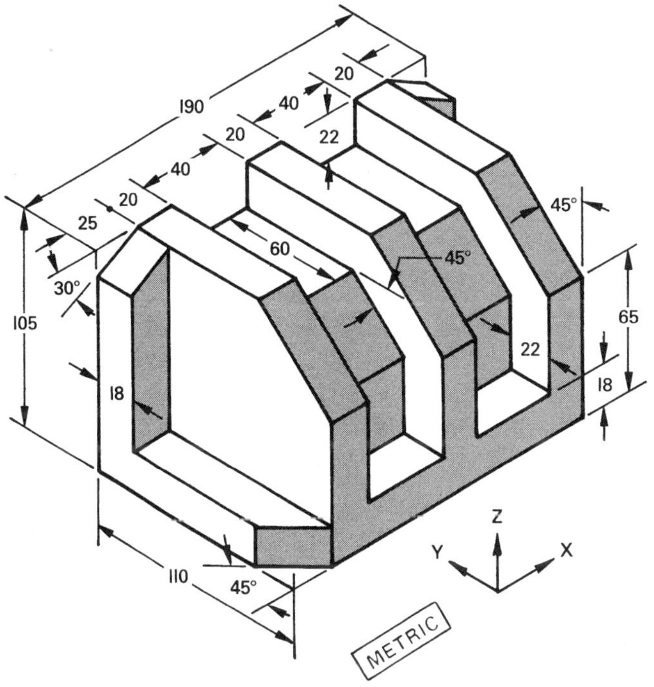

ORTHOGRAPHIC PROJECTION— BASE	REFERENCE UNIT 6–5	NAME COURSE DATE	DRAWING D–17

In the space below, draw the top, front, and right-side views of the base plate. Use third-angle orthographic projection. Allow 20 mm between views. Do not dimension. The scale is 1:2.

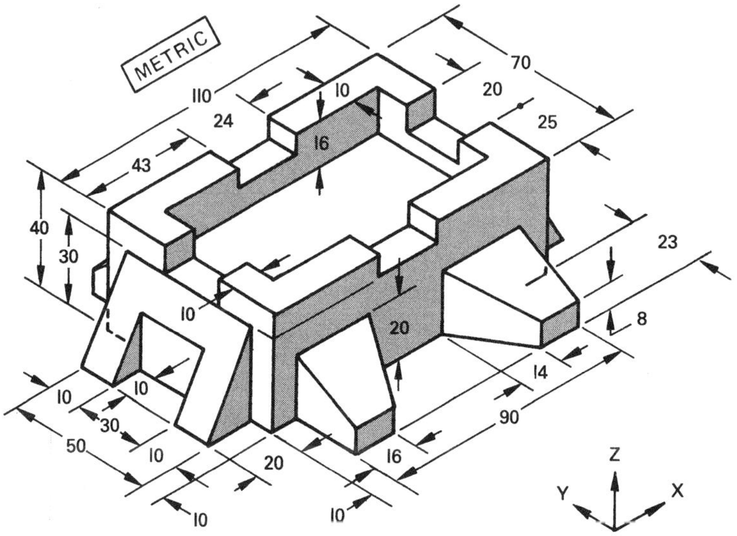

ORTHOGRAPHIC PROJECTION— BASE PLATE	REFERENCE UNIT 6–5	NAME COURSE DATE	DRAWING D–18

BRACKET

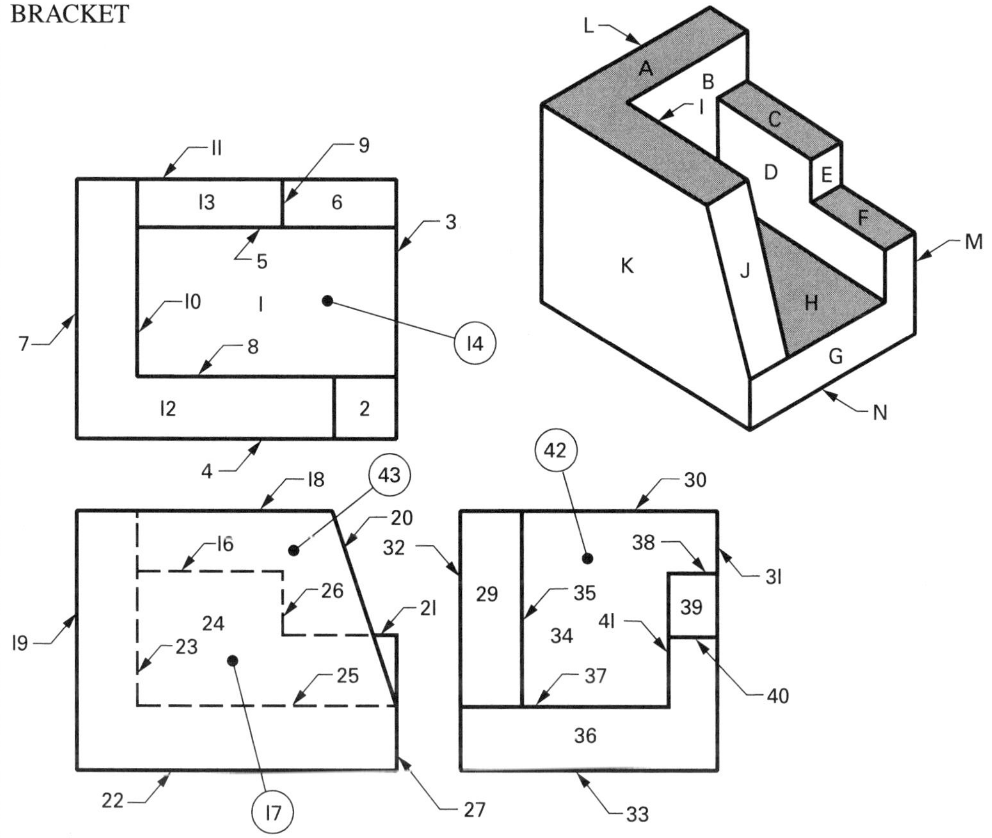

PICTORIAL DRAWING LETTERS	THIRD ANGLE ORTHOGRAPHIC DRAWING		
	TOP VIEW	FRONT VIEW	RIGHT SIDE VIEW
A			
B			
C			
D			
E			
F			
G			
H			
I			
J			
K			
L			
M			
N			

CORNER BRACKET

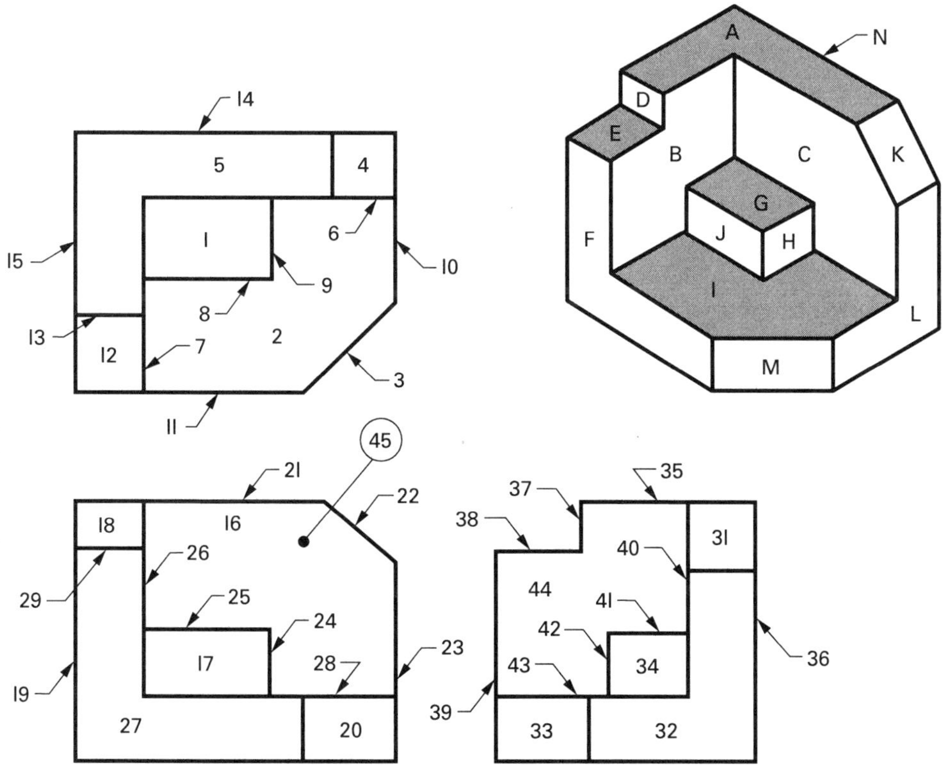

PICTORIAL DRAWING LETTERS	THIRD ANGLE ORTHOGRAPHIC DRAWING		
	TOP VIEW	FRONT VIEW	RIGHT SIDE VIEW
A			
B			
C			
D			
E			
F			
G			
H			
I			
J			
K			
L			
M			
N			

In the charts at the right, place the appropriate numbers corresponding to the letters shown on the drawings. The letters represent surfaces on the parts. The numbers enclosed by a circle refer to a hidden surface or a hidden line. In some instances, an enclosed number may refer to more than one surface.

ORTHOGRAPHIC REPRESENTATION— INTERPRETING DRAWINGS #1	REFERENCE UNIT 6–5	NAME ____ COURSE ____ DATE ____	DRAWING D–19

ANGLED STEP BRACKET

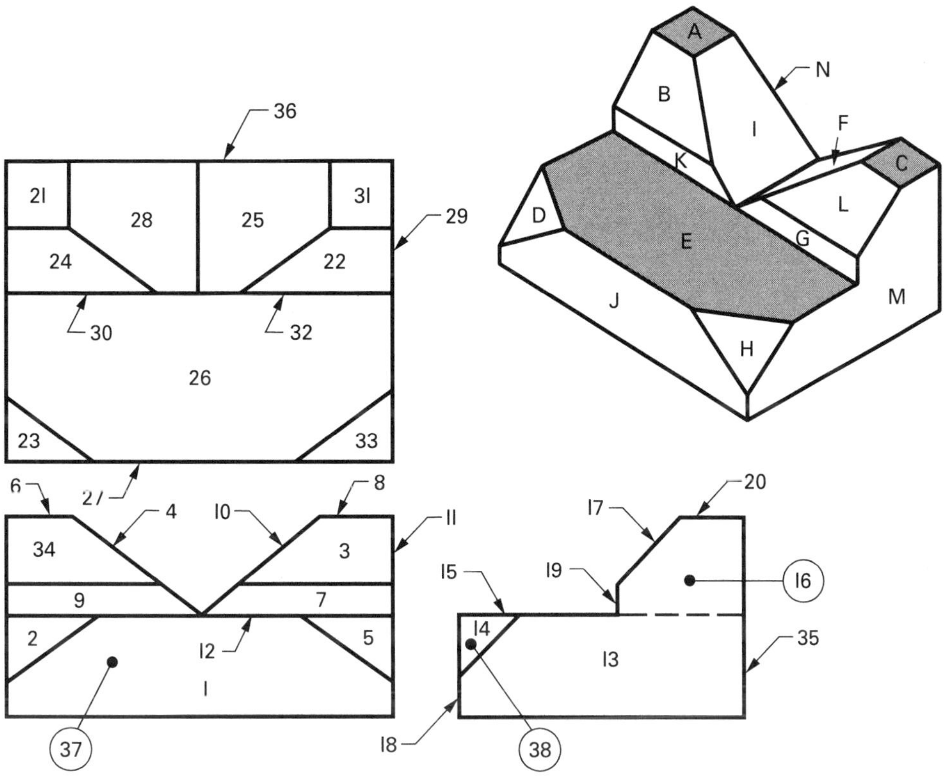

PICTORIAL DRAWING LETTERS	THIRD ANGLE ORTHOGRAPHIC DRAWING		
	TOP VIEW	FRONT VIEW	RIGHT SIDE VIEW
A			
B			
C			
D			
E			
F			
G			
H			
I			
J			
K			
L			
M			
N			

ANGLE STOP

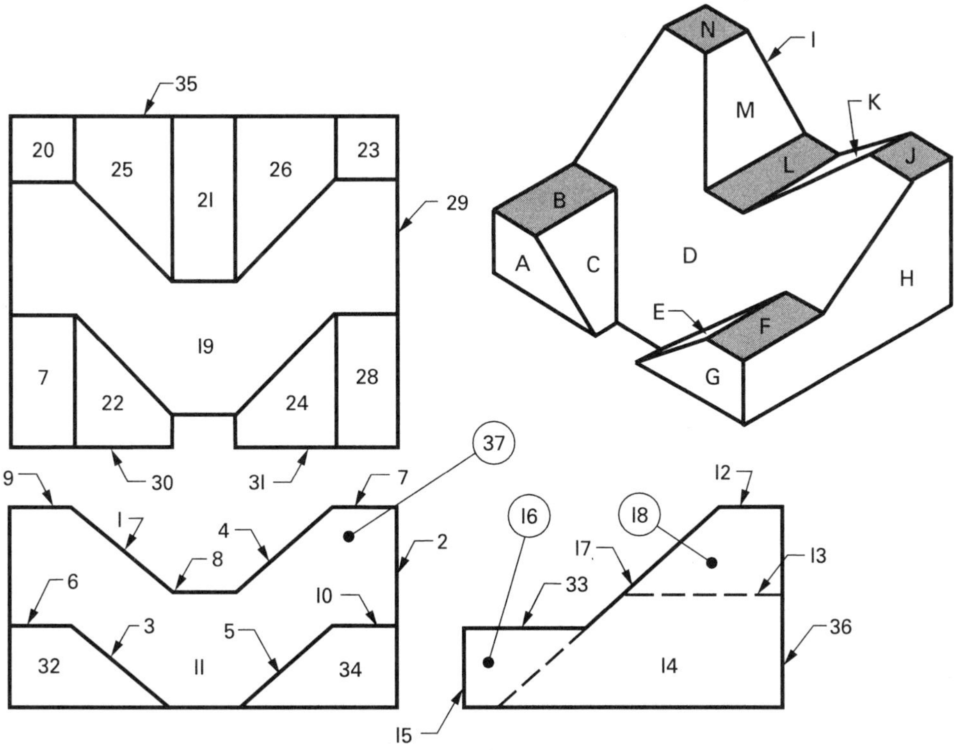

PICTORIAL DRAWING LETTERS	THIRD ANGLE ORTHOGRAPHIC DRAWING		
	TOP VIEW	FRONT VIEW	RIGHT SIDE VIEW
A			
B			
C			
D			
E			
F			
G			
H			
I			
J			
K			
L			
M			
N			

In the charts at the right, place the appropriate numbers corresponding to the letters shown on the drawings. The letters represent surfaces on the parts. The numbers enclosed by a circle refer to a hidden surface or a hidden line. In some instances, an enclosed number may refer to more than one surface.

ORTHOGRAPHIC REPRESENTATION— INTERPRETING DRAWINGS #2	REFERENCE UNIT 6–5	NAME ____ COURSE ____ DATE ____	DRAWING D–20

1

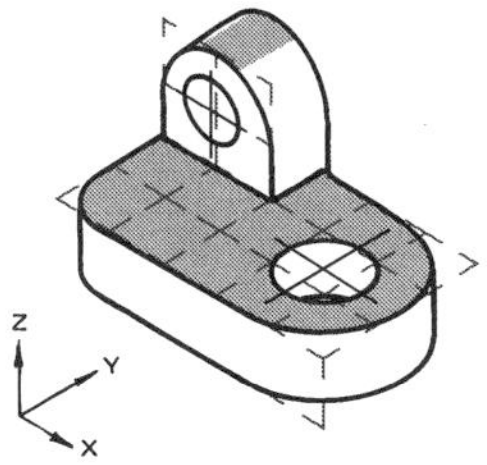

4

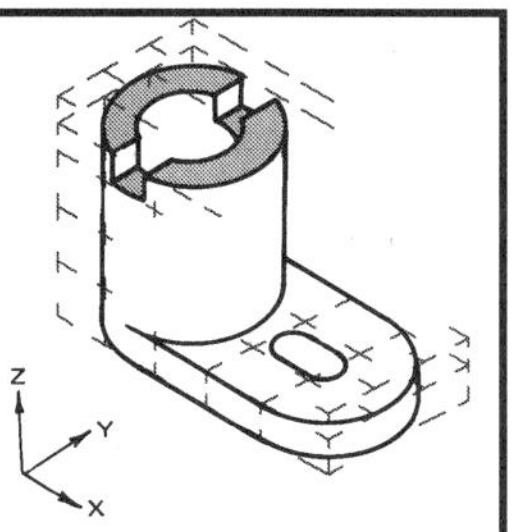

2

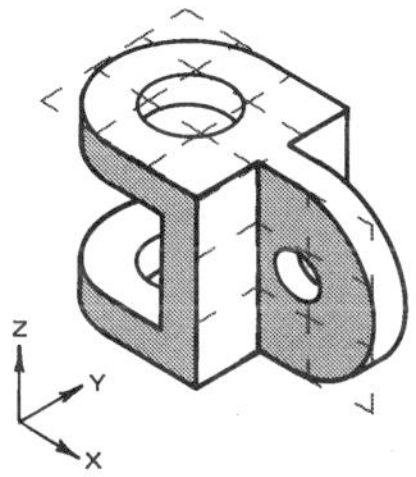

5

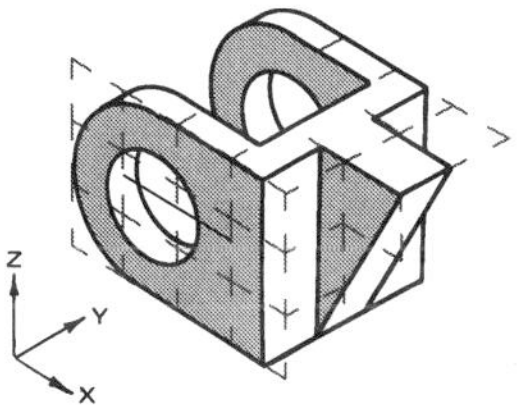

3

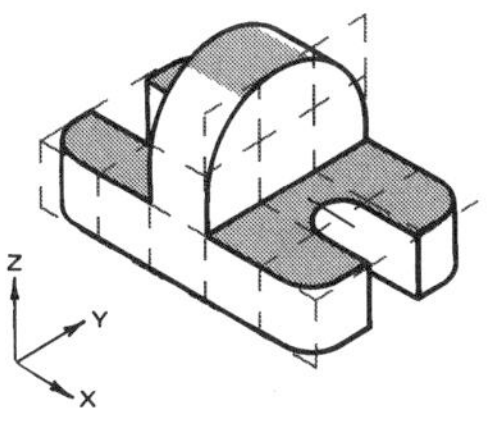

6

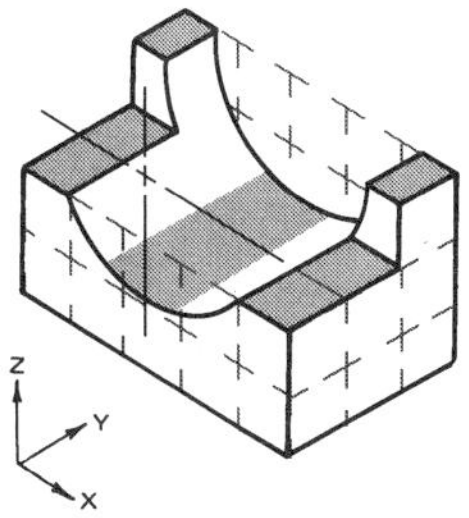

From the information shown in the figures, draw the top, front, and right-side views using third-angle projection. The bottom left corner of the front view is indicated on the drawing area. Allow 1 grid space between views.

ORTHOGRAPHIC REPRESENTATION— CIRCULAR FEATURES (1)	REFERENCE UNIT **6–6**	NAME _ COURSE _ DATE _	DRAWING **D–21**

1

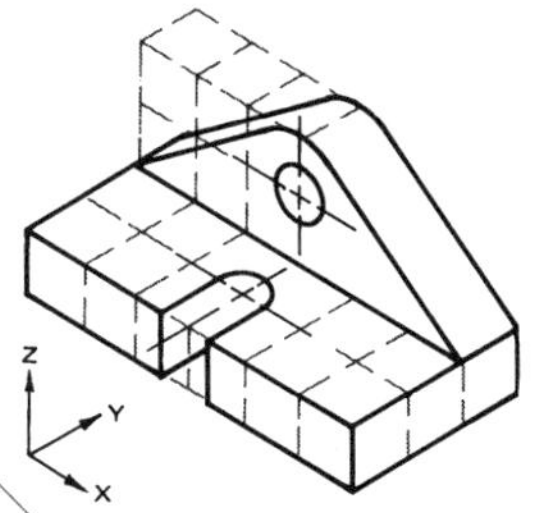

4

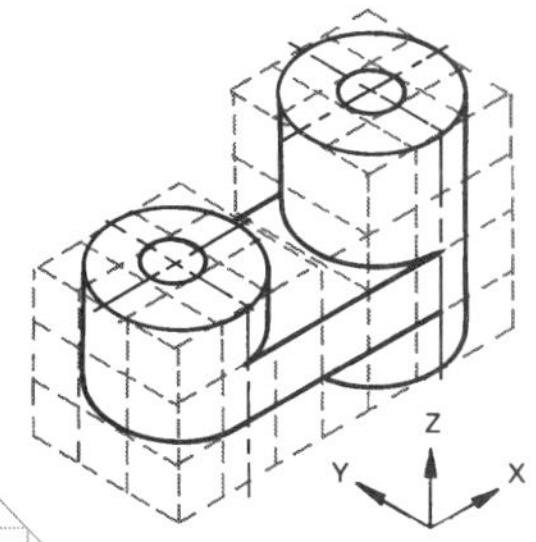

2

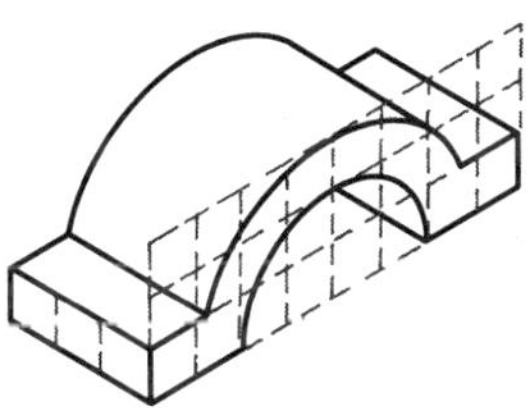

5

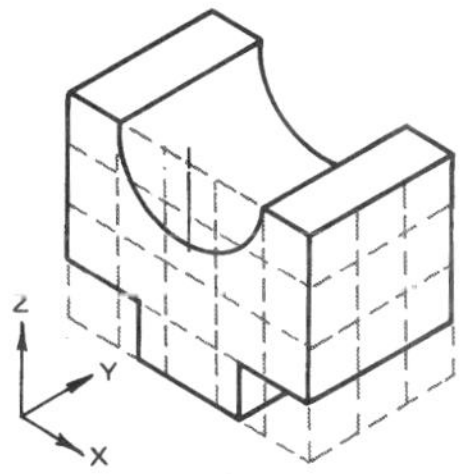

3

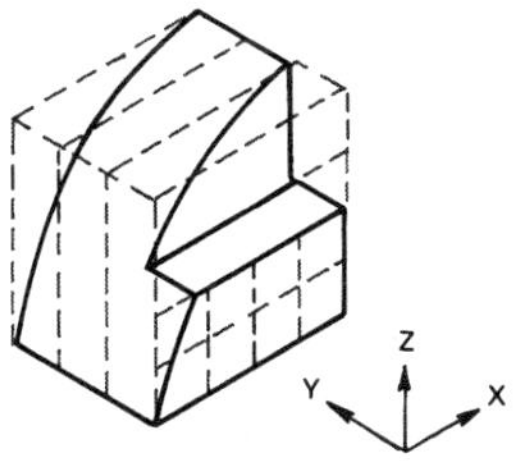

6

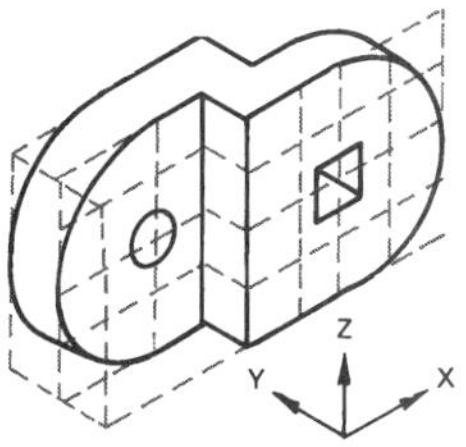

From the information shown in the figures, draw the top, front, and right-side views using third-angle projection. The bottom left corner of the front view is indicated on the drawing area. Allow 1 grid space between views.

ORTHOGRAPHIC REPRESENTATION— CIRCULAR FEATURES (2)	REFERENCE UNIT **6–6**	NAME COURSE DATE	DRAWING **D–22**

SWING BRACKET

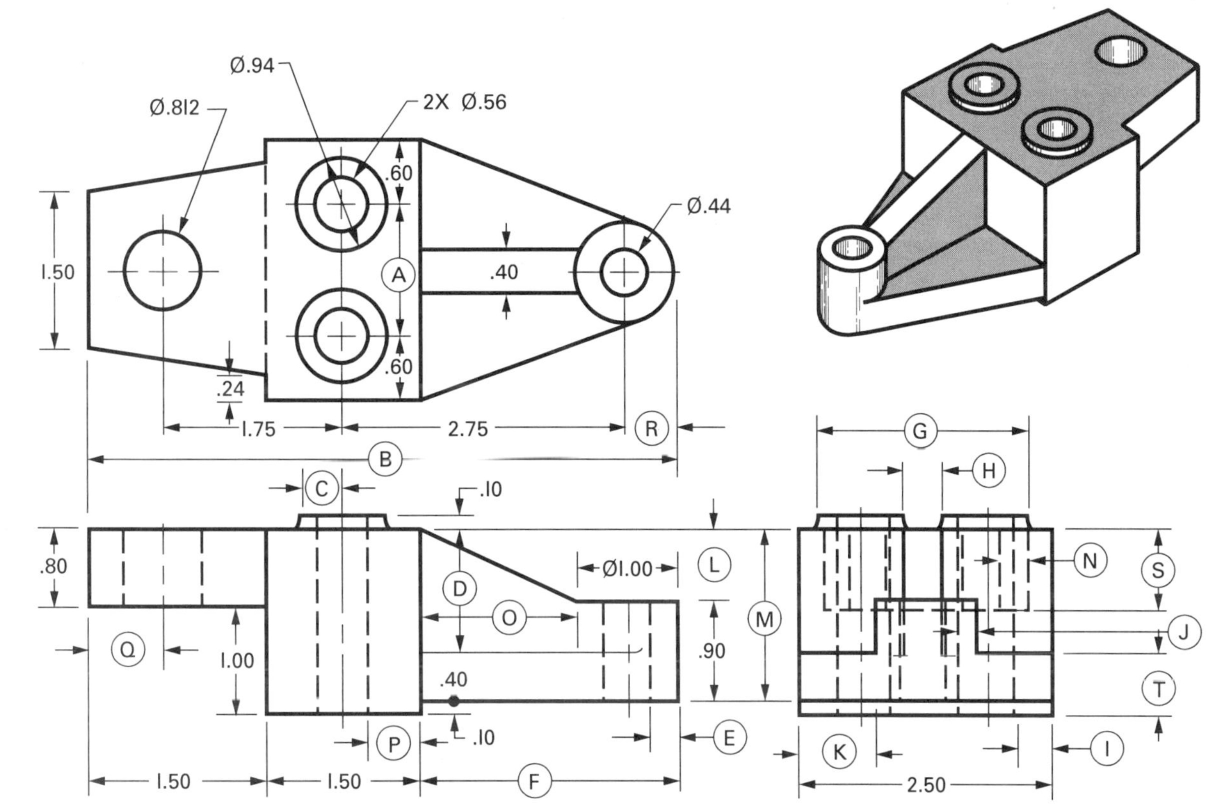

ANSWERS	
A	
B	
C	
D	
E	
F	
G	
H	
I	
J	
K	
L	
M	
N	
O	
P	
Q	
R	
S	
T	

CRADLE BRACKET

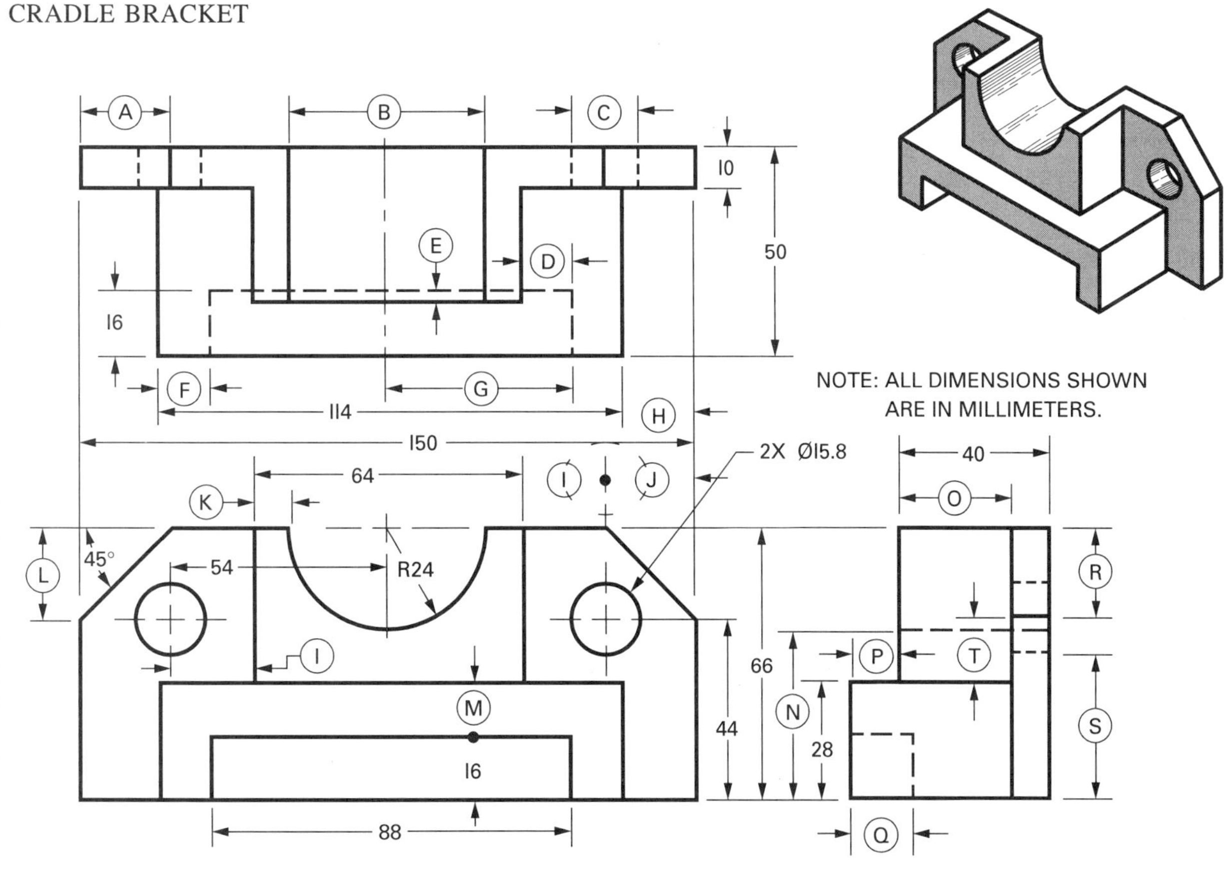

ANSWERS	
A	
B	
C	
D	
E	
F	
G	
H	
I	
J	
K	
L	
M	
N	
O	
P	
Q	
R	
S	
T	

Calculate distances A to T.

ORTHOGRAPHIC REPRESENTATION— INTERPRETING DRAWINGS #3	REFERENCE UNIT 6-6	NAME _ COURSE _ DATE _	DRAWING D-23

In the space below, draw the top and front views of the rocker arm. Use third-angle orthographic projection. Allow 15 mm between views. Do not dimension. The scale is 1:2.

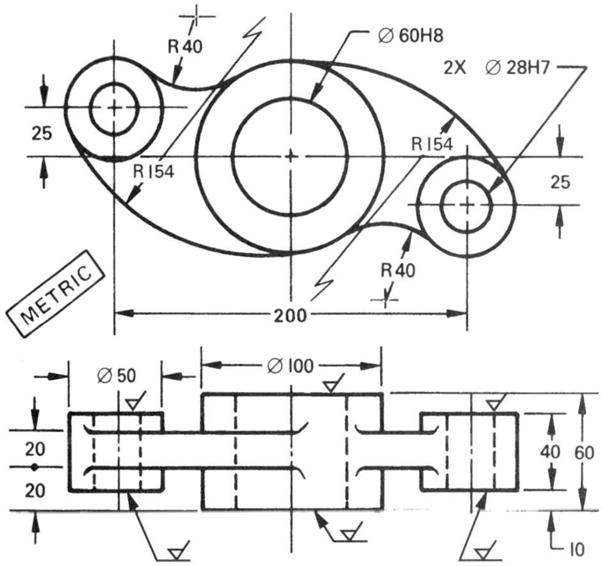

ORTHOGRAPHIC PROJECTION— ROCKER ARM	REFERENCE UNIT 6–6	NAME COURSE DATE	DRAWING D–24

1

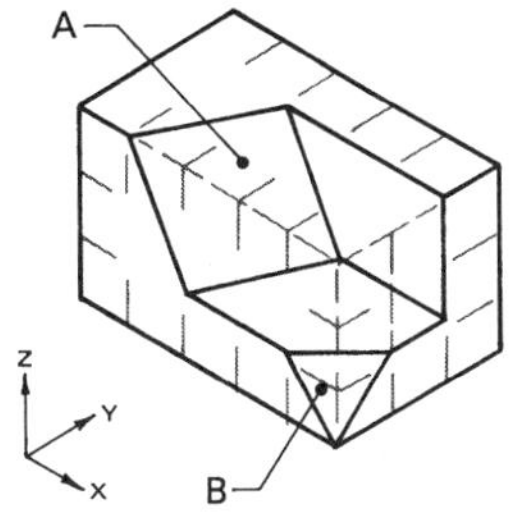

4

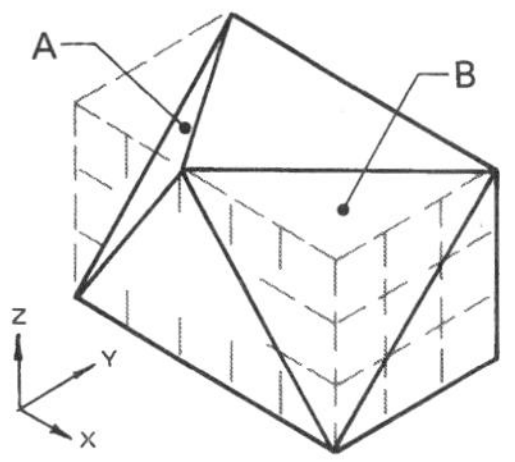

2

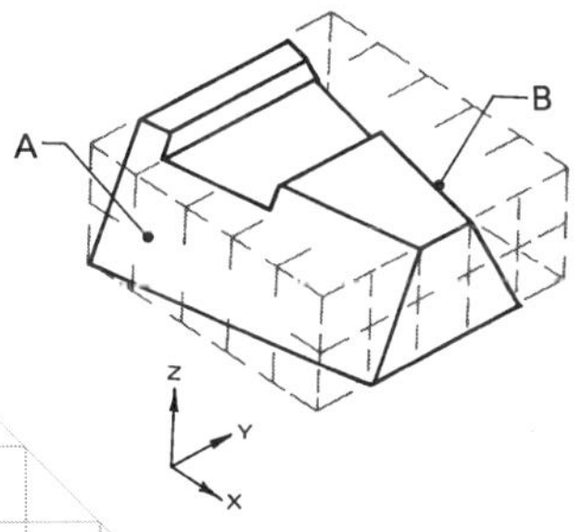

5

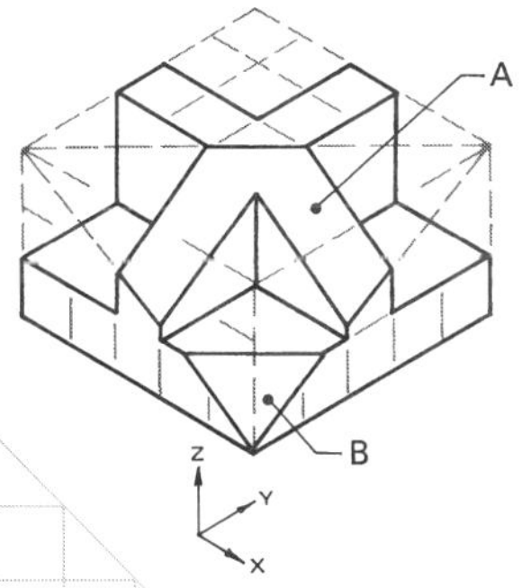

3

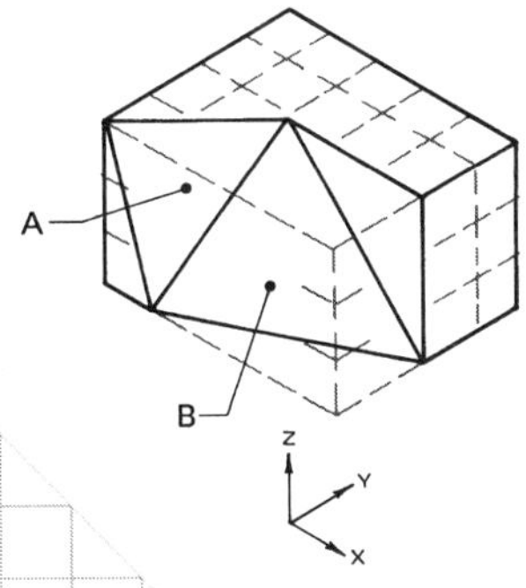

6

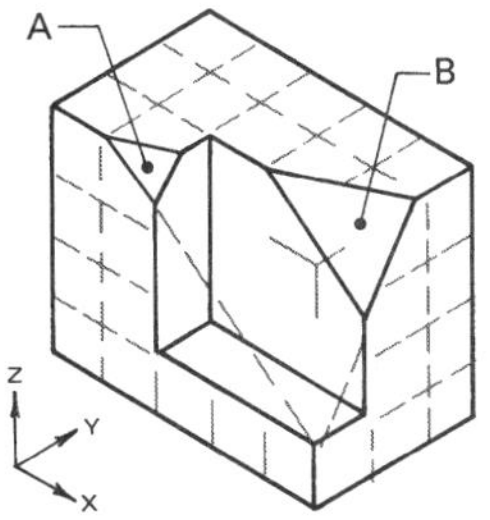

From the information shown in the figures, draw the top, front, and right-side views using third-angle projection. The bottom left-hand corner of the front view is indicated on the drawing area. Allow 1 grid space between views. Label the oblique surfaces on all three views using the letters shown in the figures.

ORTHOGRAPHIC REPRESENTATION— OBLIQUE SURFACES (1)	REFERENCE UNIT **6–7**	NAME ___ COURSE ___ DATE ___	DRAWING **D–25**

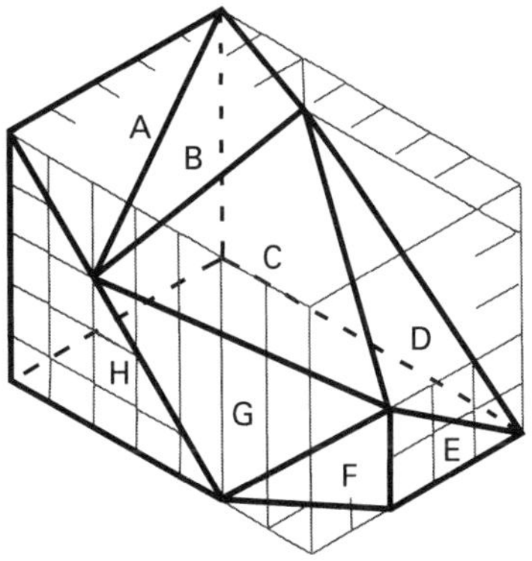

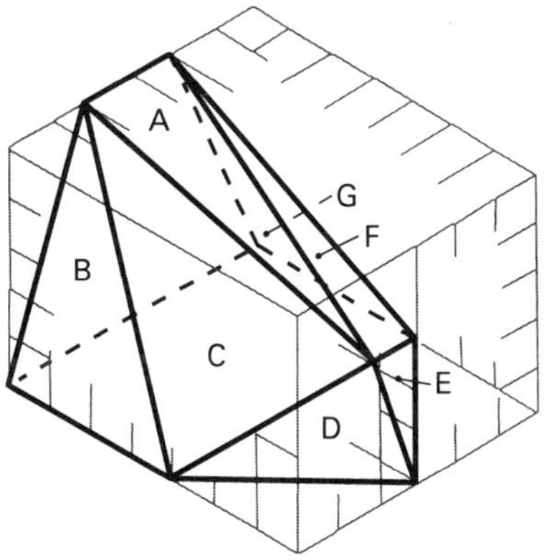

From the information shown in the figures, draw the top, front, and right-side views using third-angle projection. The bottom left-hand corner of the front view is indicated on the drawing area. Allow 1 grid space between views. Label the inclined surfaces on all three views using the letters shown in the figures.

ORTHOGRAPHIC REPRESENTATION— OBLIQUE SURFACES (2)	REFERENCE UNIT 6–7	NAME COURSE DATE	DRAWING D–26

1

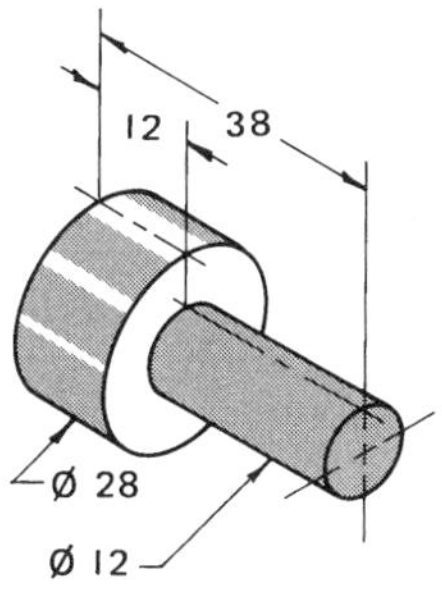

4

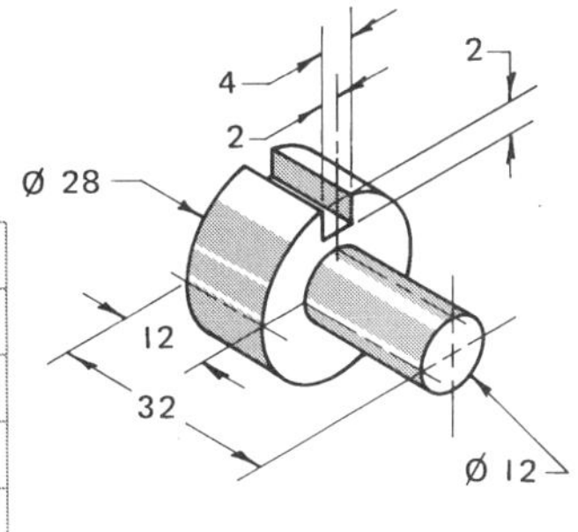

2

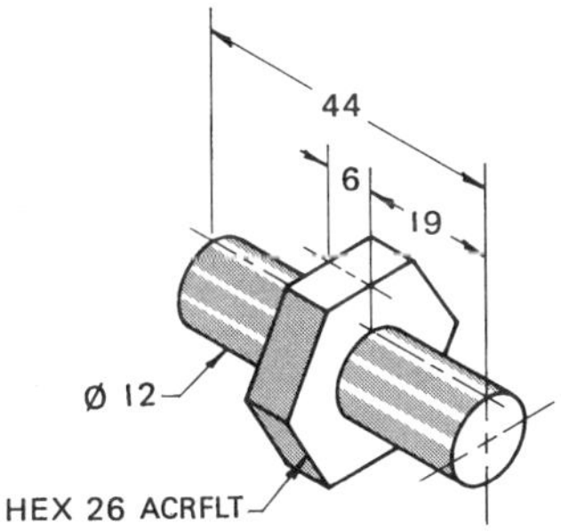

5

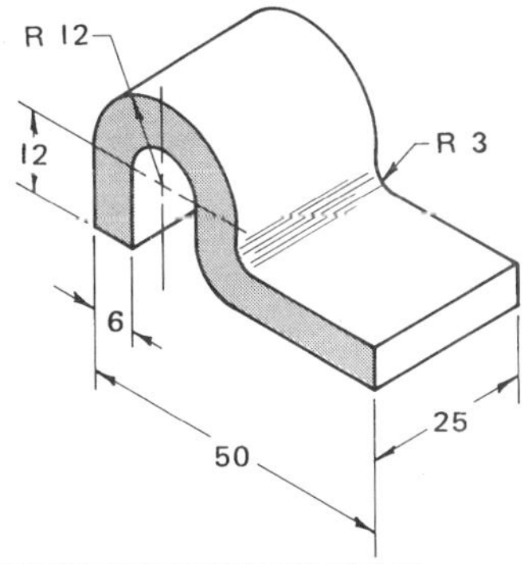

3

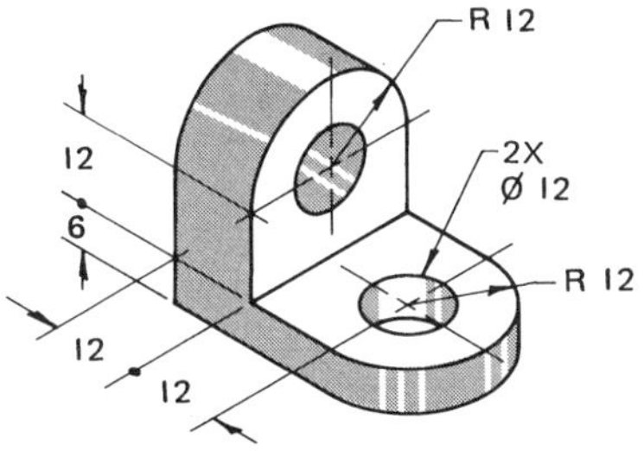

6

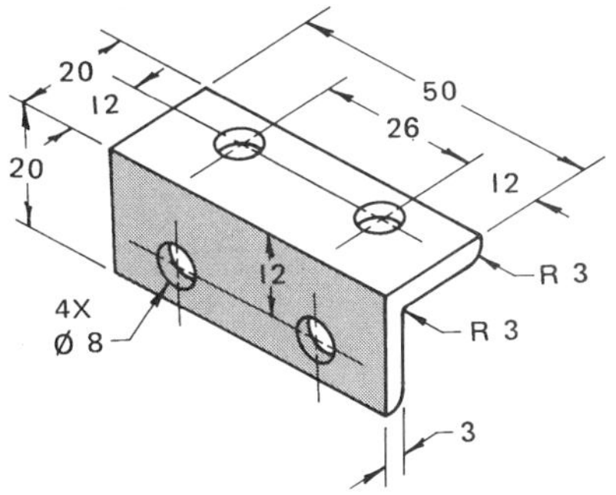

Sketch the objects shown and draw only the necessary views that will completely describe each part. The drawings need not be to scale but should be drawn in proportion to the illustrations shown. Grid size = 5 mm squares. Do not dimension.

ORTHOGRAPHIC REPRESENTATION— ONE- AND TWO-VIEW DRAWINGS	REFERENCE UNIT **6–8**	NAME COURSE DATE	DRAWING **D–27**

Draw only the views (full and partial) that will completely describe the connector shown. Dimensioning is optional. The scale is 1:1. Allow 10 mm between views. The lower center of the front view is located on the drawing area.

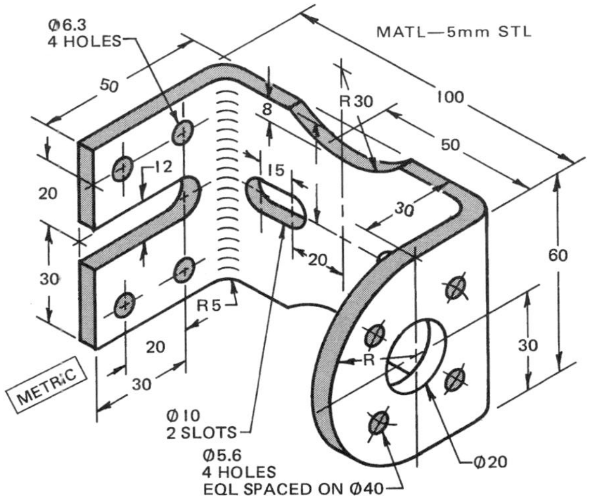

SPECIAL VIEWS—

CONNECTOR

REFERENCE UNIT **6–9**

NAME

COURSE

DATE

DRAWING **D–28**

In the space below, make a working drawing of the adjustable locking plate shown to the scale 10:1. Wherever possible, simplify the drawing by using conventional representation of common features.

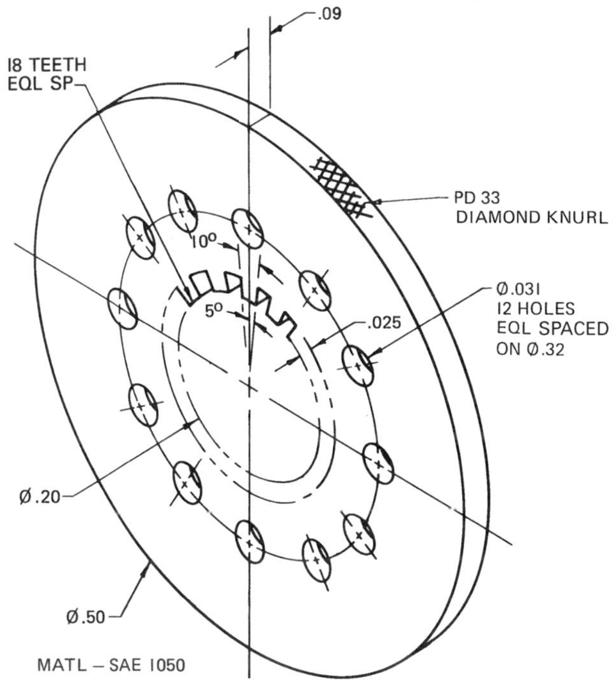

CONVENTIONAL REPRESENTATION— LOCKING PLATE	REFERENCE UNIT **6–10**	NAME COURSE DATE	DRAWING **D–29**

In the space below, draw the front and top views of the fixture base to the scale 1:1. Use two conventional breaks to shorten the length of the part. Add an enlarged view of the hexagonal feature.

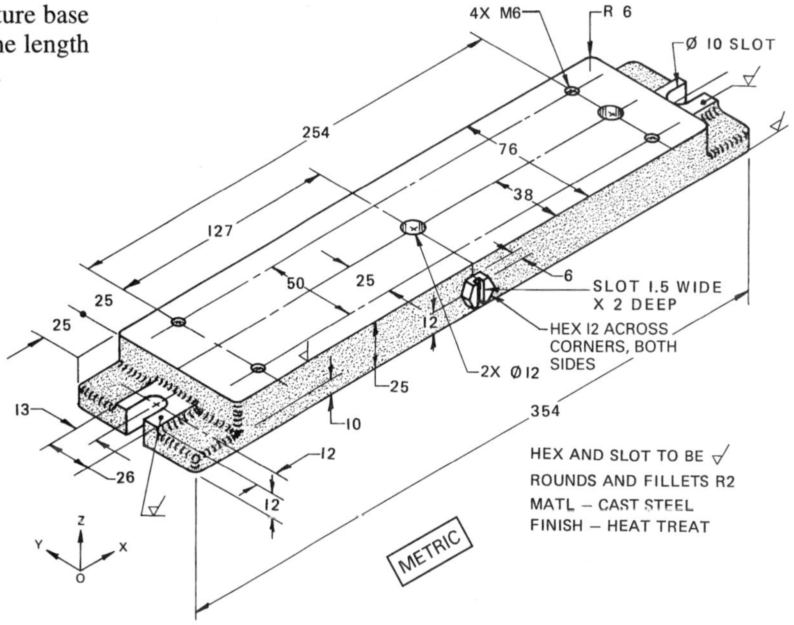

CONVENTIONAL BREAKS— FIXTURE BASE	REFERENCE UNIT 6–11	NAME COURSE	DATE	DRAWING D–30

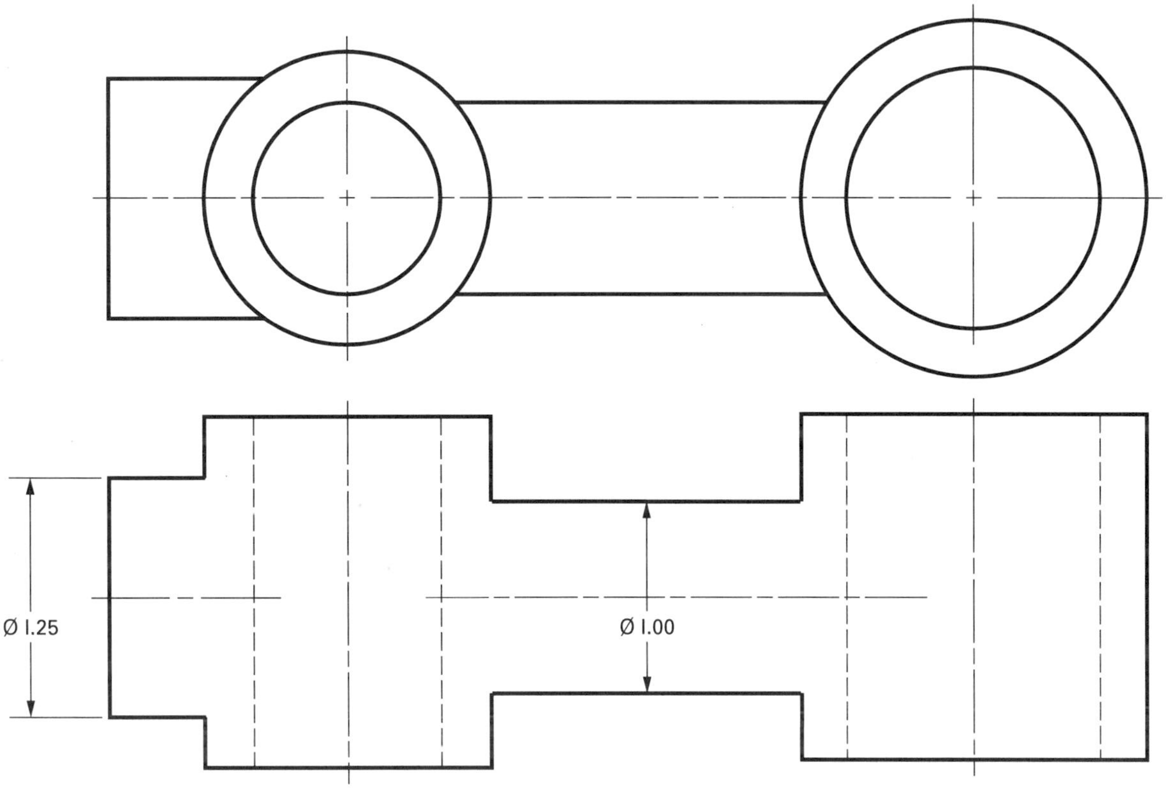

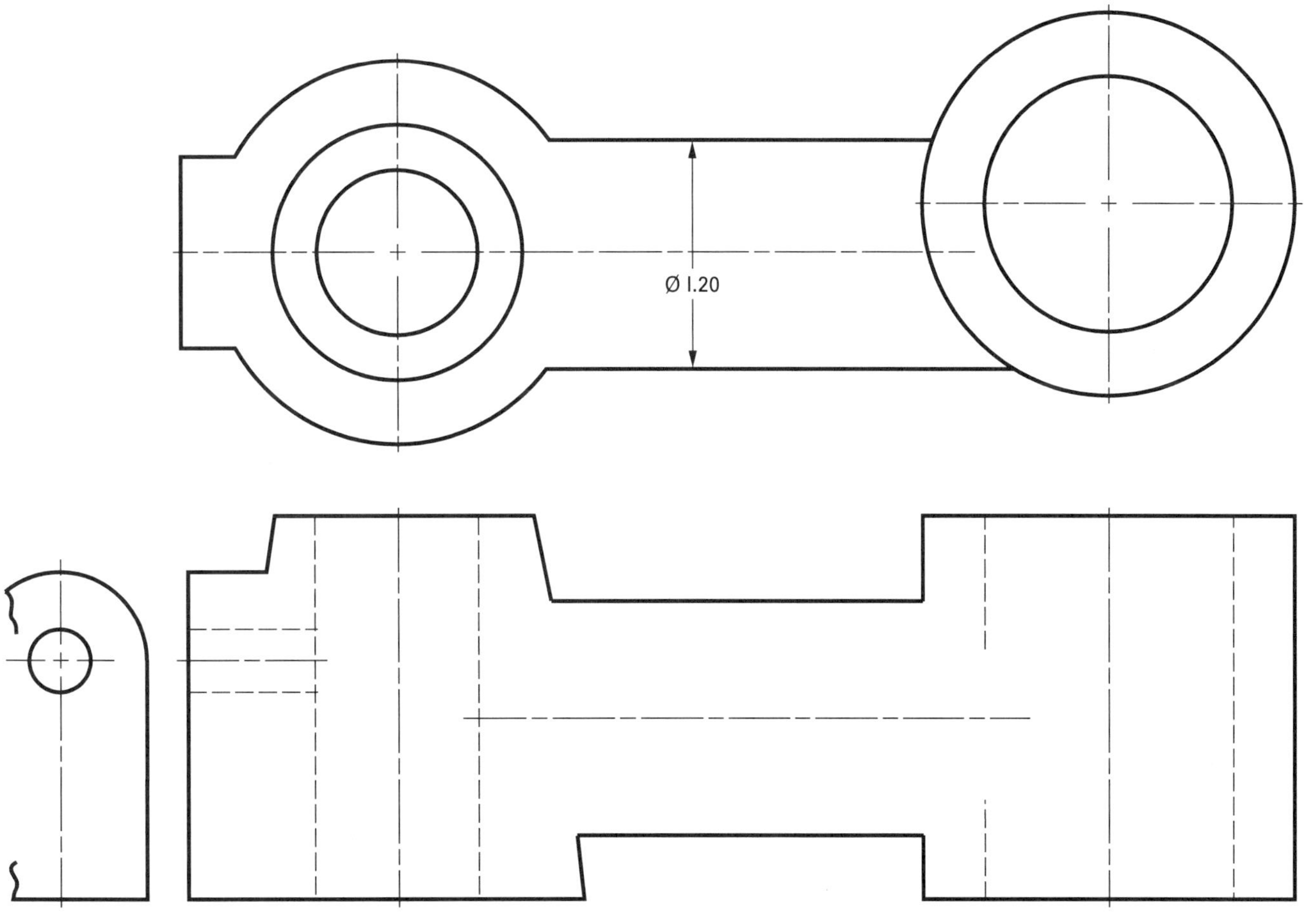

Complete the lines of intersection.

LINES OF INTERSECTION— CYLINDRICAL SURFACES	REFERENCE UNIT 6–12	NAME COURSE DATE	DRAWING D–31

Make a working drawing of the mounting bracket to the scale 1:1. All surface finishes are 63 μin. The keyed hole should have an RC6 fit with the shaft. Where required, rotate features to show their true distance from the center and edges. A revolved or removed section is recommended on a leg.

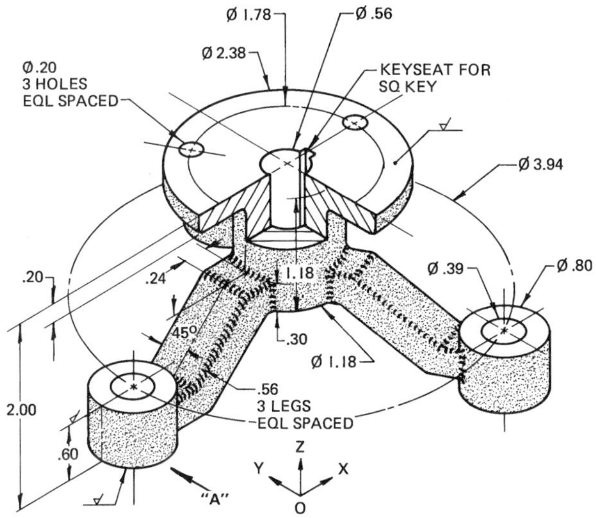

ROUNDS AND FILLETS R.12
MATL – CAST STEEL

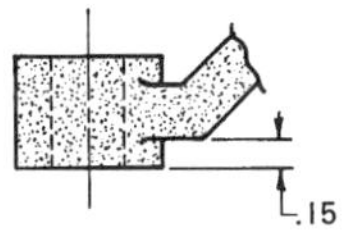

VIEW IN DIRECTION OF ARROW "A"

FORESHORTENED PROJECTION— MOUNTING BRACKET	REFERENCE UNIT 6–14	NAME COURSE DATE	DRAWING D–32

Make a working drawing of the clutch to the scale 1:1. All surface finishes are 1.6 µm. The keyed hole should have an H9d9 fit with the shaft. Where required, rotate features to show their true distance from the center and edges. A revolved section to show the shape of the ribs is recommended.

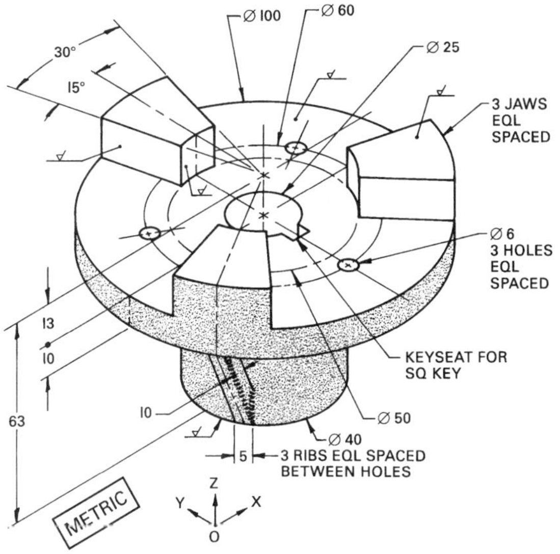

ROUNDS AND FILLETS R4
MATL-CAST STEEL

FORESHORTENED PROJECTION— CLUTCH	REFERENCE UNIT 6–14	NAME COURSE	DATE	DRAWING D–33

Make a three-view working drawing of the cutoff stop. The scale is 1:1. The T-slot surfaces should have a maximum roughness of 0.8 μm and a maximum waviness of 0.05 mm for a 25 mm length. The back surface should have a maximum roughness of 3.2 μm with no restriction on waviness.

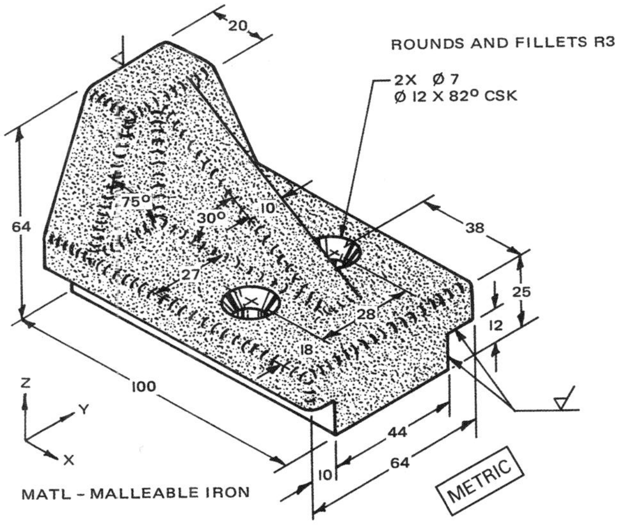

INTERSECTION OF UNFINISHED SURFACES —CUTOFF STOP	REFERENCE UNIT **6–15**	NAME COURSE DATE	DRAWING **D–34**

Make a three-view working drawing of the sparker bracket. The scale is 1:1. The back and notch should have a maximum roughness of 32 µin. and a maximum waviness of 2 µin. for a 1.00-in. length. The faces on the bosses should have a maximum roughness of 125 µin. with no restriction on waviness.

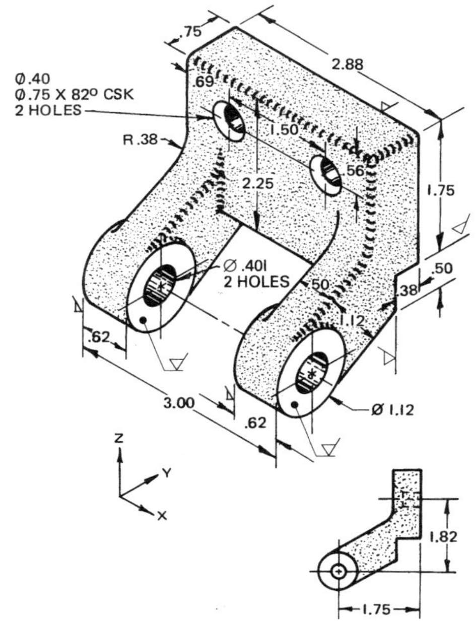

ROUNDS AND FILLETS R.12
MATL – MALLEABLE IRON

INTERSECTION OF UNFINISHED SURFACES —SPARKER BRACKET	REFERENCE UNIT 6–15	NAME COURSE DATE	DRAWING D–35

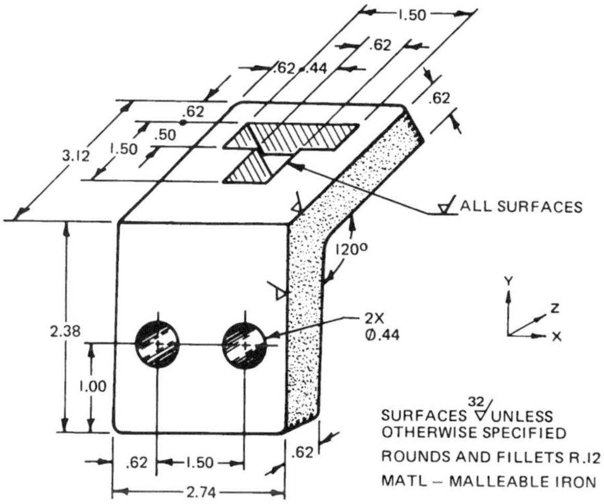

Make a detail drawing of the angle bracket. Replace the top view with an auxiliary view projected from the right-side view. The front and auxiliary views should be shown as partial views. The scale is 1:1. The bottom left corner of the side view is located on the drawing area.

PRIMARY AUXILIARY VIEWS— ANGLE BRACKET	REFERENCE UNIT **7–1**	NAME COURSE DATE	DRAWING **E–1**

Make a detail drawing of the angle plate. The front view, partial bottom view, and a partial auxiliary view are required. The scale is 1:1. The top left corner of the front view is located on the drawing area.

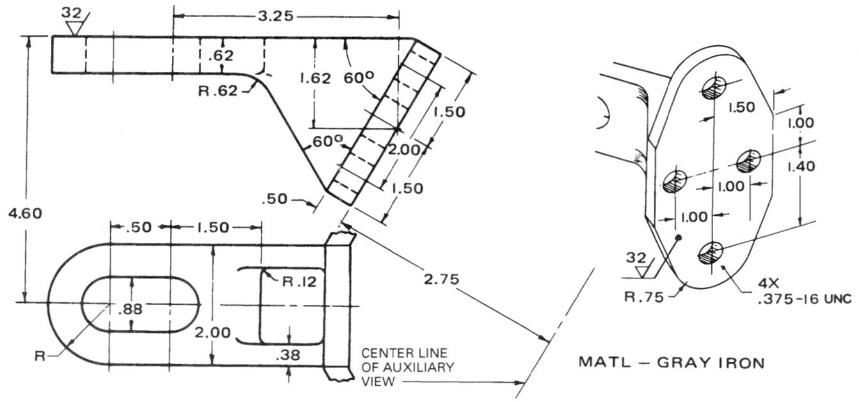

PRIMARY AUXILIARY VIEWS— ANGLE PLATE	REFERENCE UNIT 7-1	NAME COURSE DATE	DRAWING E-2

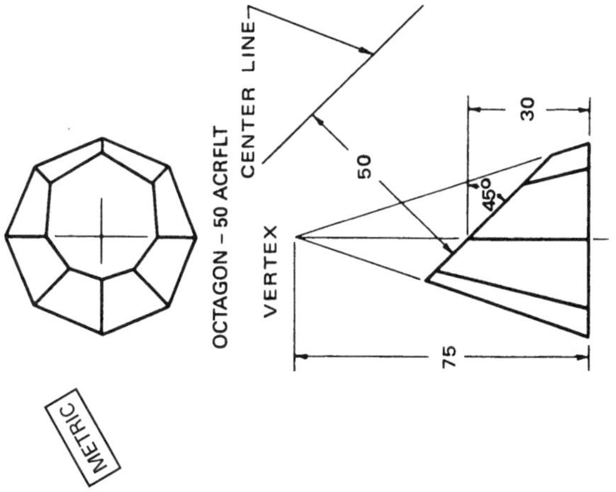

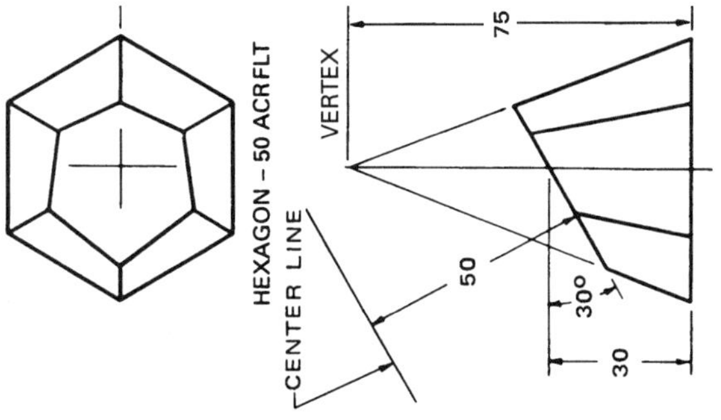

Draw the top, front, and partial auxiliary views of each of the two statue bases. The scale is 1:1. The top and front views are located on the drawing area.

PRIMARY AUXILIARY VIEWS— STATUE BASES	REFERENCE UNIT **7–1**	NAME COURSE DATE	DRAWING **E–3**

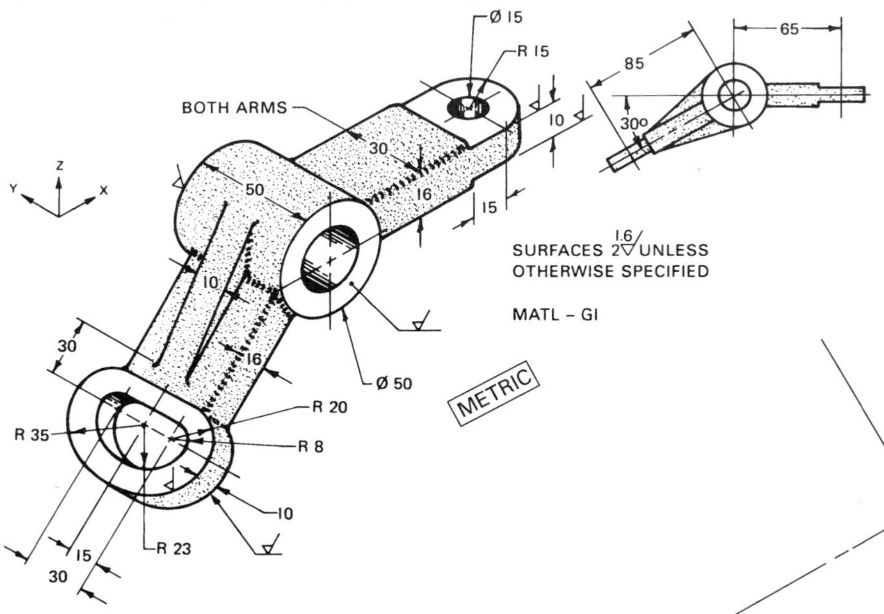

Make a detail drawing of the link. Draw the complete front view and partial top and auxiliary views. The centers of the views are located on the drawing area. The scale is 1:1.

CIRCULAR FEATURES— LINK	REFERENCE UNIT 7–2	NAME COURSE DATE	DRAWING E–4

Make a working drawing of the inclined stop. Draw partial left and right-side views and a partial auxiliary view. The centers for these views are shown on the drawing area. The scale is 1:1.

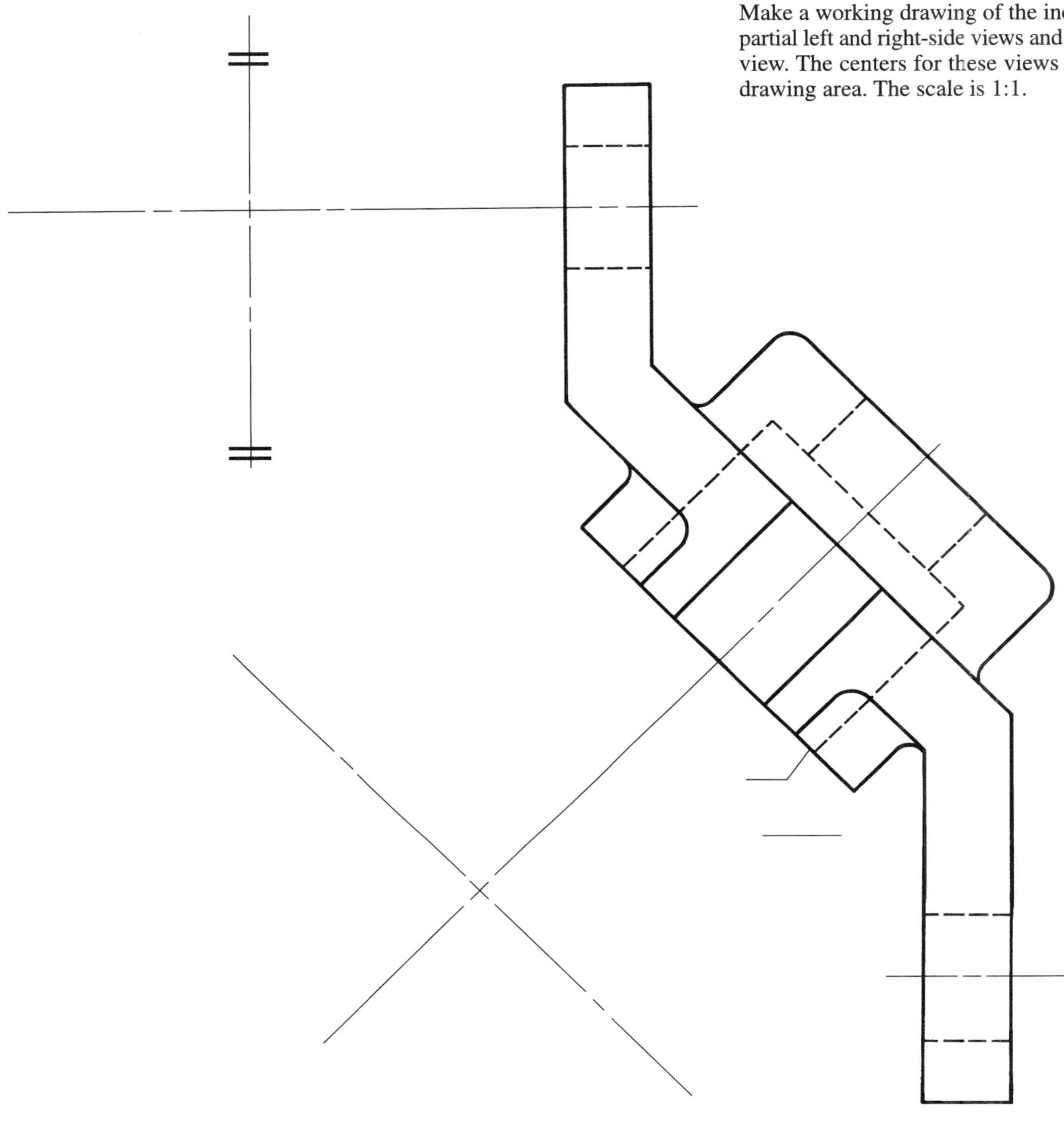

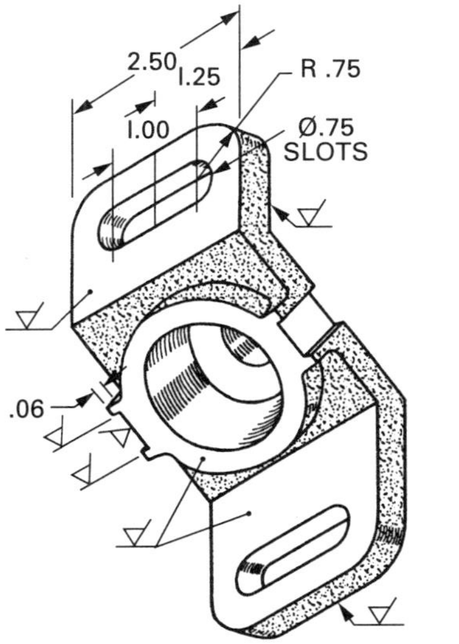

SURFACES MARKED ▽ TO BE .06 $\overset{63}{\bigtriangledown}$
MATL – MI
ROUNDS AND FILLETS R.10

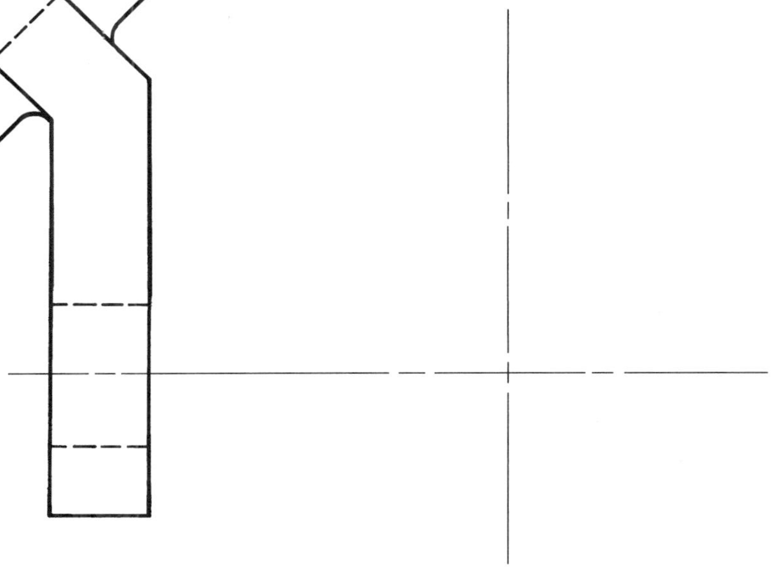

MULTIPLE AUXILIARY VIEWS— INCLINED STOP	REFERENCE UNIT 7–3	NAME ___ COURSE ___ DATE ___	DRAWING E–5

Make a working drawing of the connecting bar. Complete the front view, and add two partial auxiliary views and a bottom view. The location for these views is shown on the drawing area. The scale is 1:1.

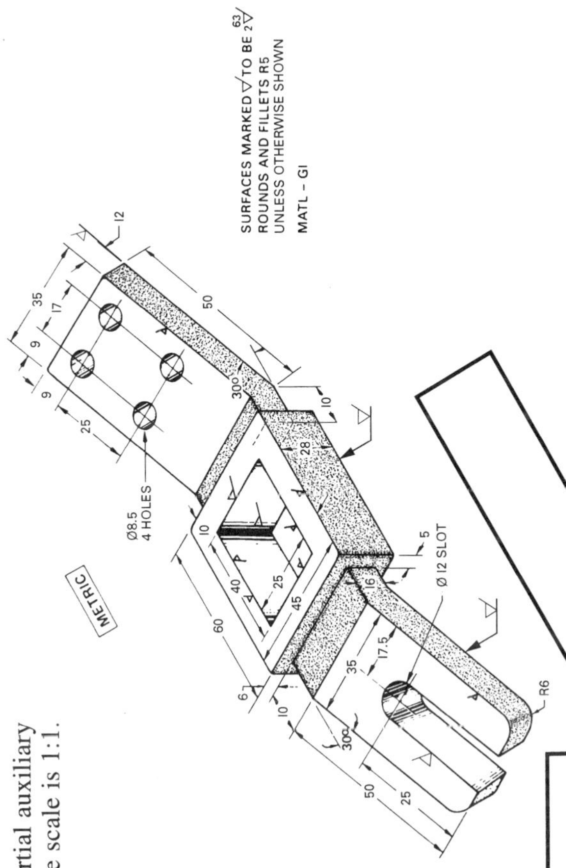

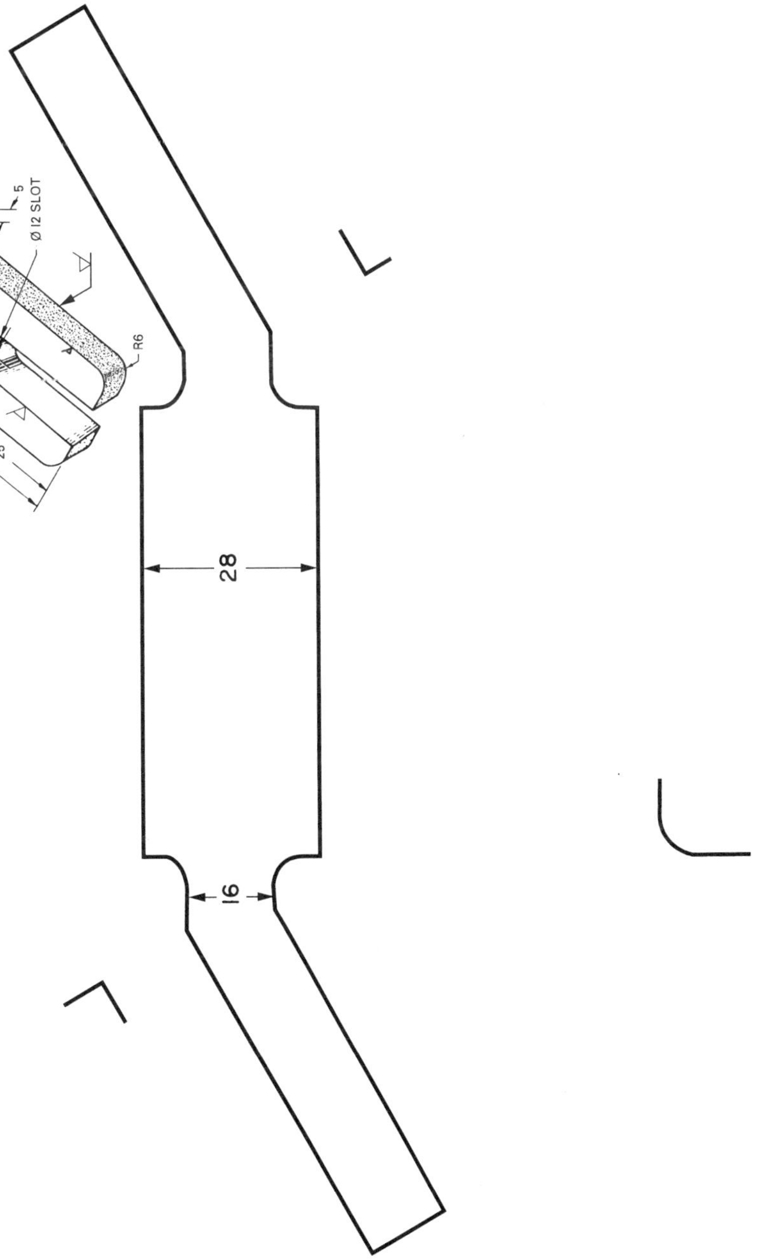

MULTIPLE AUXILIARY VIEWS— CONNECTING BAR	REFERENCE UNIT 7–3	NAME COURSE DATE	DRAWING E–6

Make a detail drawing of the angle stop. Draw complete top and front views and two partial primary auxiliary views. The centers of the four holes locate the top view, and surface A locates the other three views on the drawing area. The scale is 1:1.

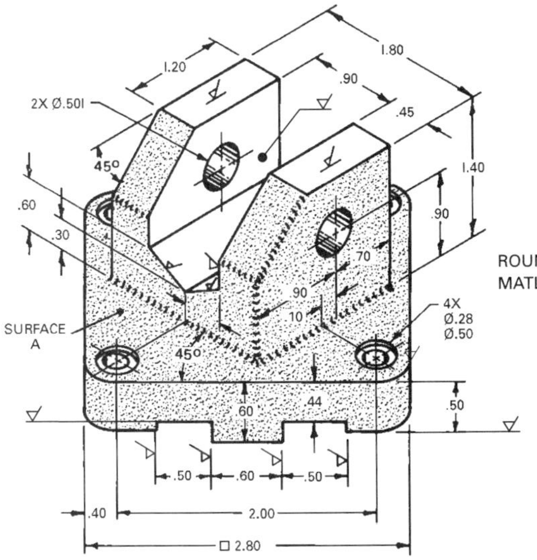

ROUNDS AND FILLETS R.16
MATL—GRAY IRON

MULTIPLE AUXILIARY VIEWS— ANGLE STOP	REFERENCE UNIT 7-3	NAME ___ COURSE ___ DATE ___	DRAWING E-7

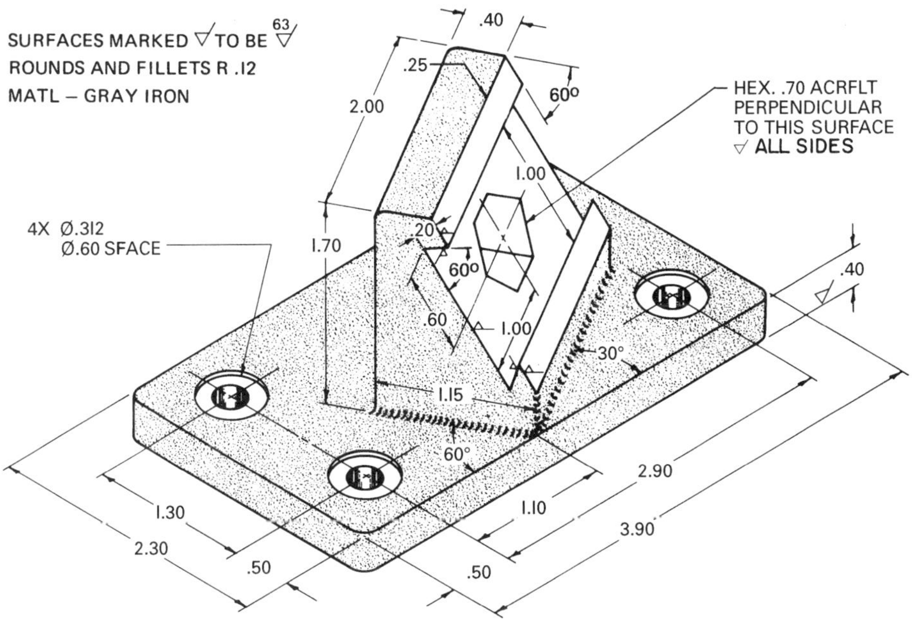

Make a detail drawing of the dovetail bracket. Draw complete top and front views and partial primary and secondary auxiliary views. The front, top, and the line representing the top surface of the base of the primary auxiliary views are located on the drawing area. The primary auxiliary view is projected from the top view. The scale is 1:1.

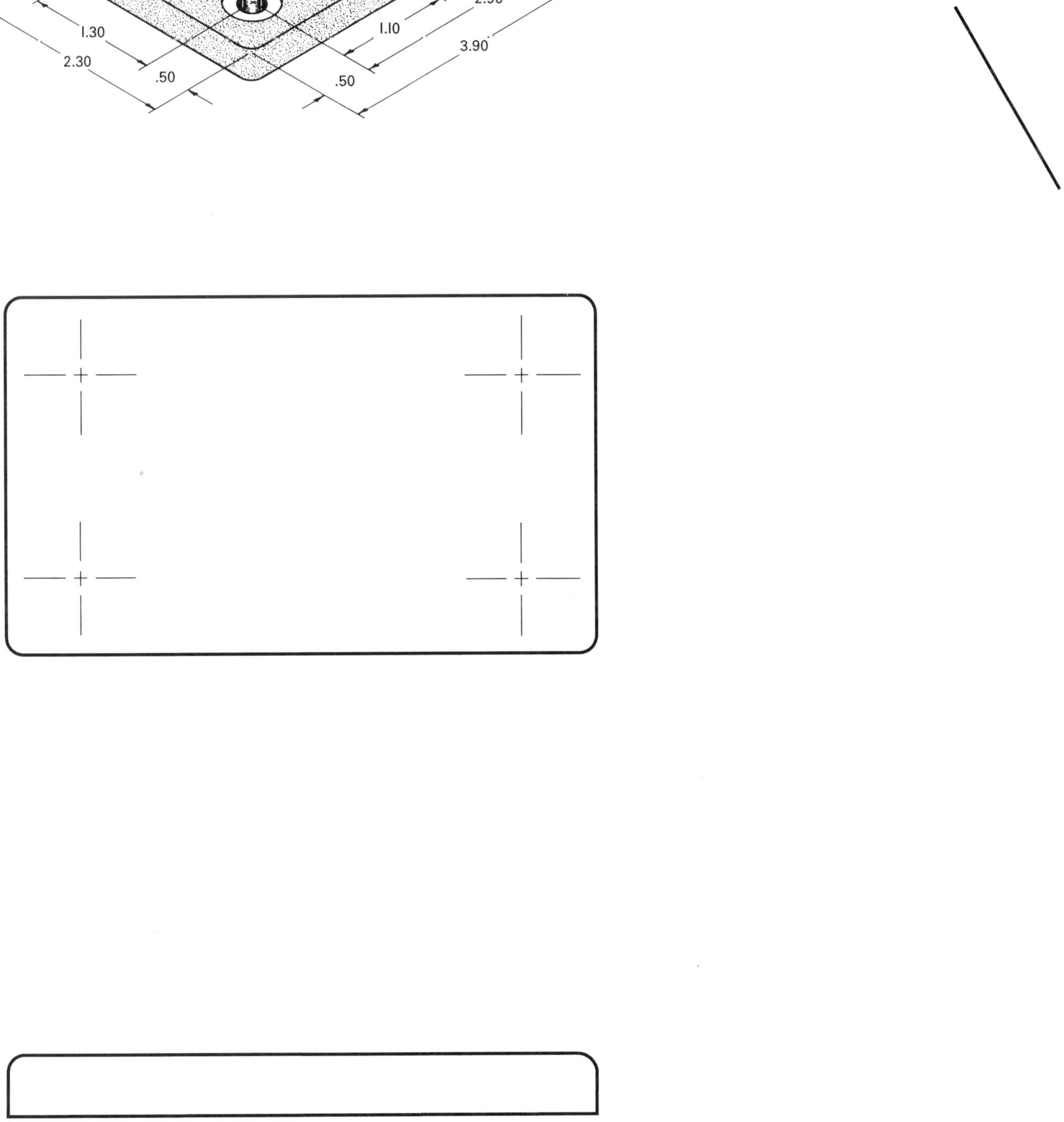

SECONDARY AUXILIARY VIEWS— DOVETAIL BRACKET	REFERENCE UNIT 7–4	NAME COURSE	DATE	DRAWING E–8

Make a working drawing of the pivot arm. Draw complete front and two primary auxiliary views, plus a partial secondary auxiliary view projected from the left primary auxiliary view. The positions for the front and primary auxiliary views are shown on the drawing area. The scale is 1:1.

METRIC

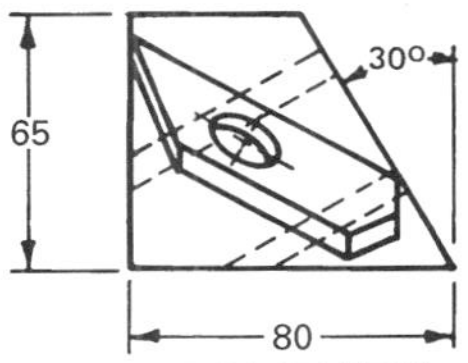

FRONT VIEW

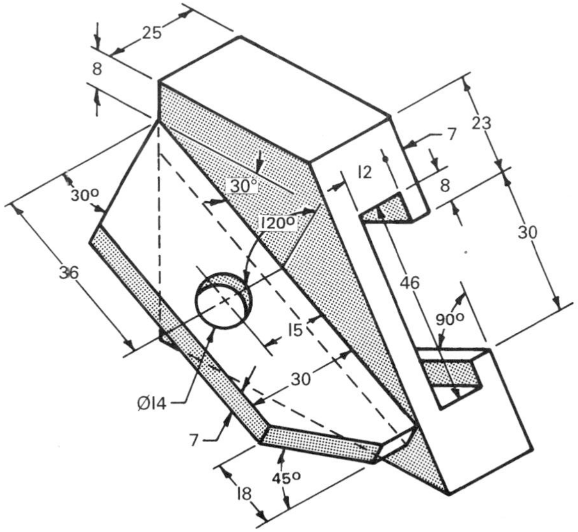

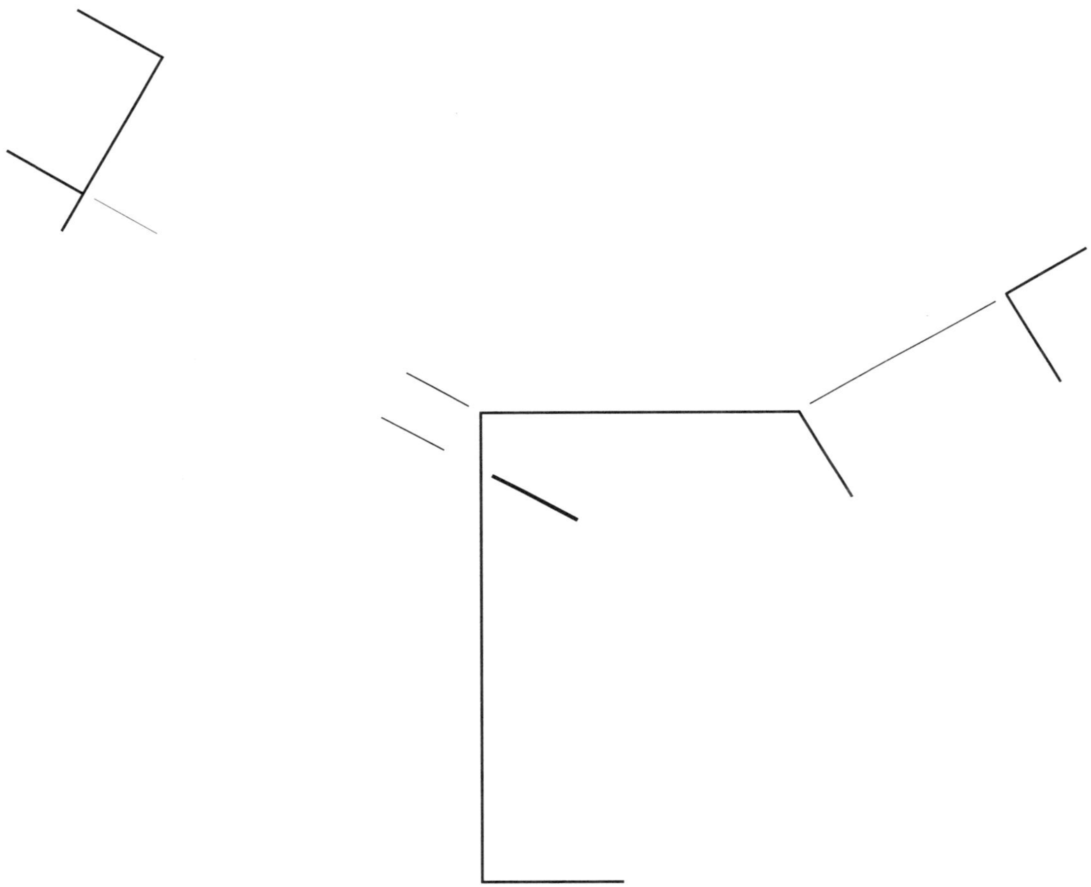

SECONDARY AUXILIARY VIEWS—

PIVOT ARM

REFERENCE UNIT	NAME	DRAWING
7–4	COURSE	DATE	E–9

REVOLUTIONS

REFERENCE UNIT **7–5**

NAME ____

COURSE ____ DATE ____

DRAWING **E–10**

Draw the other two views of parts A, B, and C when they are revolved as shown.

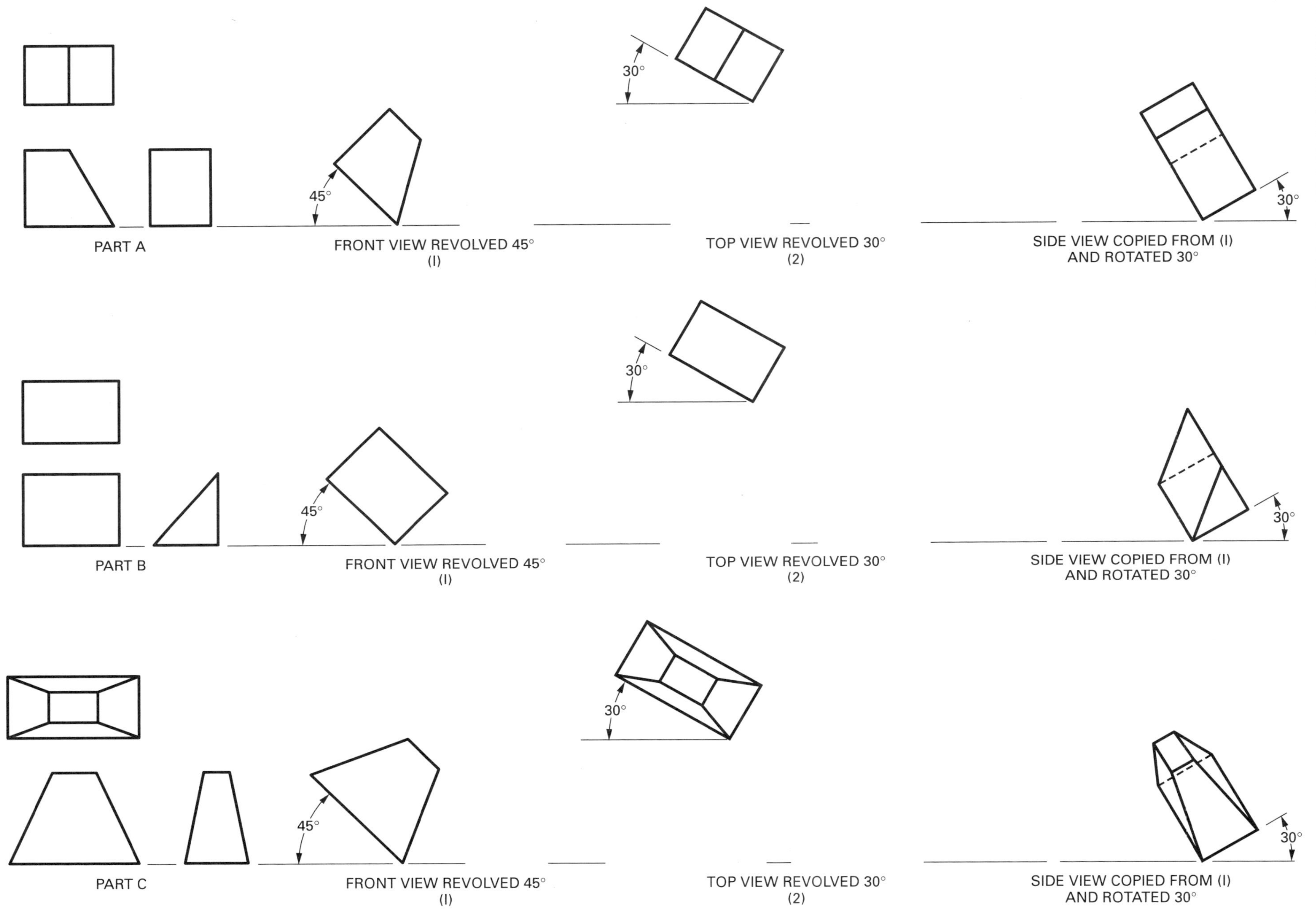

A.

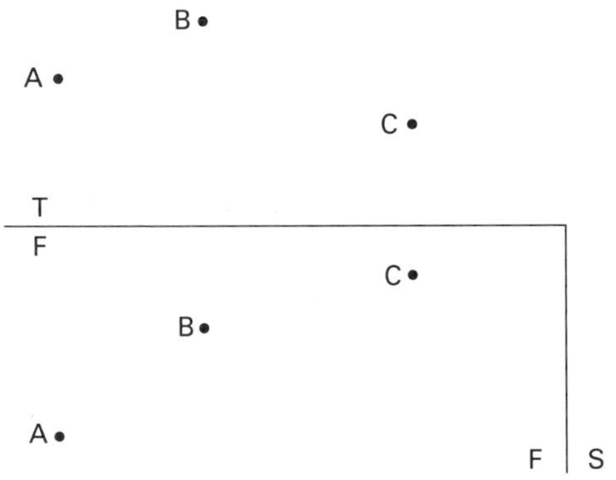

Locate the points in the side view.

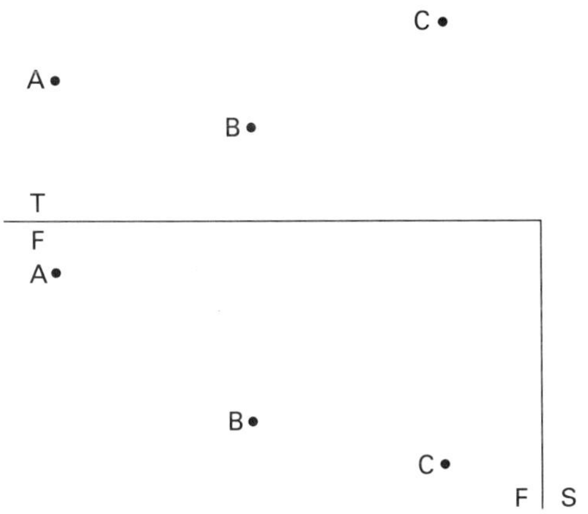

Locate the points in the side view.

B.

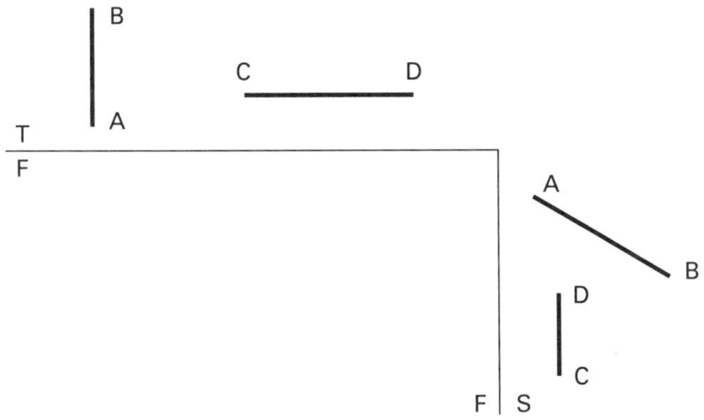

Locate the lines in the front view.

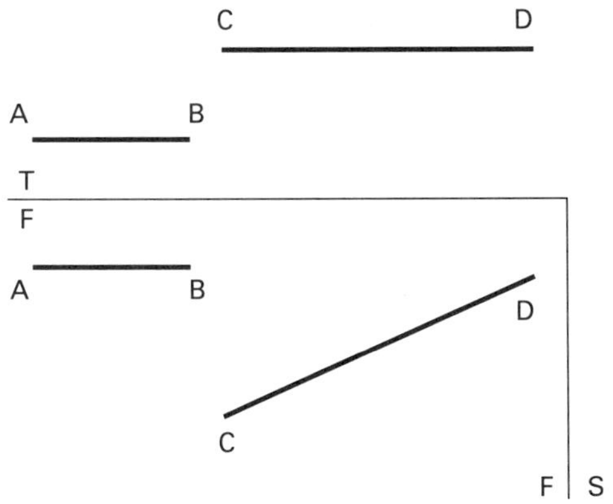

Locate the lines in the side view.

C.

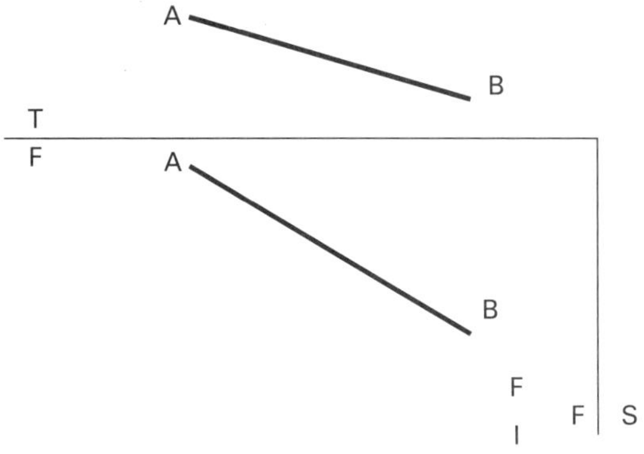

Draw side and auxiliary views. Indicate the true length of line A–B.

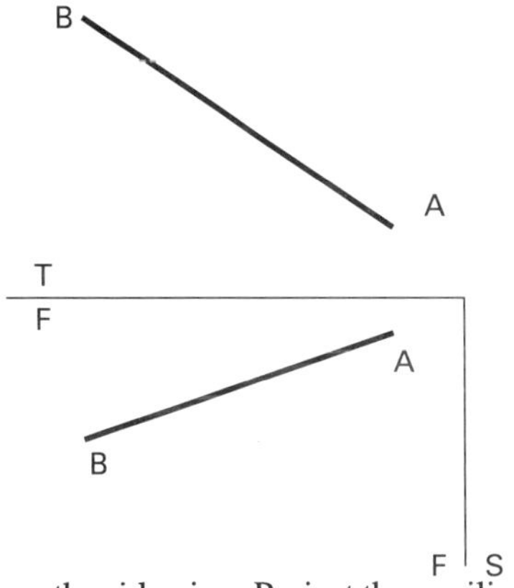

Draw the side view. Project the auxiliary view from the side view. Indicate the true length of line A–B.

D.

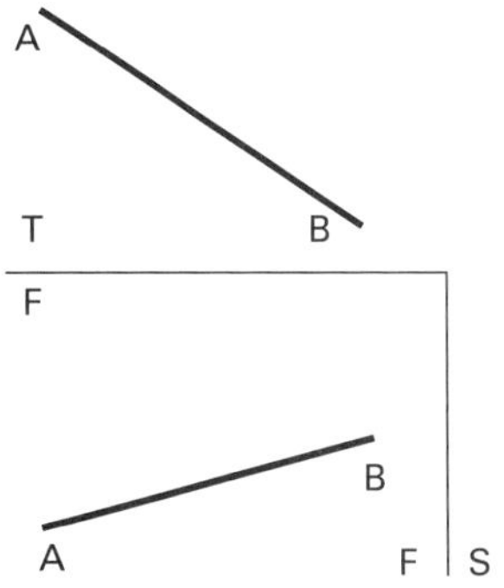

Draw the side view. Project the auxiliary view from the side view to establish the true length of line A–B.

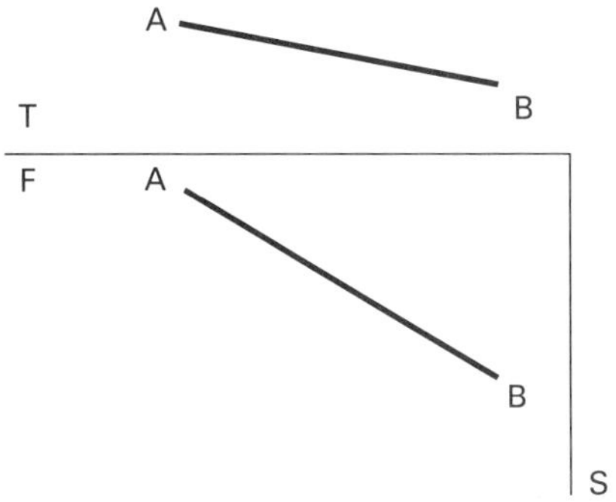

Draw the side view. Project the auxiliary view from the front view to establish the true length of line A–B. Draw a secondary auxiliary view showing the point view of line A–B.

POINTS AND LINES IN SPACE	REFERENCE UNIT 7–6	NAME: COURSE: DATE:	DRAWING E–11

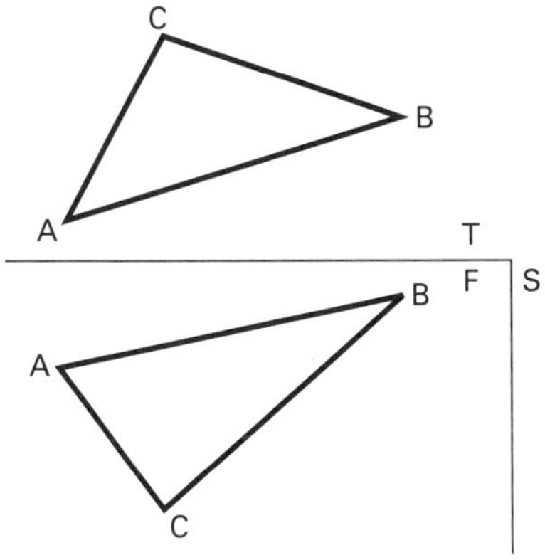

Draw the side view of triangle ABC.

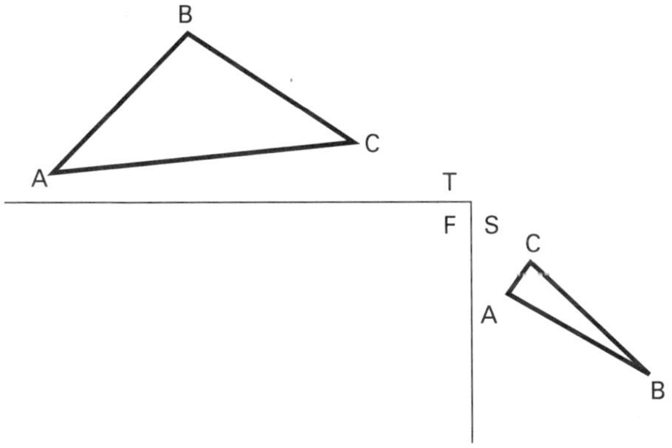

Draw the front view of triangle ABC.

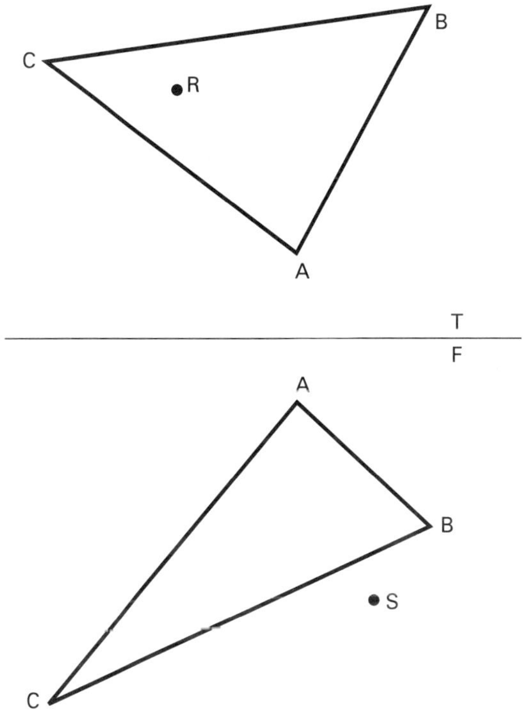

Locate the points in the other views.

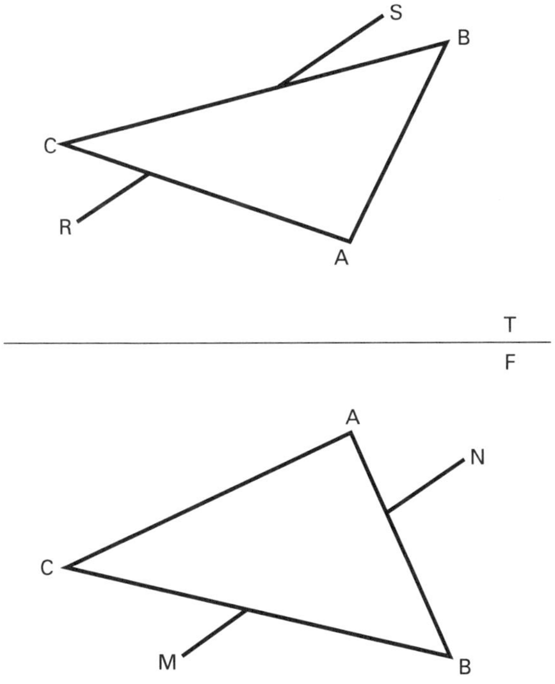

Locate the lines in the other views.

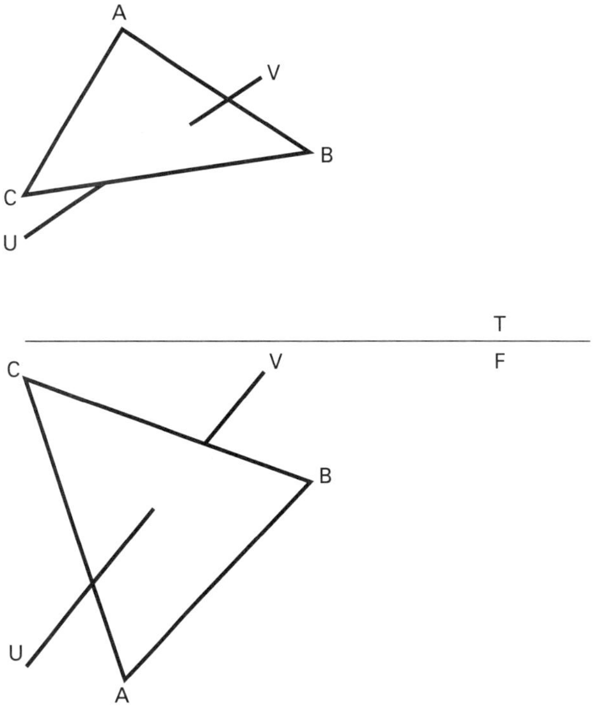

Locate the piercing point of the line and plane and complete line U–V.

PLANES IN SPACE	REFERENCE UNIT **7–7**	NAME COURSE DATE	DRAWING **E–12**

VISIBILITY OF LINES IN SPACE

REFERENCE UNIT 7–8

NAME ____

COURSE ____ DATE ____

DRAWING E–13

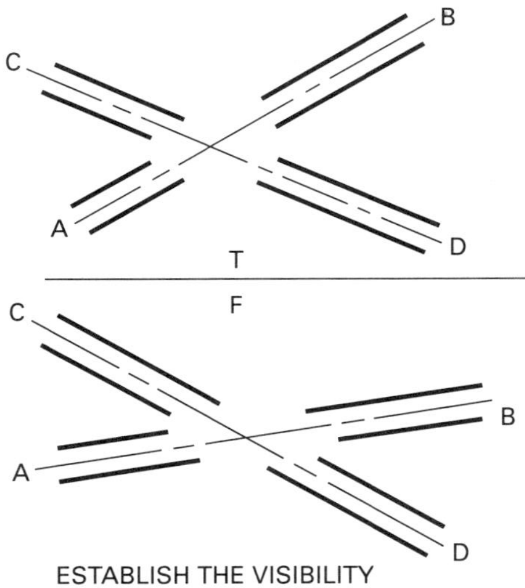

ESTABLISH THE VISIBILITY OF PIPES A-B AND C-D

1

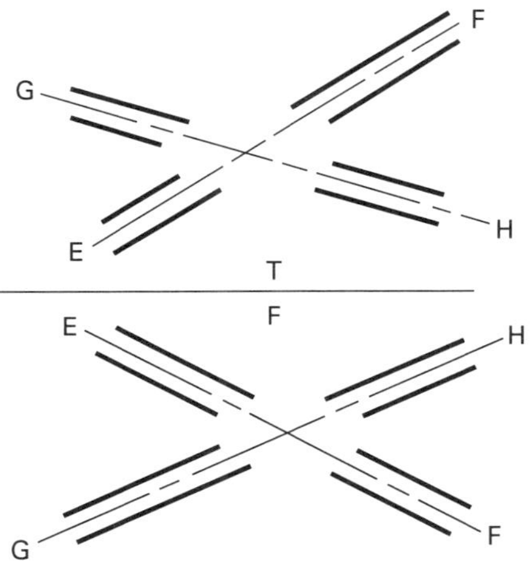

ESTABLISH THE VISIBILITY OF PIPES E-F AND G-H

2

By observation, sketch the circular pipes. Show the direction in which the pipes are sloping and which pipes are closest to the observer.

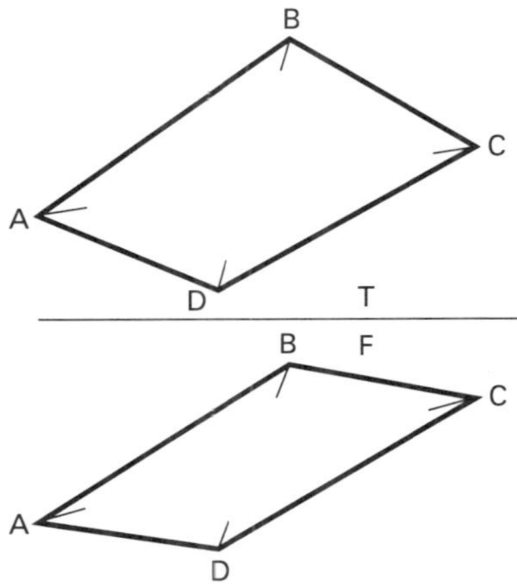

ESTABLISH THE VISIBILITY OF LINES A-C, AND B-D

3

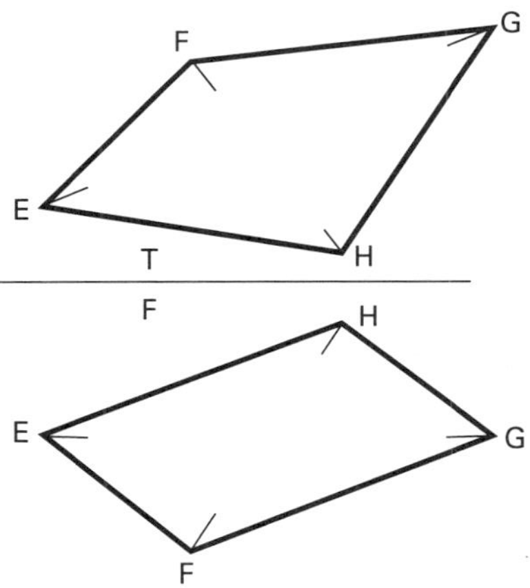

ESTABLISH THE VISIBILITY OF LINES E-G AND F-H

4

Establish the visibility of the lines and surfaces by showing which lines are seen or hidden.

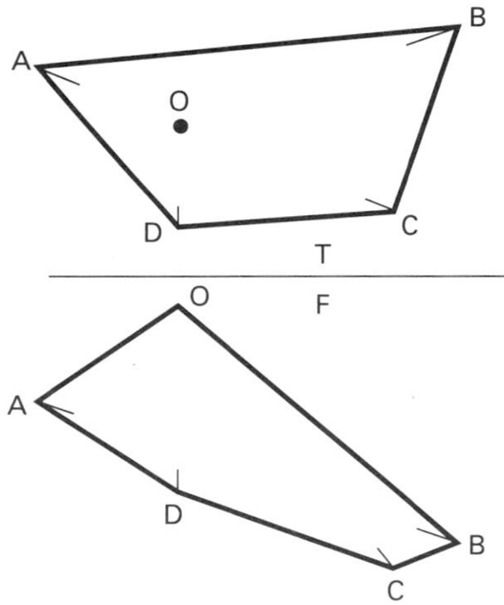

ESTABLISH THE VISIBILITY OF LINES O-A, O-B, O-C AND O-D

5

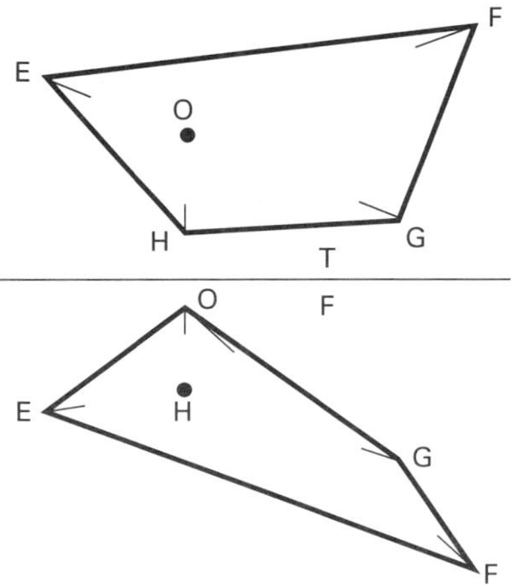

ESTABLISH THE VISIBILITY OF LINES O-E, O-F, O-G AND O-H

6

Establish the visibility of the lines and surfaces by showing which lines are seen or hidden.

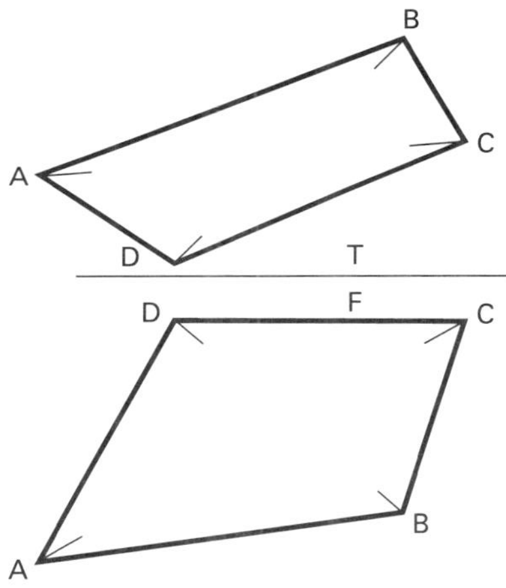

ESTABLISH THE VISIBILITY OF LINES A-C, AND B-D

7

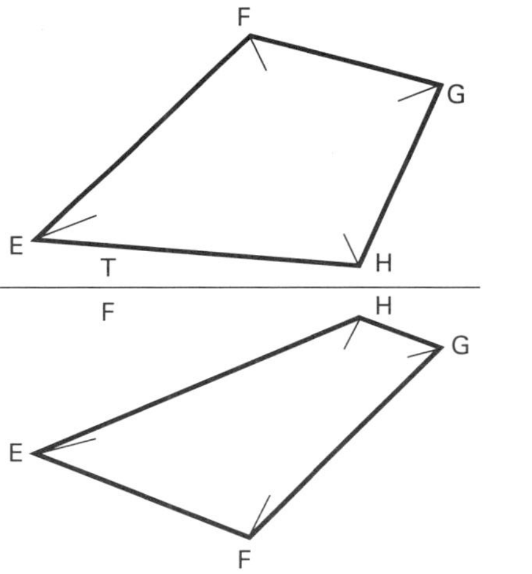

ESTABLISH THE VISIBILITY OF LINES E-G AND F-H

8

Establish the visibility of the lines and surfaces by showing which lines are seen or hidden.

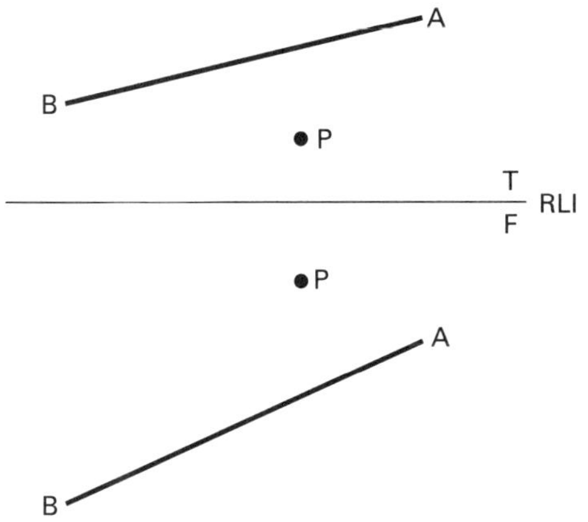

Find the distance between point P line and A–B.

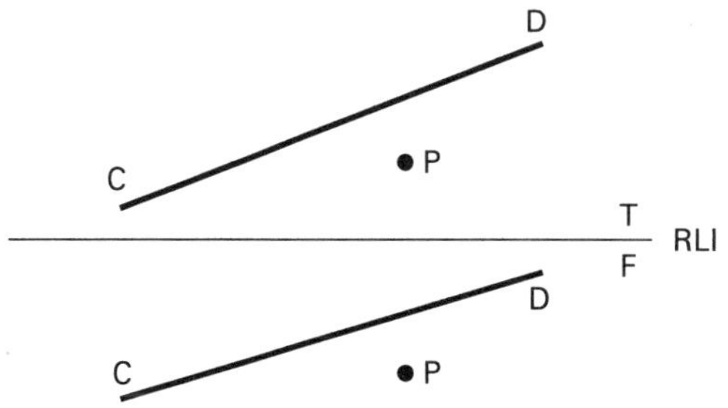

Find the distance between point P and line C–D.

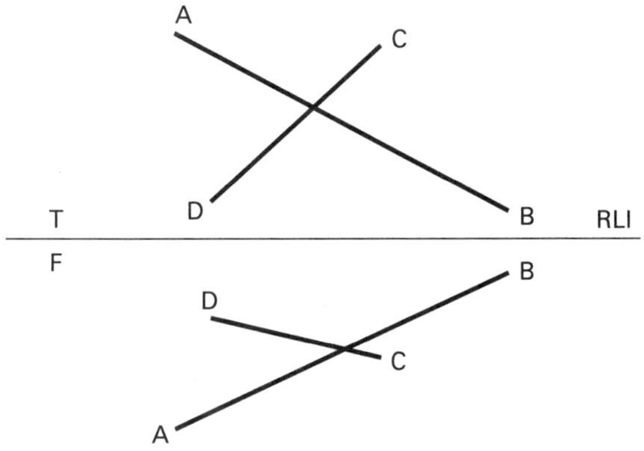

Find the shortest distance between lines A–B and C–D.

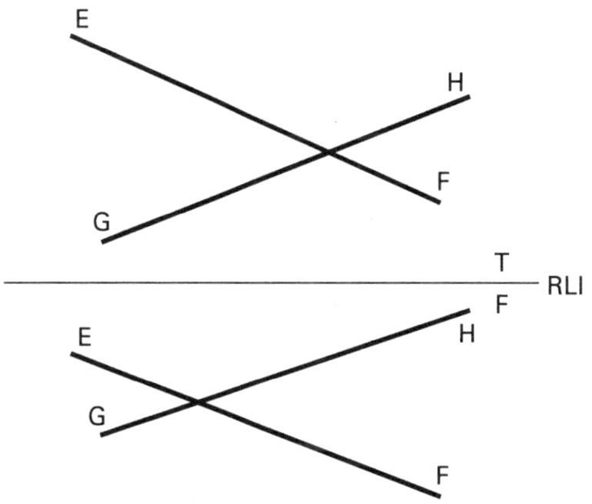

Find the shortest distance between lines E–F and G–H.

DISTANCE BETWEEN LINES AND POINTS	REFERENCE UNIT 7–9	NAME COURSE DATE	DRAWING E–14

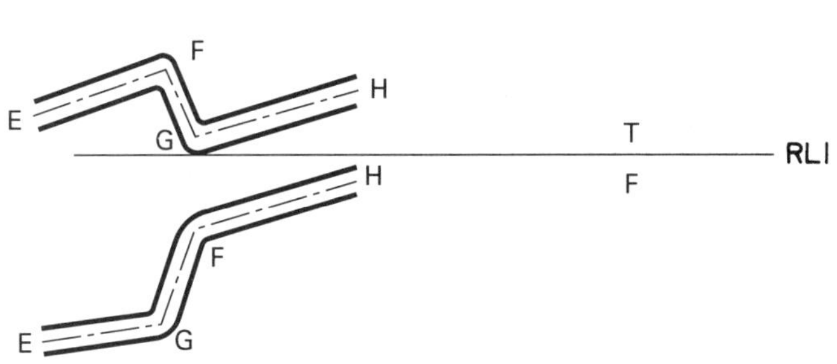

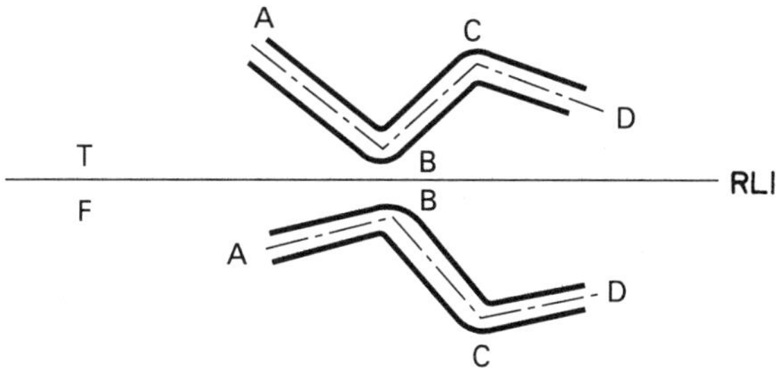

Find the true angles of EFG and FGH.

Find the true angles of ABC and BCD.

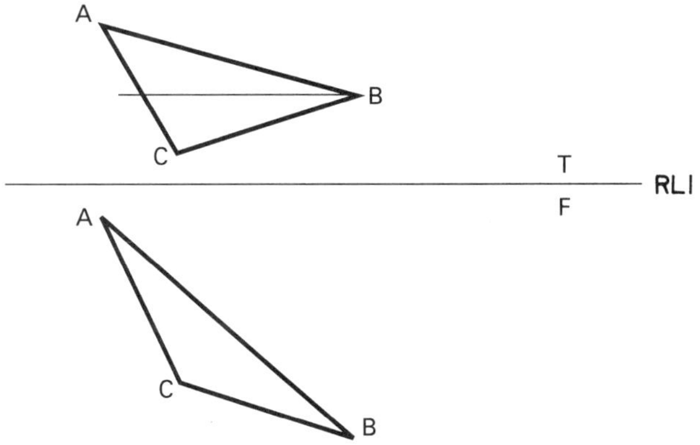

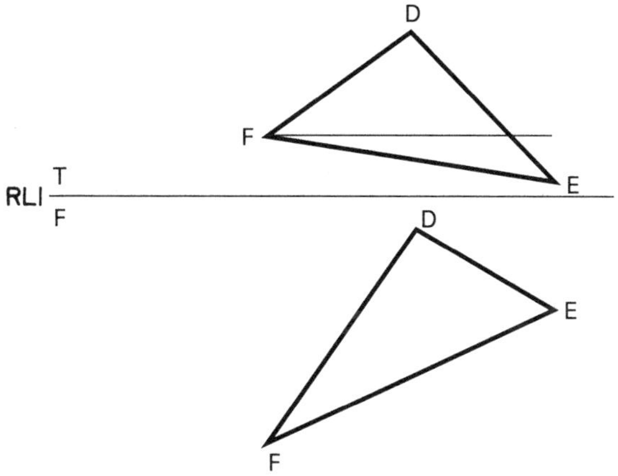

Find the true view of plane ABC.

Find the true view of plane DEF.

EDGE AND TRUE VIEW OF PLANES	REFERENCE UNIT 7–10	NAME COURSE DATE	DRAWING E–15

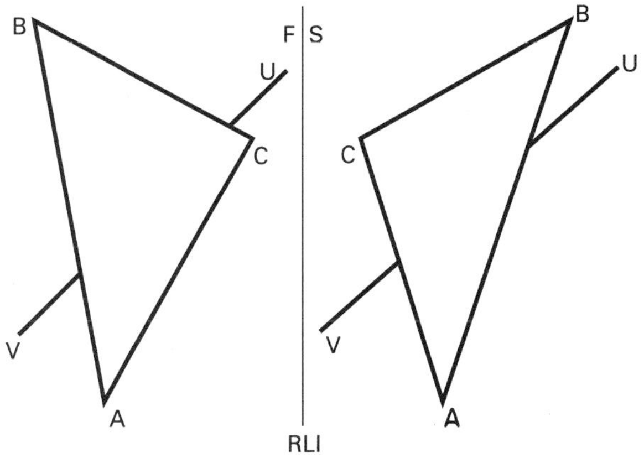

Find the angle that line U–V makes with plane ABC.

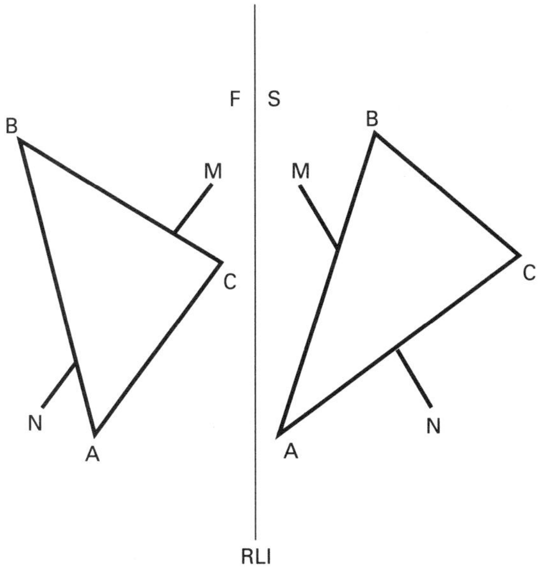

Find the angle that line M–N makes with plane ABC.

ANGLE BETWEEN LINES AND PLANES	REFERENCE UNIT 7–11	NAME COURSE DATE	DRAWING E–16

Make a three-view drawing, complete with dimensions, of the notched block. Place the dimensions with the view that shows best the shape of the part or feature. The scale is 1:1. The bottom left corner of the front view is shown on the drawing area.

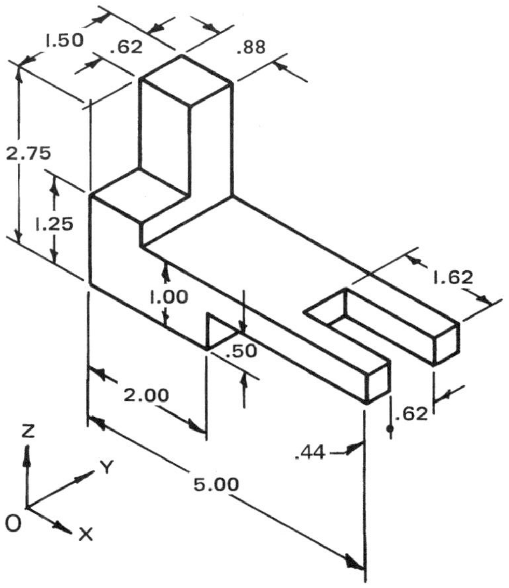

BASIC DIMENSIONING—

NOTCHED BLOCK

REFERENCE UNIT **8–1**

NAME

COURSE

DATE

DRAWING **F–1**

Make a three-view drawing, complete with dimensions of the vertical guide. Place the dimensions with the view that shows best the shape of the part or feature. The scale is 1:1. The bottom left corner of the front view is shown on the drawing area.

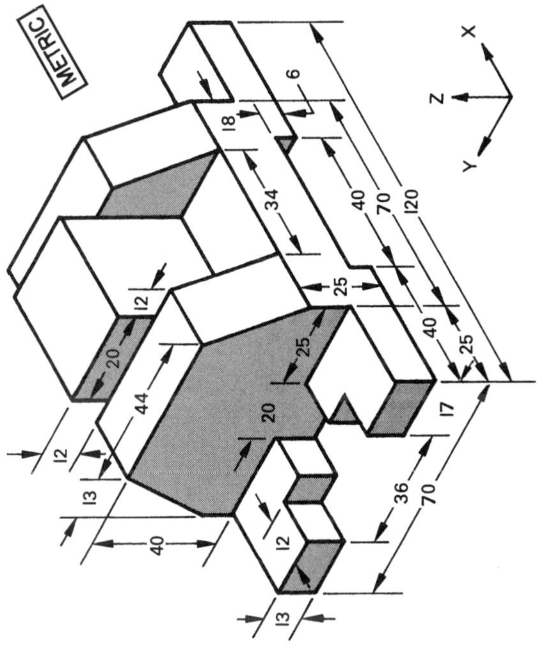

BASIC DIMENSIONING— VERTICAL GUIDE	REFERENCE UNIT **8–1**	NAME COURSE　　DATE	DRAWING **F–2**

Make a three-view drawing, complete with dimensions, of the base. Place the dimensions with the view that shows best the shape of the part or feature. The scale is 1:2. The bottom left corner of the front view is shown on the drawing area.

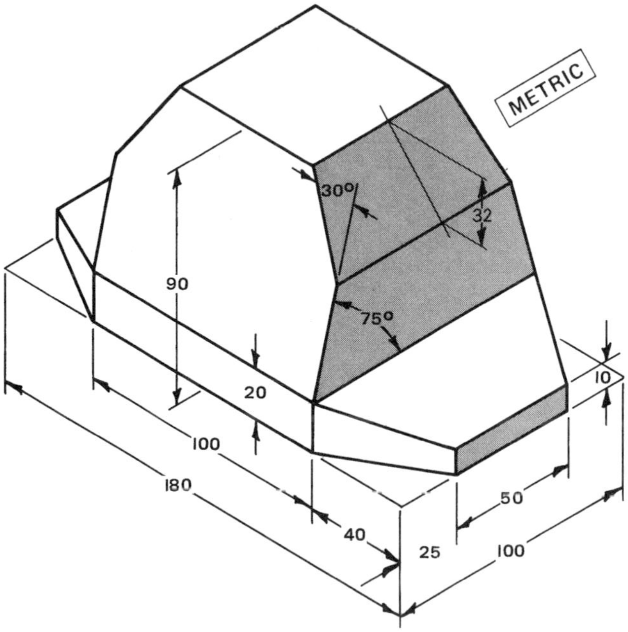

BASIC DIMENSIONING— BASE	REFERENCE UNIT 8–1	NAME COURSE DATE	DRAWING F–3

Make a one-view drawing, complete with dimensions, of the adjustable sector. The scale is 1:1.

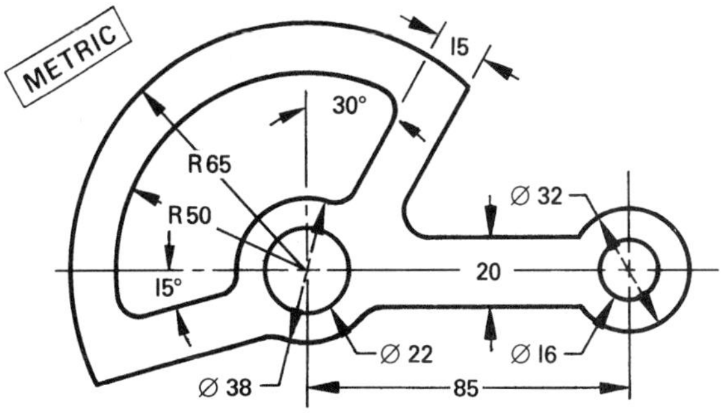

FILLETS R5

MATL SAE 1020 3 THICK

DIMENSIONING CIRCULAR FEATURES— ADJUSTABLE SECTOR	REFERENCE UNIT **8–2–D**	NAME COURSE DATE	DRAWING **F–4**

Make a three-view drawing, complete with dimensions, of the bracket. Place the dimensions with the view that shows best the shape of the part or feature. The scale is 1:1.

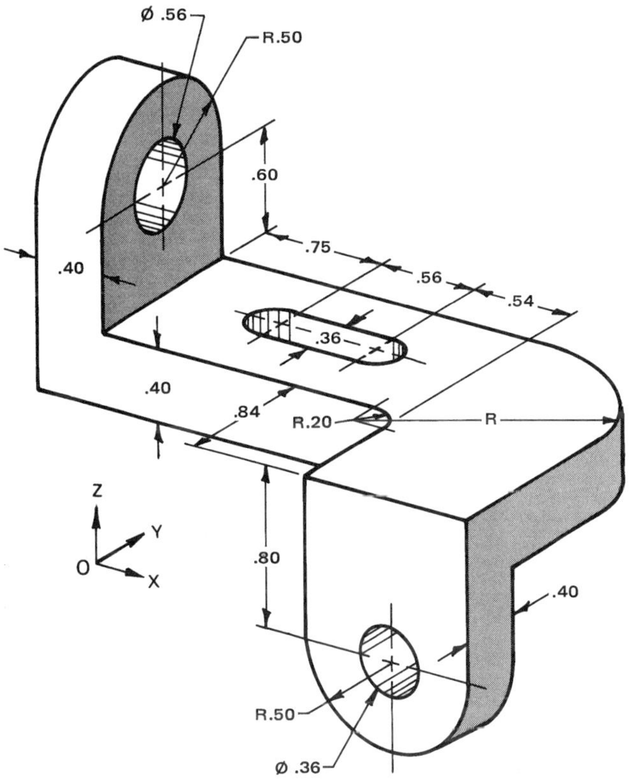

DIMENSIONING CIRCULAR FEATURES— BRACKET	REFERENCE UNIT 8–2	NAME COURSE DATE	DRAWING F–5

Redraw the handle shown to the scale 1:1. The following features are to be added and dimensioned in the locations shown.

(A) 45° × .10 in. chamfer
(B) 33DP diamond knurl for 1.20 in. starting .80 in. from left end
(C) 1:8 circular taper for 1.20 in. length on right end of Ø1.25 in.
(D) .16 × Ø.54 in. undercut on Ø.750
(E) Ø.189 × .25 in. deep, 4 holes equally spaced
(F) 30° × .10 in. chamfer (the .10 in. dimension taken horizontally along the shaft)

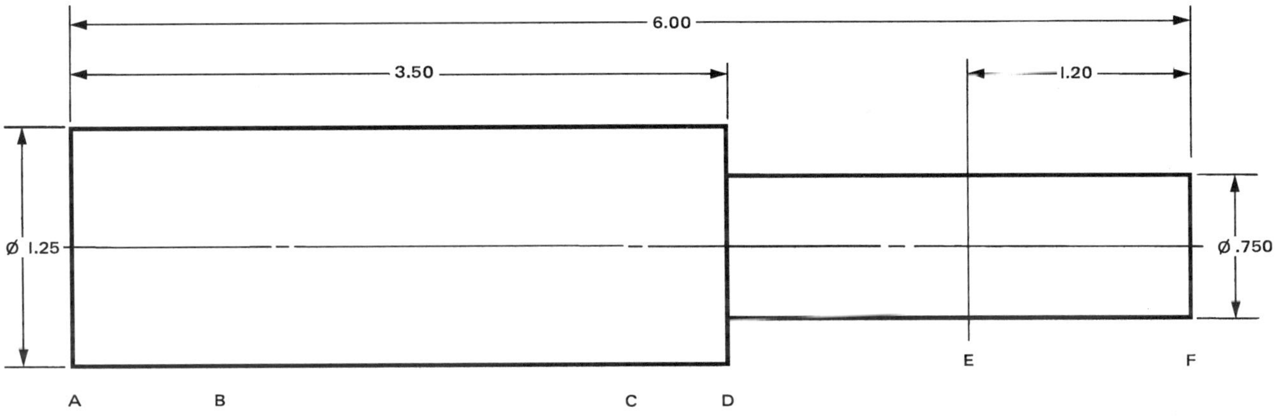

DIMENSIONING COMMON FEATURES— HANDLE	REFERENCE UNIT 8–3	NAME COURSE DATE	DRAWING F–6

Make a half-view drawing of the gasket. Add the symmetry symbol to the drawing and dimension. The scale is 1:1.

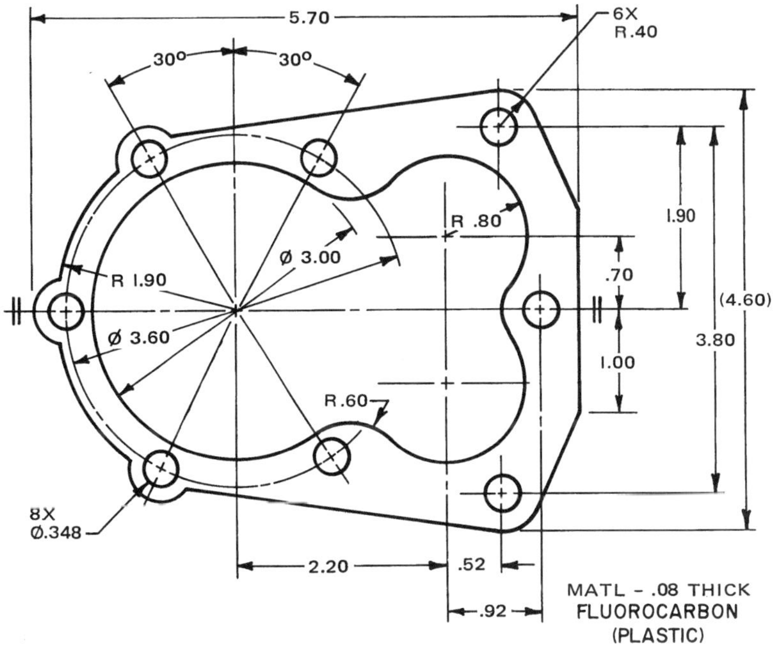

DIMENSIONING COMMON FEATURES— GASKET	REFERENCE UNIT 8–3	NAME COURSE DATE	DRAWING F–7

Make a one-view working drawing of the transmission cover to the scale 1:1. The arrowless dimensioning shown should be replaced with rectangular coordinate dimensioning and have the following dimensional changes:

(a) holes E and D are located from left and bottom edges
(b) holes A and C are located from center of hole D
(c) hole B is located from center of hole E

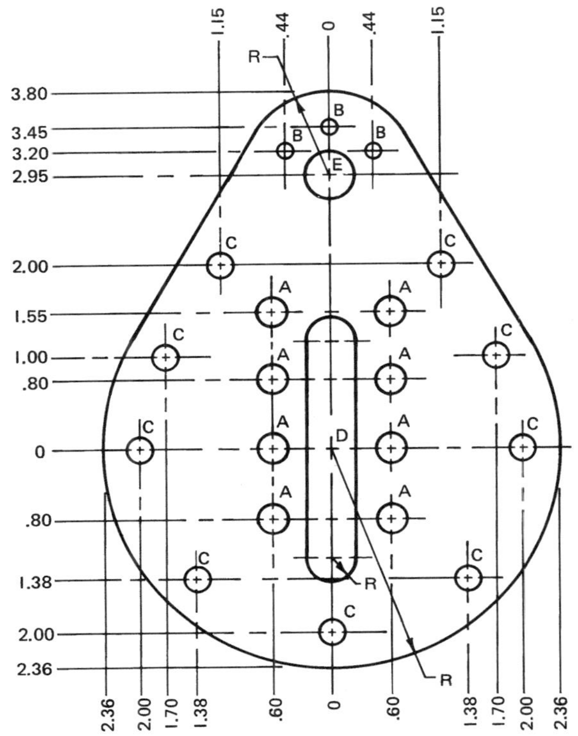

MATL—SAE1008
.12 THICK

HOLE	SIZE
A	.30
B	.16
C	.24
D	.40 × 2.80
E	.50

DIMENSIONING METHODS—
TRANSMISSION COVER

REFERENCE UNIT **8–4**

NAME

COURSE

DATE

DRAWING **F–8**

Redraw the adapter plate four times to the scale 1:1. Different methods of dimensioning should be used for each drawing: rectangular coordinate, polar, arrowless, and tabular.

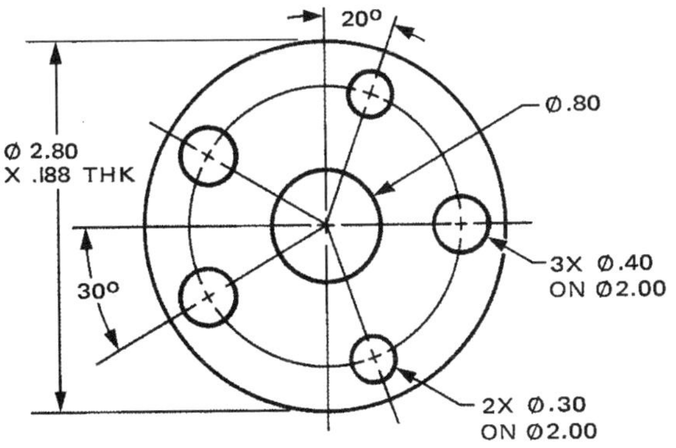

TABULAR

ARROWLESS

RECTANGULAR COORDINATE

POLAR

DIMENSIONING METHODS— ADAPTER PLATE	REFERENCE UNIT **8–4**	NAME ___ COURSE ___ DATE ___	DRAWING **F–9**

LIMITS AND TOLERANCES

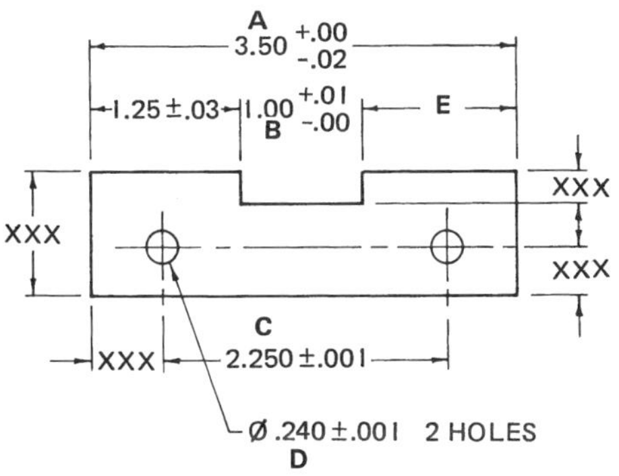

Complete this chart from the information given above.

		A	B	C	D	E
BASIC SIZE						X
TOLERANCE						X
LIMITS OF SIZE	MAX					
	MIN					

LIMITS AND TOLERANCES

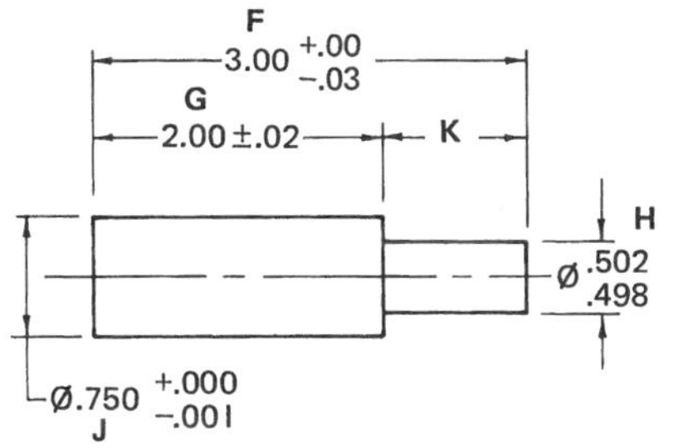

Complete this chart from the information given above.

		F	G	H	J	K
BASIC SIZE						X
TOLERANCE						X
LIMITS OF SIZE	MAX					
	MIN					

LIMITS AND TOLERANCES

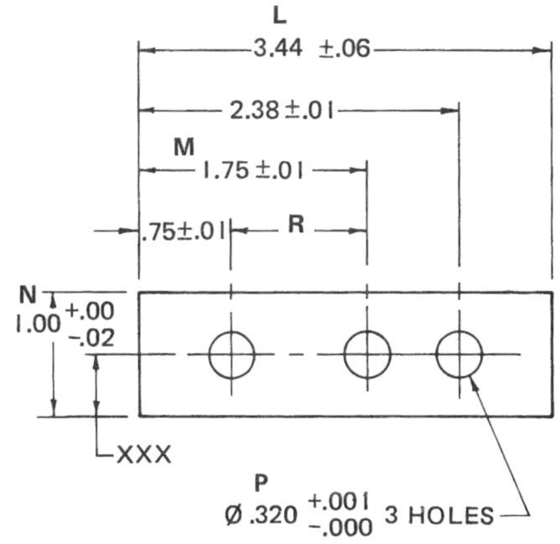

Complete this chart from the information given above.

		L	M	N	P	R
BASIC SIZE						X
TOLERANCE						X
LIMITS OF SIZE	MAX					
	MIN					

INCH LIMITS AND TOLERANCES

LIMITS AND TOLERANCES

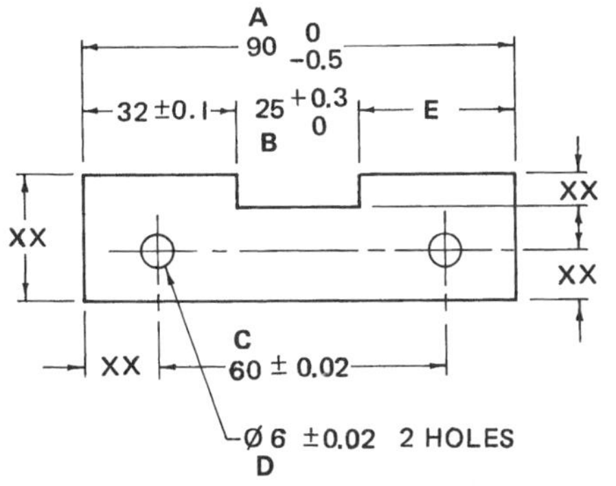

Complete this chart from the information given above.

		A	B	C	D	E
BASIC SIZE						X
TOLERANCE						X
LIMITS OF SIZE	MAX					
	MIN					

LIMITS AND TOLERANCES

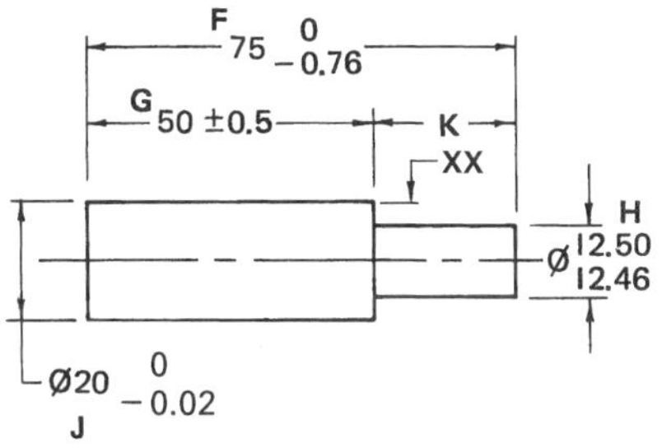

Complete this chart from the information given above.

		F	G	H	J	K
BASIC SIZE						X
TOLERANCE						X
LIMITS OF SIZE	MAX					
	MIN					

LIMITS AND TOLERANCES

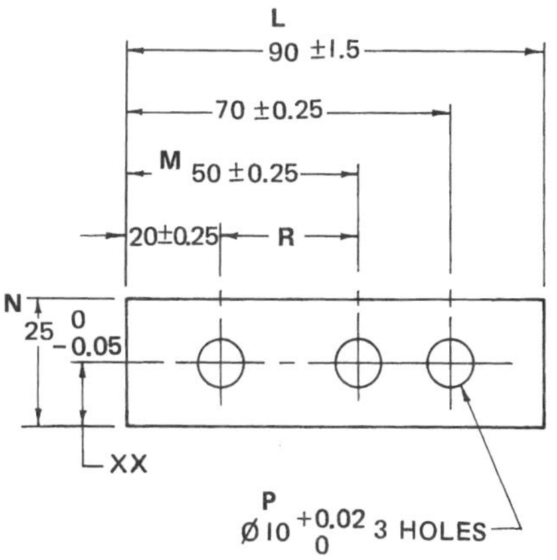

Complete this chart from the information given above.

		L	M	N	P	R
BASIC SIZE						X
TOLERANCE						X
LIMITS OF SIZE	MAX					
	MIN					

MILLIMETER (METRIC) LIMITS AND TOLERANCES

Calculate the sizes and tolerances.

LIMITS AND TOLERANCES	REFERENCE UNIT 8–5	NAME COURSE DATE	DRAWING F–10

CLEARANCE FITS (RUNNING OR SLIDING)	TRANSITION FITS (LOCATIONAL)	INTERFERENCE FITS (FORCE OR SHRINK)

CLEARANCE FITS (RUNNING OR SLIDING)

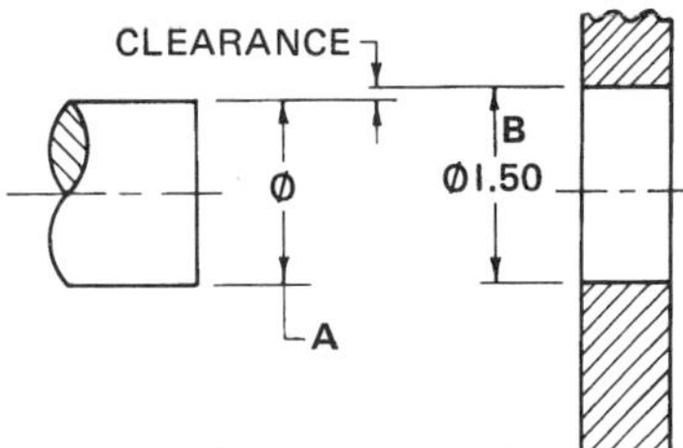

RC2 SLIDING FIT

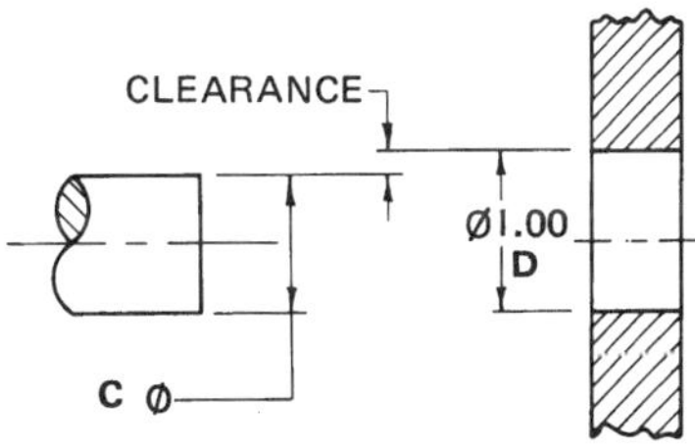

RC5 RUNNING FIT

Complete this chart using the proper limit and fit tables.

FIT	ØA	ØB	ØC	ØD
RC2				
RC5				

TRANSITION FITS (LOCATIONAL)

LC2 LOCATIONAL CLEARANCE

LT3 LOCATIONAL TRANSITION

LN2 LOCATIONAL INTERFERENCE

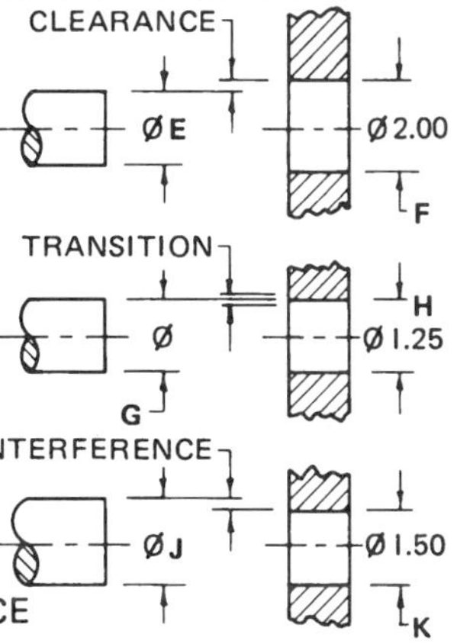

Complete this chart using the proper limit and fit tables.

DIA	LC2	LT3	LN2
ØE			
ØF			
ØG			
ØH			
ØJ			
ØK			

INTERFERENCE FITS (FORCE OR SHRINK)

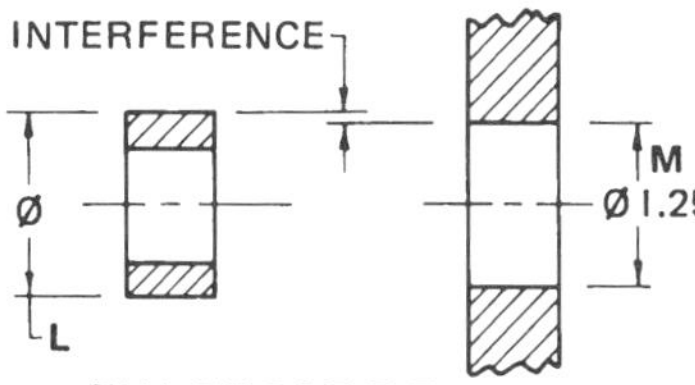

FNI FORCE FIT

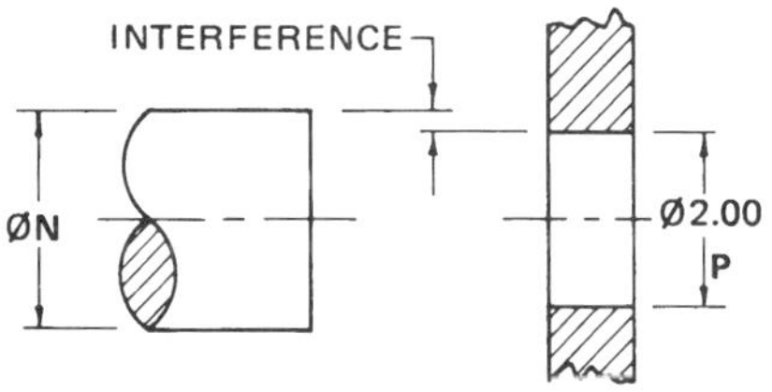

FN4 SHRINK FIT

Complete this chart using the proper limit and fit tables.

FIT	ØL	ØM	ØN	ØP
FNI				
FN4				

CLEARANCE FITS

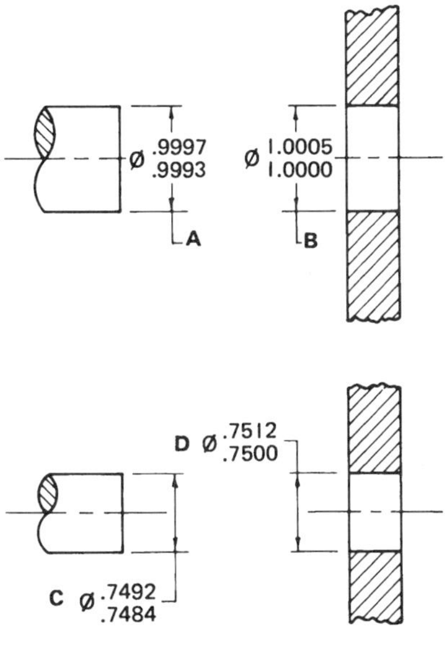

Complete this chart from the information given above.

	A	B	C	D
TOLERANCE ON HOLE				
TOLERANCE ON SHAFT				
MINIMUM CLEARANCE				
MAXIMUM CLEARANCE				

INTERFERENCE FITS

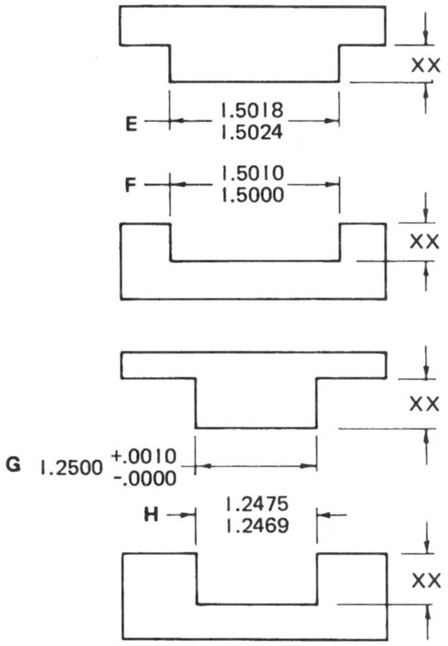

Complete this chart from the information given above.

	E	F	G	H
TOLERANCE ON PART				
TOLERANCE ON SLOT				
MINIMUM INTERFERENCE				
MAXIMUM INTERFERENCE				

CLEARANCE AND INTERFERENCE FITS

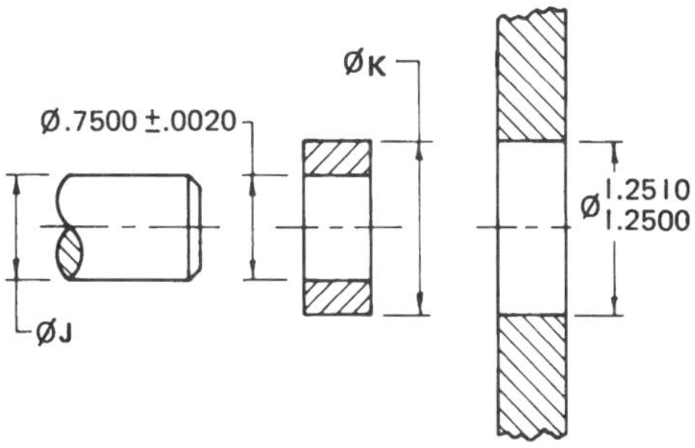

QI. DIMENSION SHAFT (J) TO HAVE A TOLERANCE OF .0012 AND A MINIMUM CLEARANCE OF .0025.

Q2. DIMENSION BUSHING (K) TO HAVE A TOLERANCE OF .0010 AND A MAXIMUM INTERFERENCE OF .0016.

Q3. DIMENSION SHAFT (J) TO HAVE A TOLERANCE OF .0012 AND A MINIMUM CLEARANCE OF .0014.

Q4. DIMENSION BUSHING (K) TO HAVE A TOLERANCE OF .0008 AND A MAXIMUM INTERFERENCE OF .0022.

Complete this chart from the information given above.

	J	K
QI		
Q2		
Q3		
Q4		

Using the table of fits in the appendix, calculate the missing dimensions.

FITS AND ALLOWANCES— INCH	REFERENCE UNIT 8-6	NAME COURSE DATE	DRAWING F-11

CLEARANCE FITS (RUNNING OR SLIDING)	TRANSITION FITS (LOCATIONAL)	INTERFERENCE FITS (FORCE OR SHRINK)

CLEARANCE FITS (RUNNING OR SLIDING)

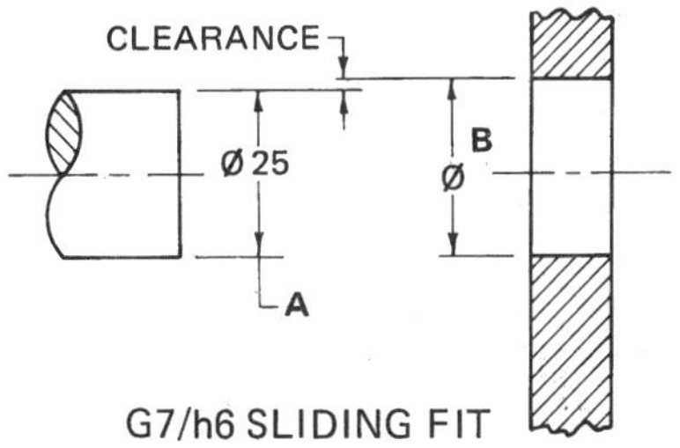

G7/h6 SLIDING FIT

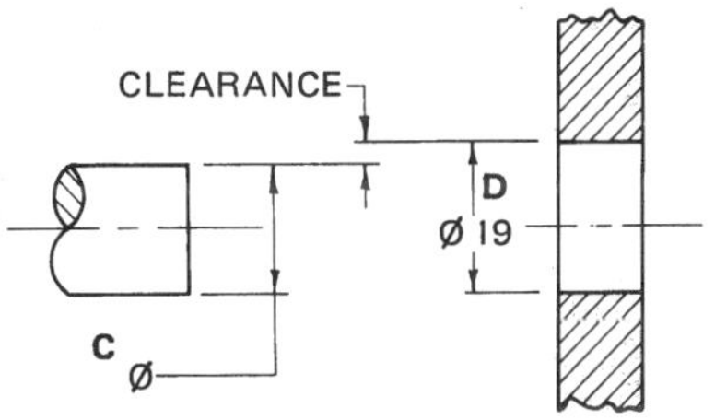

H9/d9 RUNNING FIT

Complete this chart using the proper limit and fit tables.

FIT	ØA	ØB	ØC	ØD
G7/h6				
H9/d9				

TRANSITION FITS (LOCATIONAL)

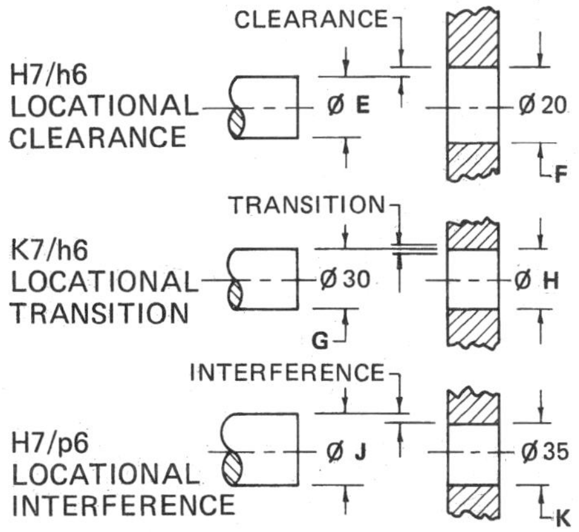

Complete this chart using the proper limit and fit tables.

DIA	H7/h6	K7/h6	H7/p6
ØE			
ØF			
ØG			
ØH			
ØJ			
ØK			

INTERFERENCE FITS (FORCE OR SHRINK)

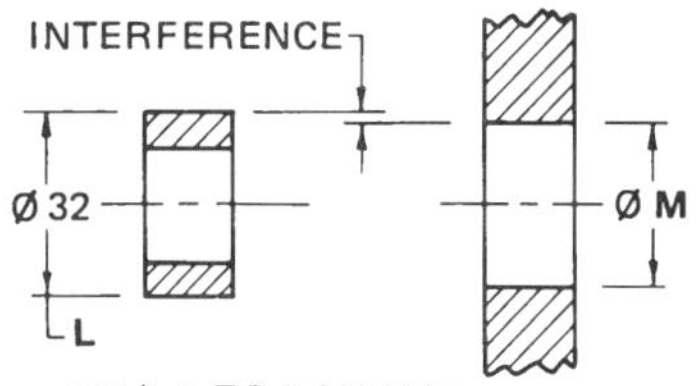

U7/h6 FORCE FIT

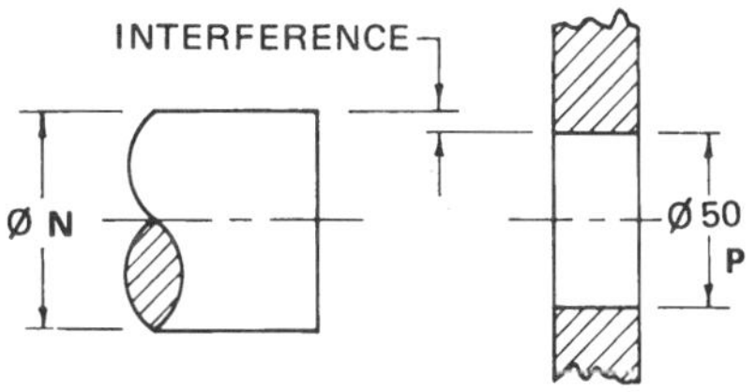

H7/s6 SHRINK FIT

Complete this chart using the proper limit and fit tables.

FIT	ØL	ØM	ØN	ØP
U7/h6				
U7/s6				

CLEARANCE FITS

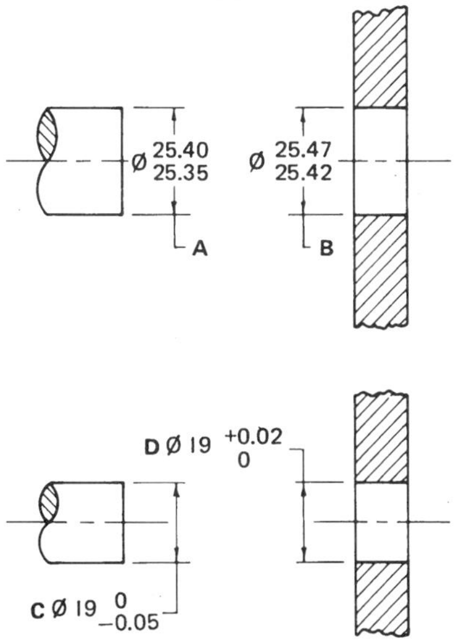

Complete this chart from the information given above.

	A	B	C	D
TOLERANCE ON HOLE				
TOLERANCE ON SHAFT				
MINIMUM CLEARANCE				
MAXIMUM CLEARANCE				

INTERFERENCE FITS

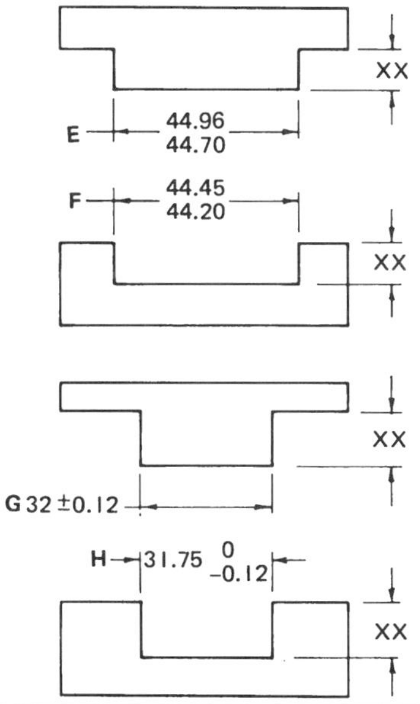

Complete this chart from the information given above.

	E	F	G	H
TOLERANCE ON PART				
TOLERANCE ON SLOT				
MINIMUM INTERFERENCE				
MAXIMUM INTERFERENCE				

CLEARANCE AND INTERFERENCE FITS

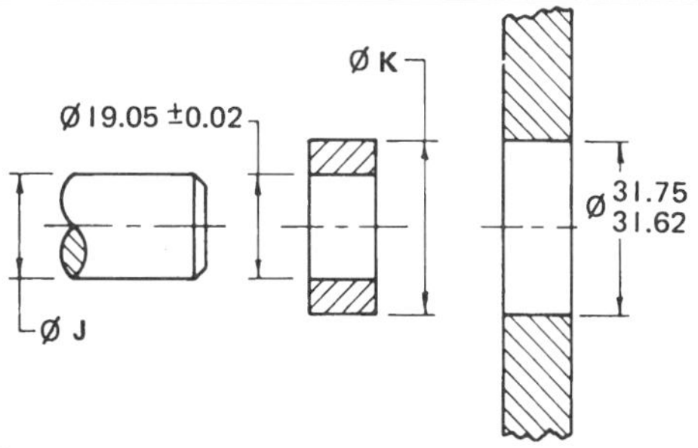

Q1. DIMENSION SHAFT (J) TO HAVE A TOLERANCE OF 0.05 AND A MINIMUM CLEARANCE OF 0.02.

Q2. DIMENSION BUSHING (K) TO HAVE A TOLERANCE OF 0.07 AND A MAXIMUM INTERFERENCE OF 0.22.

Q3. DIMENSION SHAFT (J) TO HAVE A TOLERANCE OF 0.02 AND A MINIMUM CLEARANCE OF 0.05.

Q4. DIMENSION BUSHING (K) TO HAVE A TOLERANCE OF 0.07 AND A MAXIMUM INTERFERENCE OF 0.25.

Complete this chart from the information given above.

	J	K
Q1		
Q2		
Q3		
Q4		

Using the tables of fits in the appendix, calculate the missing dimensions.

FITS AND ALLOWANCES— METRIC	REFERENCE UNIT 8-6	NAME COURSE DATE	DRAWING F-12

INCH FITS

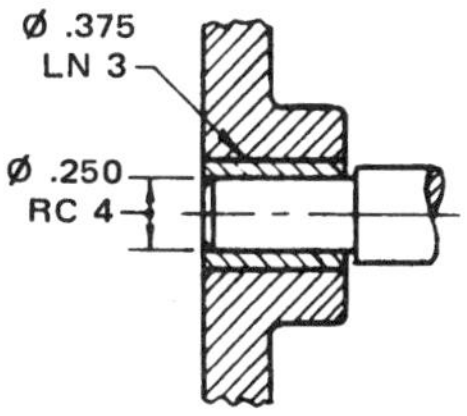

(A) SHAFT IN BUSHED HOLE

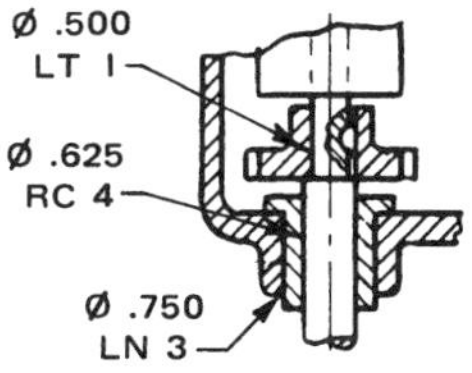

(B) GEAR AND SHAFT IN BUSHED BEARING

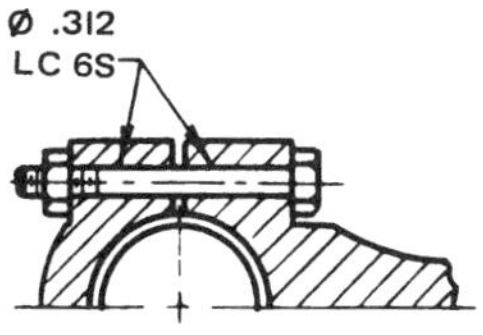

(C) CONNECTING-ROD BOLT

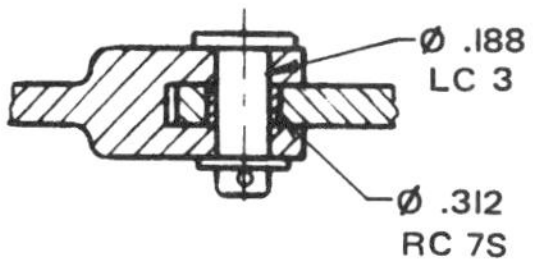

(D) LINK PIN (SHAFT BASIS FITS)

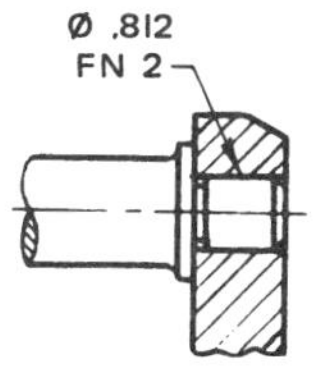

(E) CRANK PIN IN CAST IRON

DESIGN SKETCH	BASIC DIAMETER SIZE IN.	SYMBOL	BASIS	FEATURE	LIMITS OF SIZE		CLEARANCE OR INTERFERENCE	
					MAX	MIN	MAX	MIN
A	.375		HOLE	HOLE				
				SHAFT				
A	.250		HOLE	HOLE				
				SHAFT				
B	.500		HOLE	HOLE				
				SHAFT				
B	.625		HOLE	HOLE				
				SHAFT				
B	.750		HOLE	HOLE				
				SHAFT				
C	.312		SHAFT	HOLE				
				SHAFT				
D	.188		HOLE	HOLE				
				SHAFT				
D	.312		SHAFT	HOLE				
				SHAFT				
E	.812		HOLE	HOLE				
				SHAFT				

Using the inch tables of fits in the appendix, calculate the missing dimensions.

FITS— INCH	REFERENCE UNIT 8-6	NAME COURSE DATE	DRAWING F-13

METRIC FITS

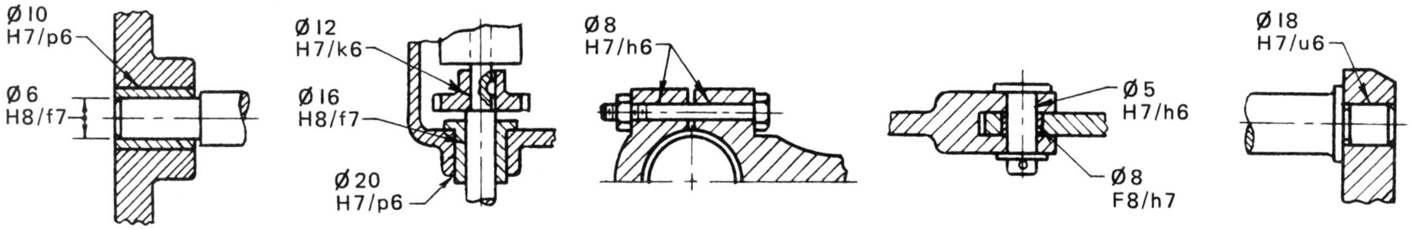

(A) SHAFT IN BUSHED HOLE (B) GEAR AND SHAFT IN BUSHED BEARING (C) CONNECTING-ROD BOLT (D) LINK PIN (SHAFT BASIS FITS) (E) CRANK PIN IN CAST IRON

DESIGN SKETCH	BASIC DIAMETER SIZE mm	SYMBOL	BASIS	FEATURE	LIMITS OF SIZE		CLEARANCE OR INTERFERENCE	
					MAX	MIN	MAX	MIN
A	10		HOLE	HOLE				
				SHAFT				
A	6		HOLE	HOLE				
				SHAFT				
B	12		HOLE	HOLE				
				SHAFT				
B	16		HOLE	HOLE				
				SHAFT				
B	20		HOLE	HOLE				
				SHAFT				
C	8		SHAFT	HOLE				
				SHAFT				
D	5		HOLE	HOLE				
				SHAFT				
D	8		SHAFT	HOLE				
				SHAFT				
E	18		HOLE	HOLE				
				SHAFT				

Using the metric tables of fits in the appendix, calculate the missing dimensions.

FITS— METRIC	REFERENCE UNIT 8–6	NAME COURSE	DATE	DRAWING F–14

Redraw the spindle to the scale 2:1. Scale the part for sizes. The drawing can be dimensioned in either inches or millimeters. Other considerations are listed below:

(a) "A" diameter to have an LC3 (inch) or H7/h6 (metric) fit
(b) "B" diameter requires a 96 diamond knurl or its equivalent
(c) "C" diameter to have an LT3 (inch) or H7/k6 (metric) fit
(d) "D" diameter to be a minimum relief (undercut)
(e) "E" to be a standard No. 605 Woodruff key in center of segment and the diameter to be controlled by an RC3 (inch) or H7/g6 (metric) fit
(f) "F" to be undercut for a standard retaining ring and controlled form to manufacturer's specifications

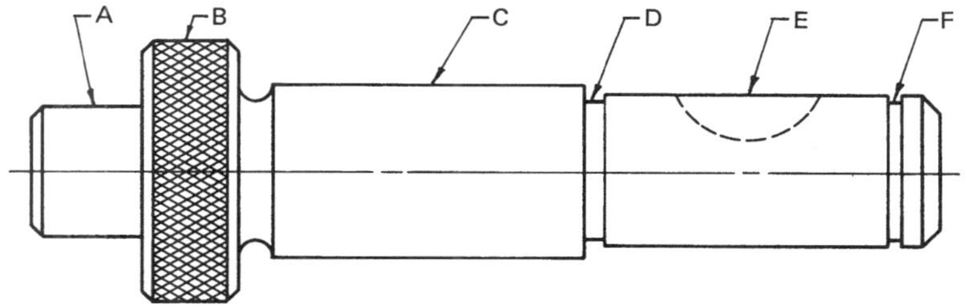

FITS AND ALLOWANCES—
SPINDLE

REFERENCE UNIT **8–6**

NAME

COURSE

DATE

DRAWING **F–15**

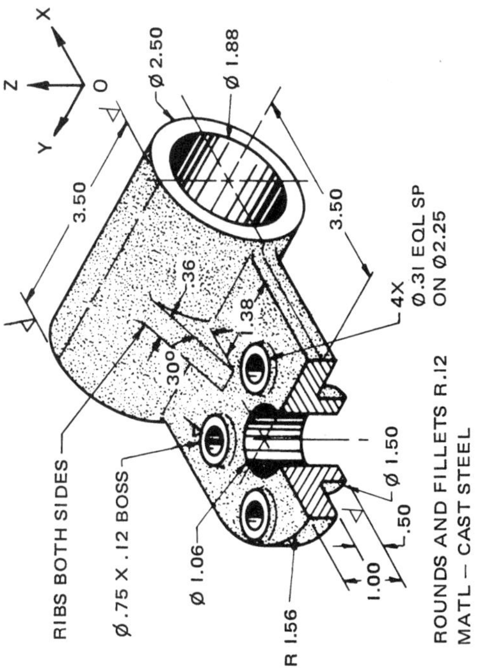

Make a two-view working drawing of the link. The amount of material to be removed from the end surfaces of the hub is .09 in. and .06 in. on the bosses and bottom of the vertical hub, respectively. The two large holes are to have an LN3 fit for journal bearings. The scale is 1:1. The location of the Ø1.88 hole is shown on the drawing area.

SURFACE TEXTURE— LINK	REFERENCE UNIT 8–7–A	NAME COURSE DATE	DRAWING F–16

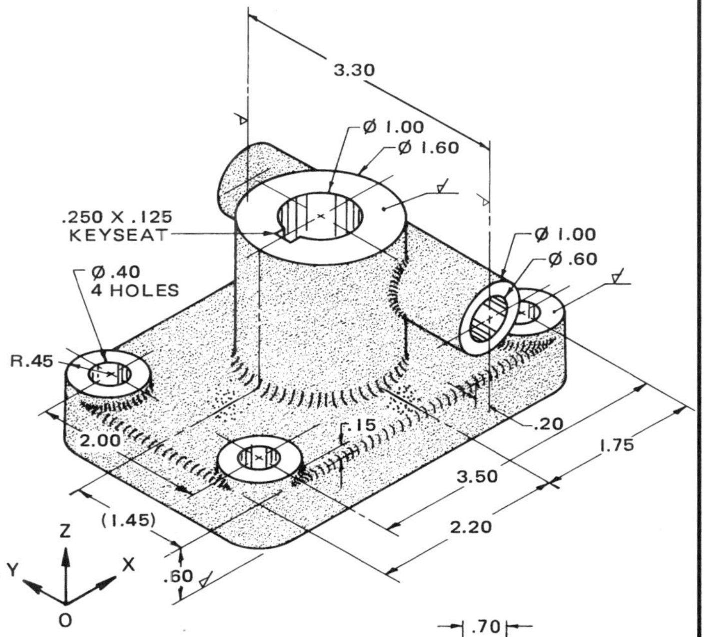

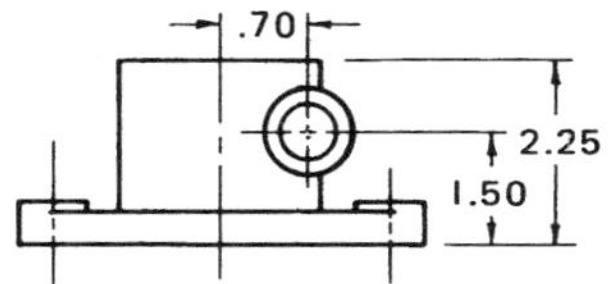

FRONT VIEW

ROUNDS AND FILLETS R.10
MATL — GRAY IRON

Draw the front and top views of the column bracket. The scale is 1:1. The following information is to be added to the drawing:

(a) The bottom of the base is to have a maximum roughness value of 125 µin. and a machining allowance of .06 in.
(b) The tops of the bosses are to have a maximum roughness value of 250 µin. and a machining allowance of .04 in.
(c) The end surfaces of the hubs supporting the shafts are to have maximum and minimum roughness values of 63 and 32 µin., respectively, and a machining allowance of .04 in.
(d) The large hole is to be dimensioned for an RC4 fit. The small hole is to be dimensioned for an LN3 fit for plain bearings.

SURFACE TEXTURE—
COLUMN BRACKET

REFERENCE UNIT **8–7–C**

NAME
COURSE
DATE

DRAWING **F–17**

Make a two-view working drawing of the shaft base. Draw the front view in full section. The scale is 1:1. Surfaces shown with ∇ are to have a surface texture rating of 125 µin. and a machining allowance of .06 in. Use symbolic dimensioning whenever possible.

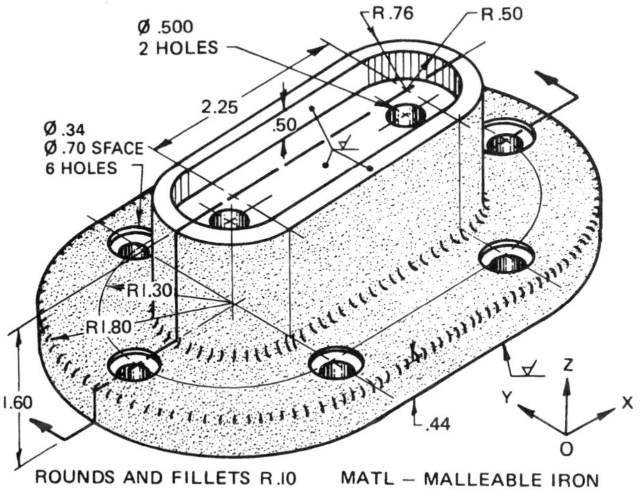

FULL SECTIONS— SHAFT BASE	REFERENCE UNIT 9–1	NAME COURSE DATE	DRAWING G–1

Make a three-view working drawing of the slide bracket. Show the front view in full section. The scale is 1:1. Surfaces shown with ▽ should have a surface texture rating of 63 μin. and a machining allowance of .06 in. Use limit dimensions for the holes showing a fit. The locations for the views are shown on the drawing area.

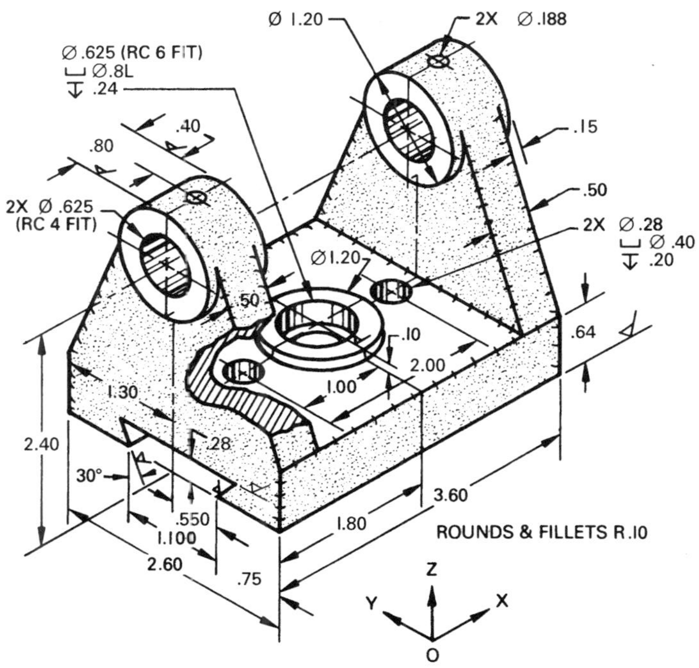

FULL SECTIONS— SLIDE BRACKET	REFERENCE UNIT 9–1	NAME COURSE DATE	DRAWING G–2

Make a working drawing of the casing showing the top, front, and four sectional views. The location of these views is shown on the drawing area. The scale is 1:1. Use symbolic dimensioning wherever possible.

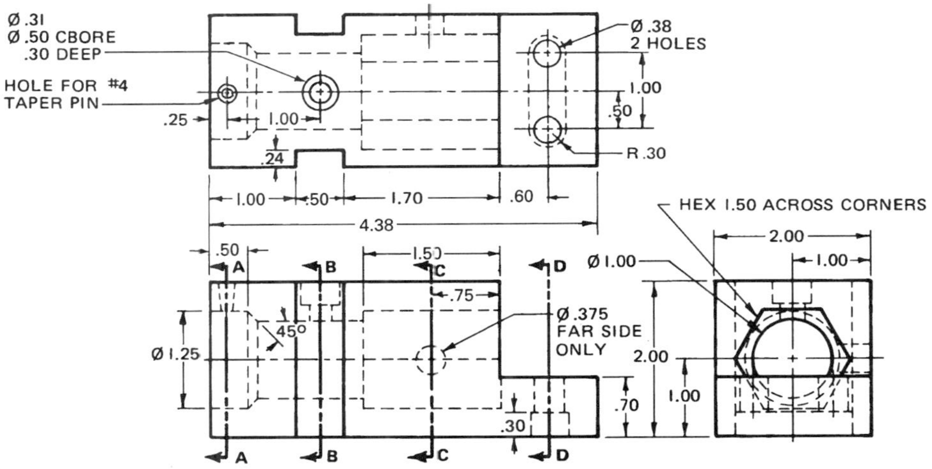

A-A

B-B

C-C

D-D

TWO OR MORE SECTIONS—

CASING

REFERENCE UNIT **9–2**

NAME

COURSE

DATE

DRAWING **G–3**

Make a working drawing of the housing showing the front, top, and four sectional views. The location of these views is shown on the drawing area. The scale is 1:1.

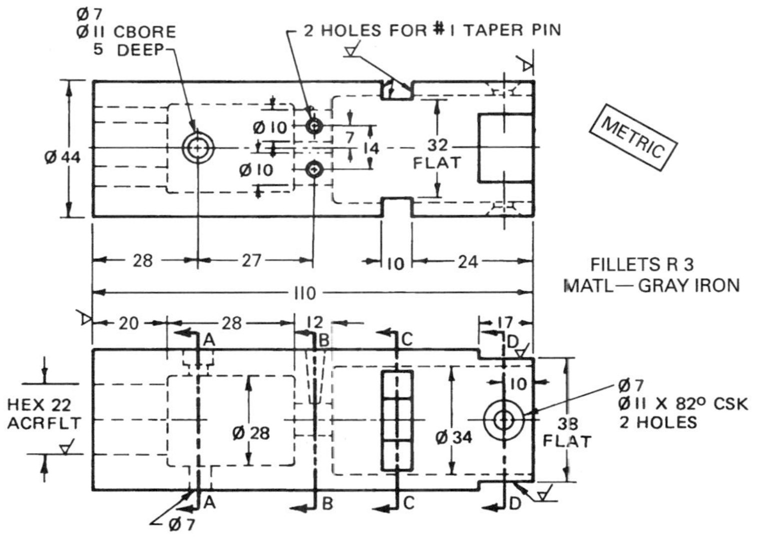

A-A

B-B

C-C

D-D

TWO OR MORE SECTIONS—

HOUSING

REFERENCE UNIT 9–2

NAME

COURSE

DATE

DRAWING G–4

Make a one-view half-section drawing of the step pulley. Add a partial view to show the hole and key detail. The scale is 1:1.

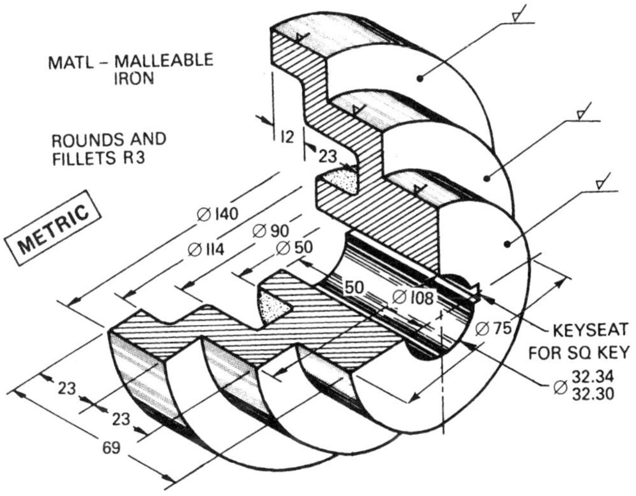

HALF-SECTIONS— STEP PULLEY	REFERENCE UNIT 9–3	NAME COURSE DATE	DRAWING G–5

Make a one-view half-section drawing of the step-V pulley. Add a partial view to show the hole and key detail. The scale is 1:1.

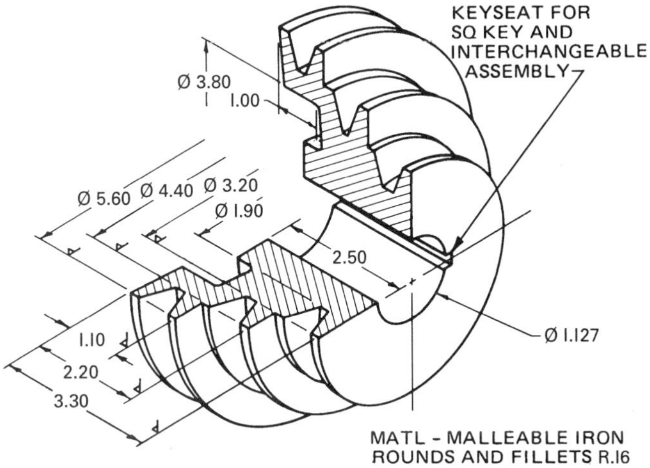

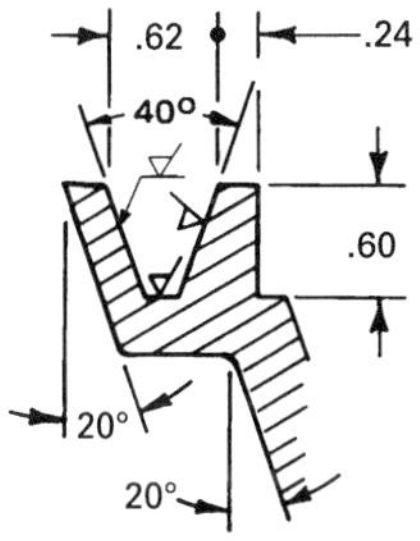

HALF-SECTIONS— STEP-V PULLEY	REFERENCE UNIT 9–3	NAME COURSE DATE	DRAWING G–6

Make a detailed half-section drawing of the pipe plug to the scale 1:1. Use symbolic dimensioning wherever possible and add undercut sizes.

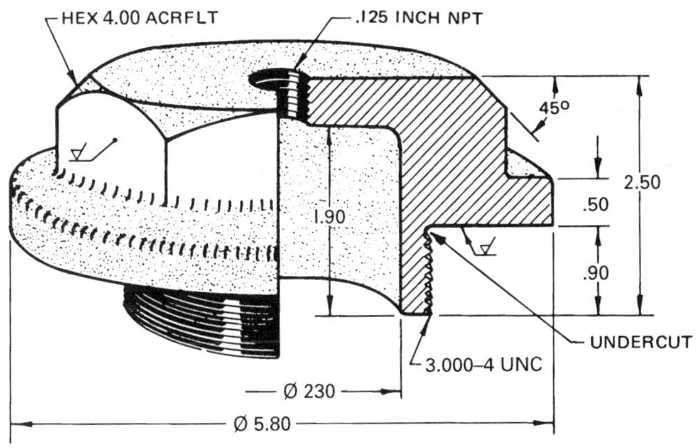

ROUNDS AND FILLETS R .10 MATL – MALLEABLE IRON

THREADS IN SECTION— PIPE PLUG	REFERENCE UNIT 9–4	NAME COURSE DATE	DRAWING G–7

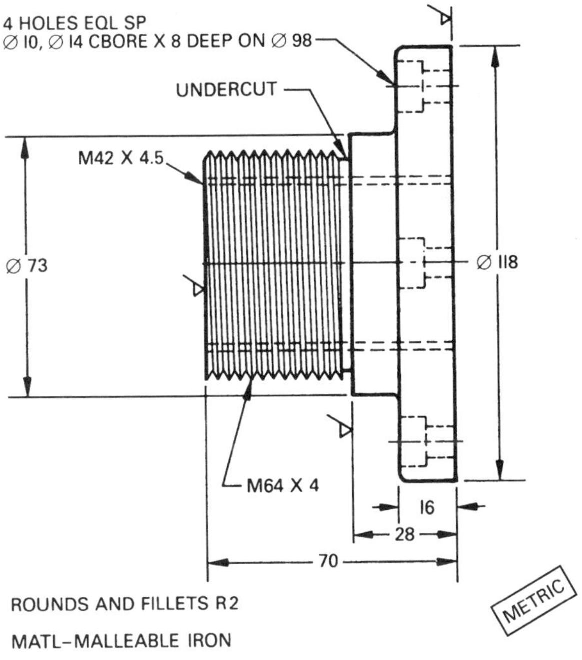

Make a detail drawing of the end plate to the scale 1:1. Determine the number of views and the best type of section that will clearly describe the part. Use symbolic dimensioning wherever possible and add undercut sizes.

THREADS IN SECTION— END PLATE	REFERENCE UNIT **9–4**	NAME COURSE DATE	DRAWING **G–8**

Make a one-view section assembly drawing of the flanged connection. Include an item list and identify the parts on the assembly. The scale is 1:1.

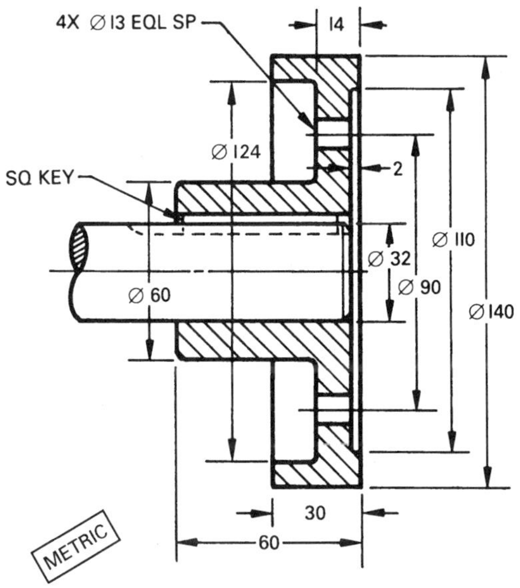

FLANGES HELD TOGETHER BY M12 X 1.75 X 45 LG HEX HD BOLTS WITH LOCKWASHERS

2mm NEOPRENE GASKET BETWEEN FLANGES

QTY	ITEM	MATL	DESCRIPTION	PT. NO.

ASSEMBLIES IN SECTION— FLANGED CONNECTION	REFERENCE UNIT 9–5	NAME ___ COURSE ___ DATE ___	DRAWING G–9

FASTEN ASSEMBLY TO A 6mm STEEL PLATE BY FOUR M 10 X 40 mm LG HEX HD BOLTS, NUTS AND LOCK WASHERS. SHOW THE STEEL PLATE IN PHANTOM LINES.

ROUNDS AND FILLETS R 3

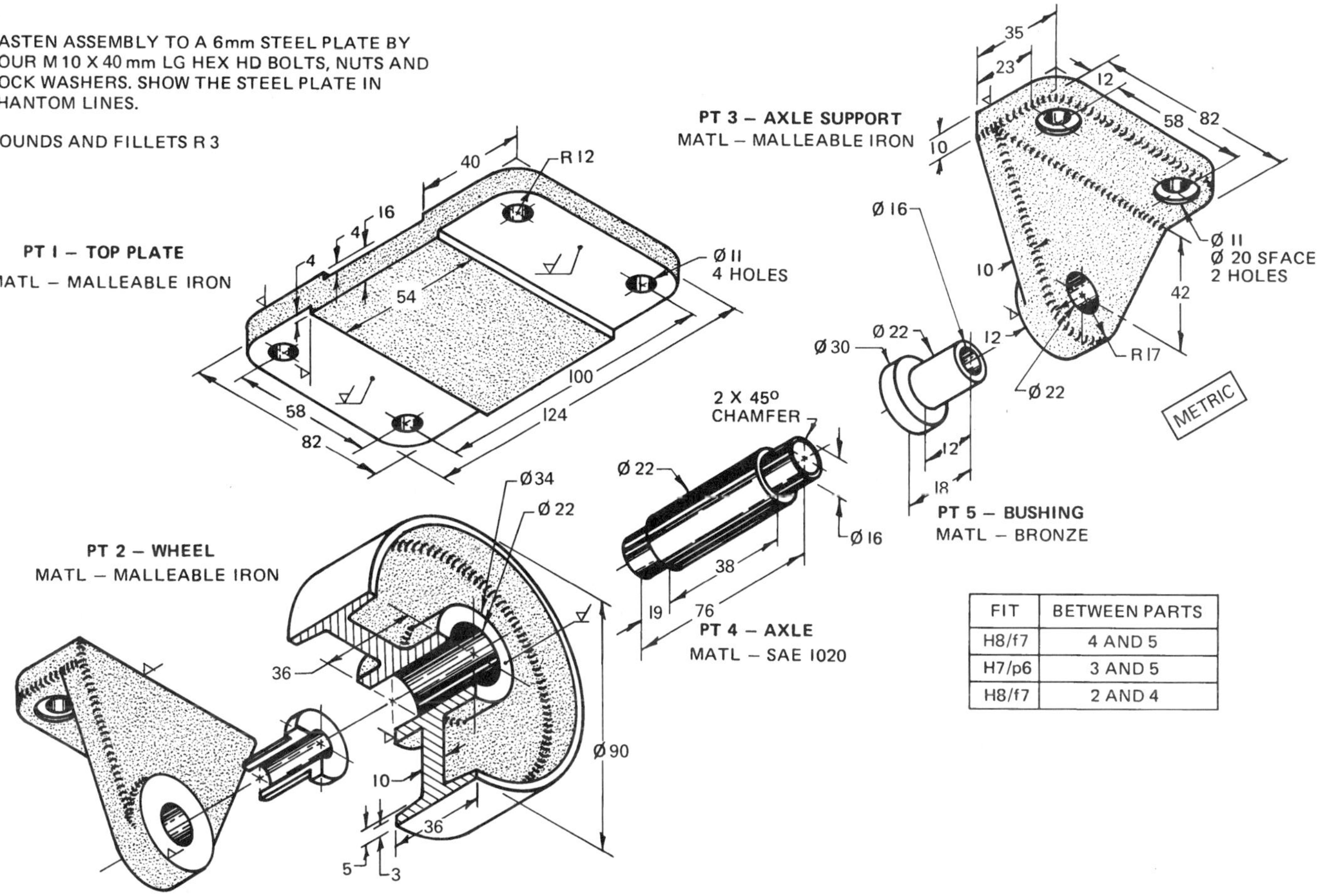

FIT	BETWEEN PARTS
H8/f7	4 AND 5
H7/p6	3 AND 5
H8/f7	2 AND 4

Make a one-view section assembly drawing of the caster. Identify the parts on the assembly. Assume that this drawing will be used in a catalog and add appropriate dimensions and information for the potential purchaser. Only the bolt and nut on the right side need be shown. The scale is 1:1. The location of Part 4 is shown on the drawing area. Prepare an item list on a separate sheet.

ASSEMBLIES IN SECTION— CASTER ASSEMBLY	REFERENCE UNIT 9–5	NAME COURSE DATE	DRAWING G–10

L–L

M–M

N–N

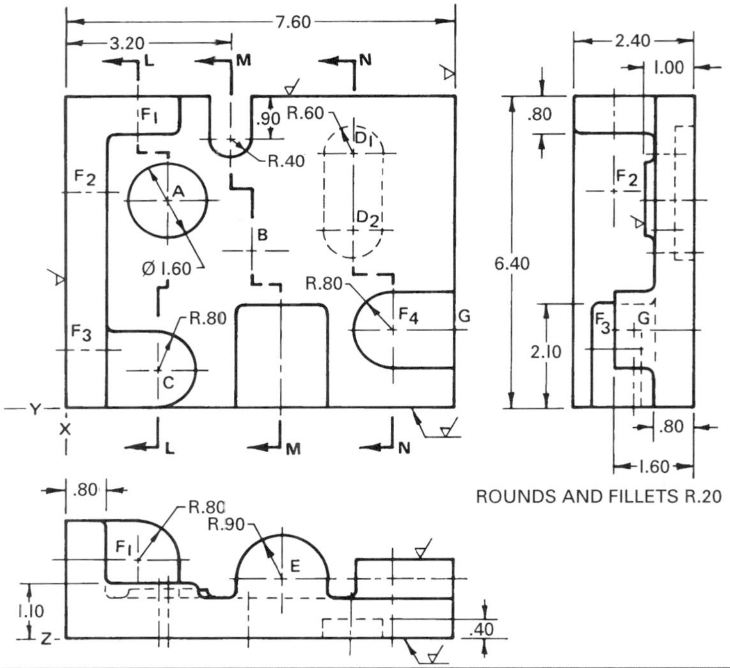

HOLE	HOLE SIZE	LOCATION		
		X	Y	Z
A	.750 –10 UNC	2.00	4.00	
B	Ø.44 Ø.81CSK × 82°	3.60	3.30	
C	Ø.44 ⌴ Ø.81 ↧.25	1.80	.80	
D1	Ø.500	5.60	5.20	
D2	Ø.500	5.60	3.60	
E	.750 –10 UNC × 1.20 DP	4.20		1.20
F1	Ø.80	1.40		1.60
F2	Ø.80		4.40	1.60
F3	Ø.80		1.20	1.60
F4	Ø.80	6.40	1.60	
G	Ø.188 THRU		1.60	1.20

Draw the top and front views and replace the side view with sections L–L, M–M, and N–N. Their locations are shown on the drawing area. The scale is 1:2.

OFFSET SECTIONS—

BASE PLATE

REFERENCE UNIT **9–6**

NAME

COURSE

DATE

DRAWING **G–11**

G-G

H-H

J-J

HOLE	SIZE	LOCATION X	LOCATION Y	LOCATION Z
A_1	Ø12	16	9	
A_2	Ø12	100	9	
A_3	Ø12	30	92	
A_4	Ø12	87	92	
B_1	Ø8	38	32	
B_2	Ø8	80	32	
C_1	M6 X 12 DP	12	50	
C_2	M6	104	52	
D	Ø6 CBORE Ø12 X 6 DP	58	70	
E	Ø10 X 12 DP	58		11
F_1	Ø6		32	20
F_2	Ø6		70	20

MATL – MALLEABLE IRON
ROUNDS AND FILLETS R3

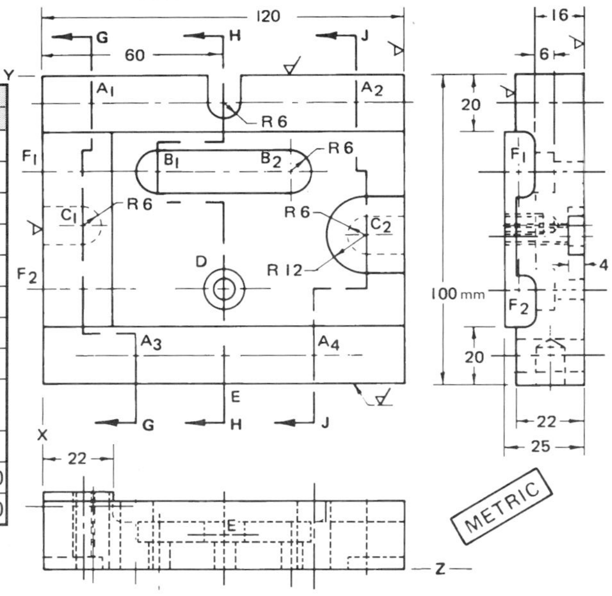

Draw the top and front views and replace the side view with sections G-G, H-H, and J-J. Their locations are shown on the drawing area. The scale is 1:1.

OFFSET SECTIONS—

MOUNTING PLATE

REFERENCE UNIT **9–6**

NAME

COURSE

DATE

DRAWING **G–12**

Make a three-view working drawing of the two-post column base to the scale of 1:4. Draw the front and side views in full section.

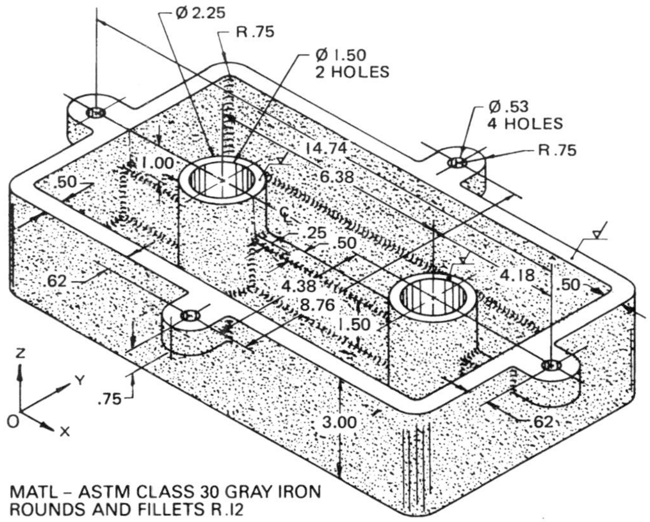

MATL – ASTM CLASS 30 GRAY IRON
ROUNDS AND FILLETS R.12

RIBS, HOLES, AND LUGS— TWO-POST COLUMN BASE	REFERENCE UNIT 9–7	NAME COURSE DATE	DRAWING G–13

Make a three-view working drawing of the shaft support to the scale 1:2. Show the front view as an offset section and the right-side view as a full-section taken through the center of the part.

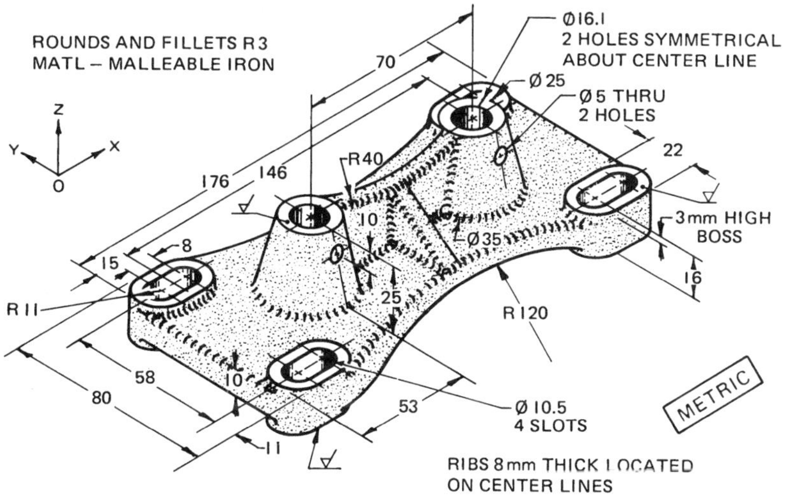

RIBS, HOLES, AND LUGS— SHAFT SUPPORT	REFERENCE UNIT 9–7	NAME COURSE	DATE	DRAWING G–14

Make a two-view working drawing of the connector to the scale 1:1. Show a revolved section of the arm on the top view. The machined surfaces are to have a surface texture rating of 1.6 and a machining allowance of 2 mm.

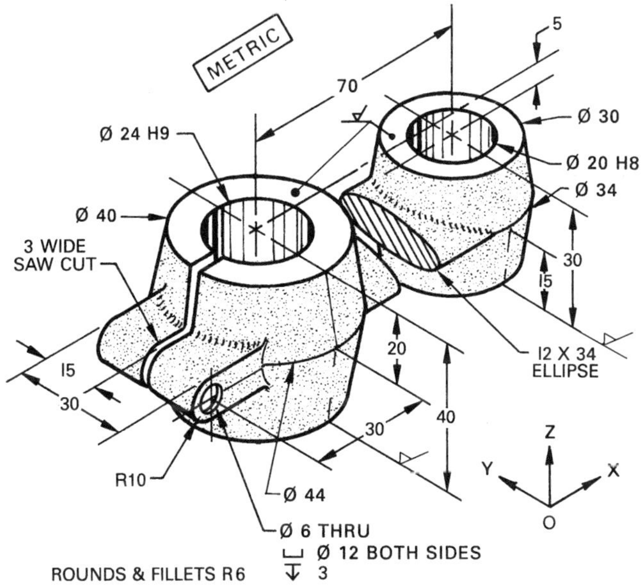

REVOLVED SECTION— CONNECTOR	REFERENCE UNIT 9–8	NAME COURSE DATE	DRAWING G–15

Make a working drawing of the chisel shown. Show either revolved or removed sections taken at planes A through D. The scale is 1:1. A conventional break on the octagonal portion is required to shorten the length. The location of the front view is shown on the drawing area.

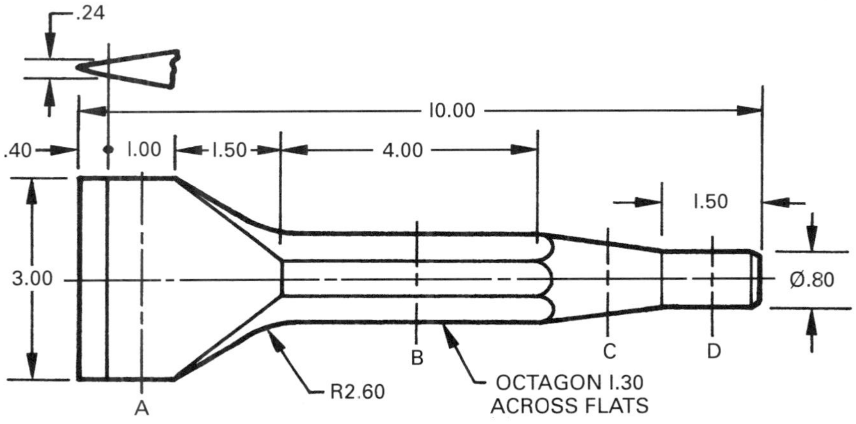

REVOLVED SECTION—

CHISEL

REFERENCE UNIT **9–8**

NAME

COURSE

DATE

DRAWING **G–16**

Make a two-view working drawing of the offset handwheel. Draw the side view in full-section and show a revolved view of the spoke in the front view. The scale is 1:1. The front view is located on the drawing area.

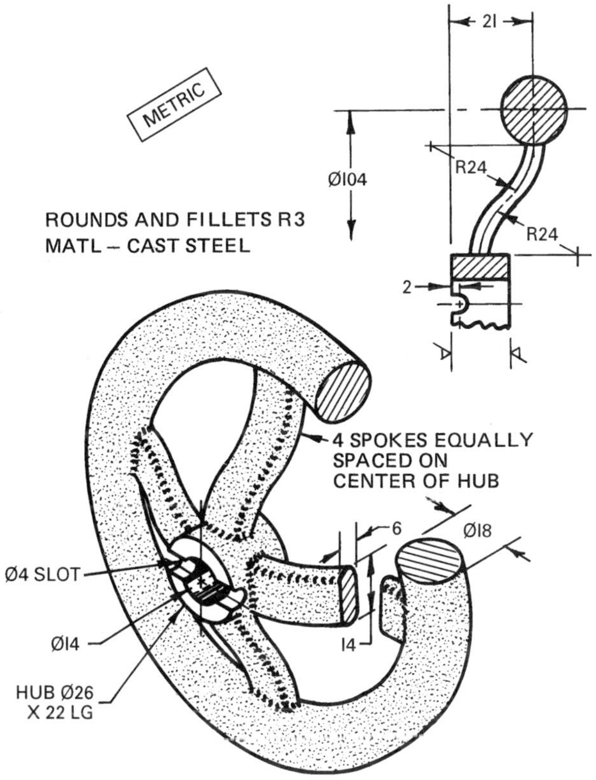

SPOKES AND ARMS— OFFSET HANDWHEEL	REFERENCE UNIT **9–9**	NAME COURSE DATE	DRAWING **G–17**

Draw the front and top views of the hold-down bracket to the scale 1:1. Use a broken-out section to show the large hole, retaining ring slot and boss detail on the front view. Dimension only the hole and retaining ring slot sizes. The bottom right corner of the front view is located on the drawing area.

ROUNDS AND FILLETS R3
MATL– MALLEABLE IRON

METRIC

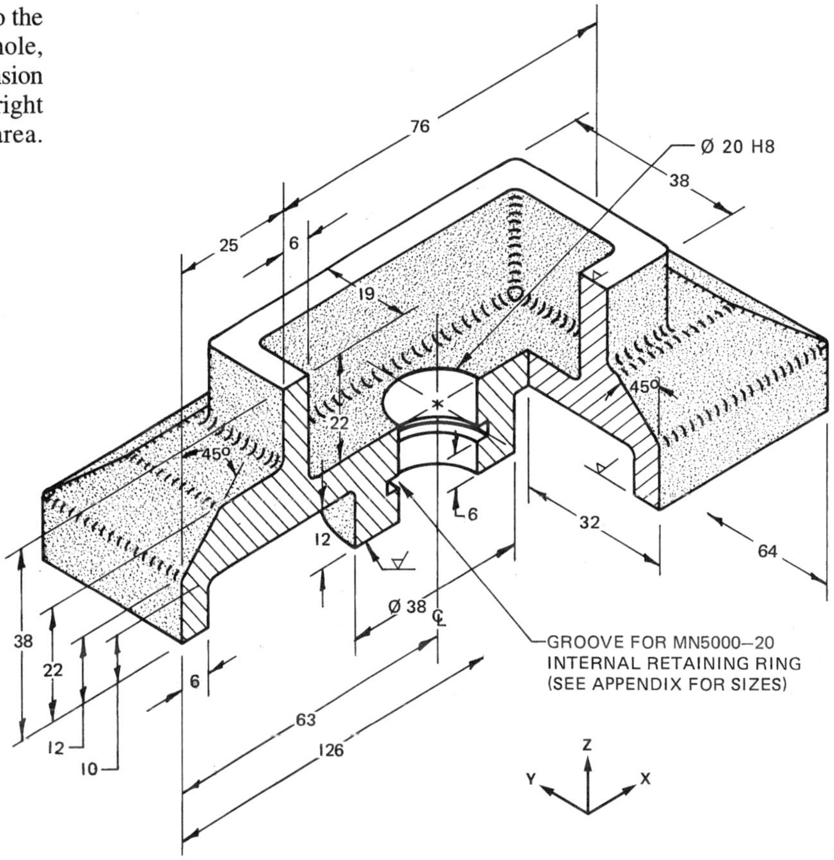

BROKEN-OUT SECTION— HOLD-DOWN BRACKET	REFERENCE UNIT **9–10**	NAME COURSE DATE	DRAWING **G–18**

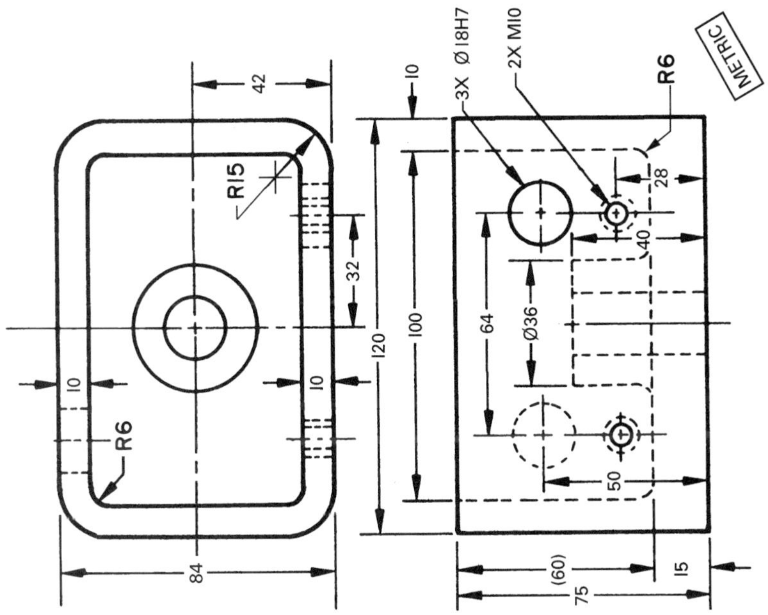

Make a two-view working drawing of the bearing housing. Show the front view as a phantom section. The scale is 1:1.

PHANTOM SECTION— BEARING HOUSING	REFERENCE UNIT 9–11	NAME COURSE DATE	DRAWING G–19

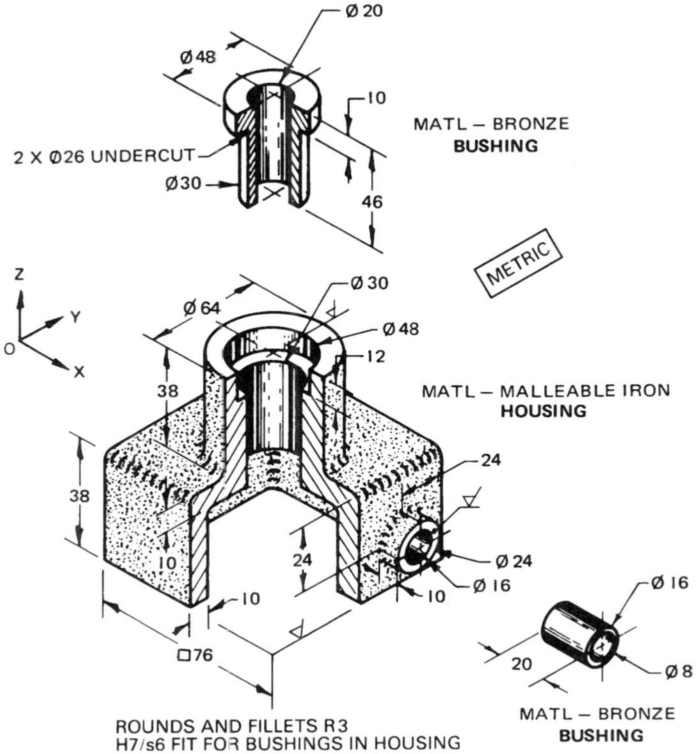

Make a two-view assembly drawing of the housing shown. The front view should be drawn as a phantom section. Show only the hole and shaft sizes for the fits depicted. The scale is 1:1. The bottom of the front view and the center of the top view are located on the drawing area.

PHANTOM SECTION— HOUSING	REFERENCE UNIT **9–11**	NAME ______ COURSE ______ DATE ______	DRAWING **G–20**

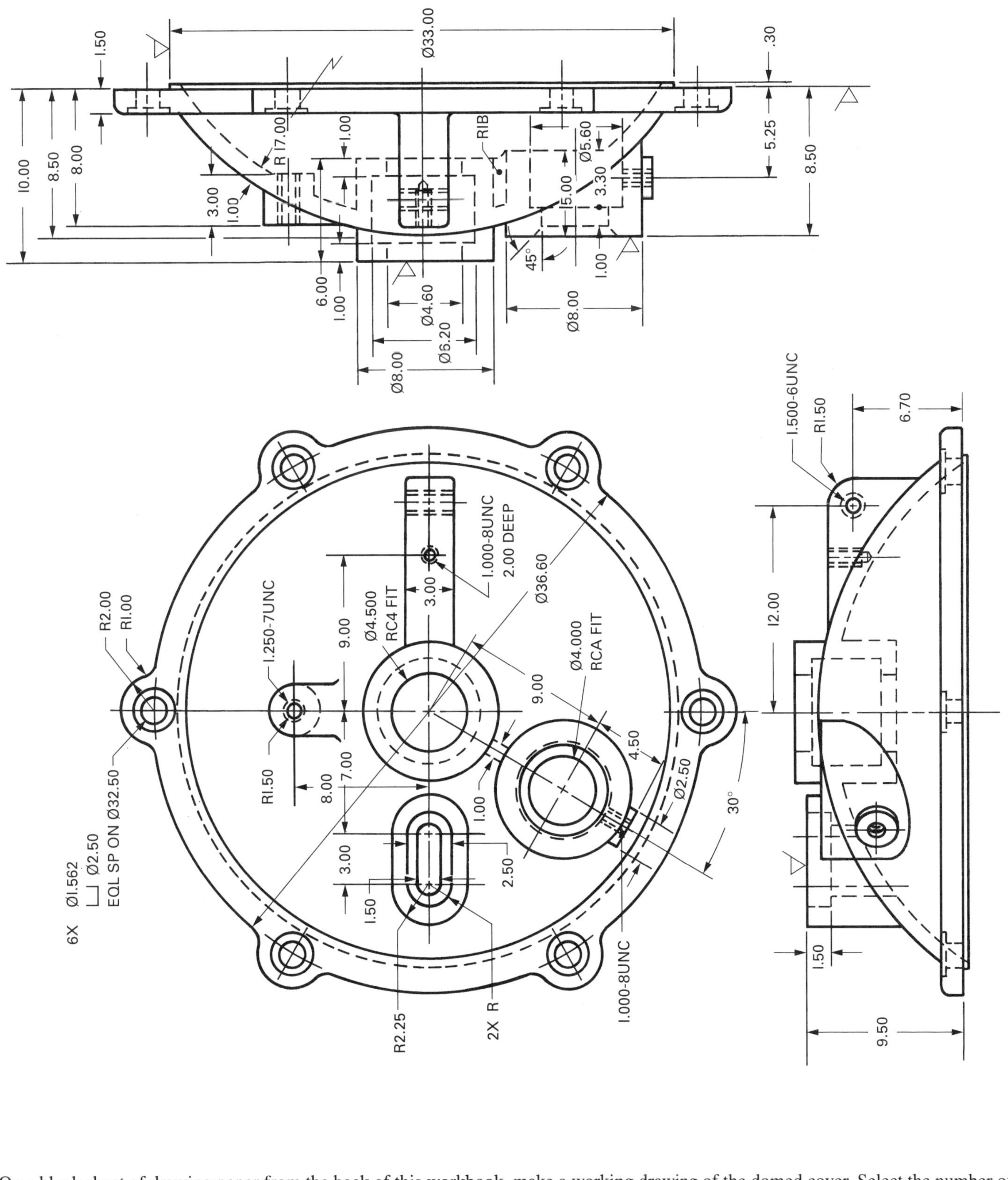

On a blank sheet of drawing paper from the back of this workbook, make a working drawing of the domed cover. Select the number of views, scale, and appropriate sectional views that will improve the clarity of the drawing.

SECTION REVIEW— DOMED COVER	REFERENCE UNIT 9–12	NAME COURSE	DATE	DRAWING G–21

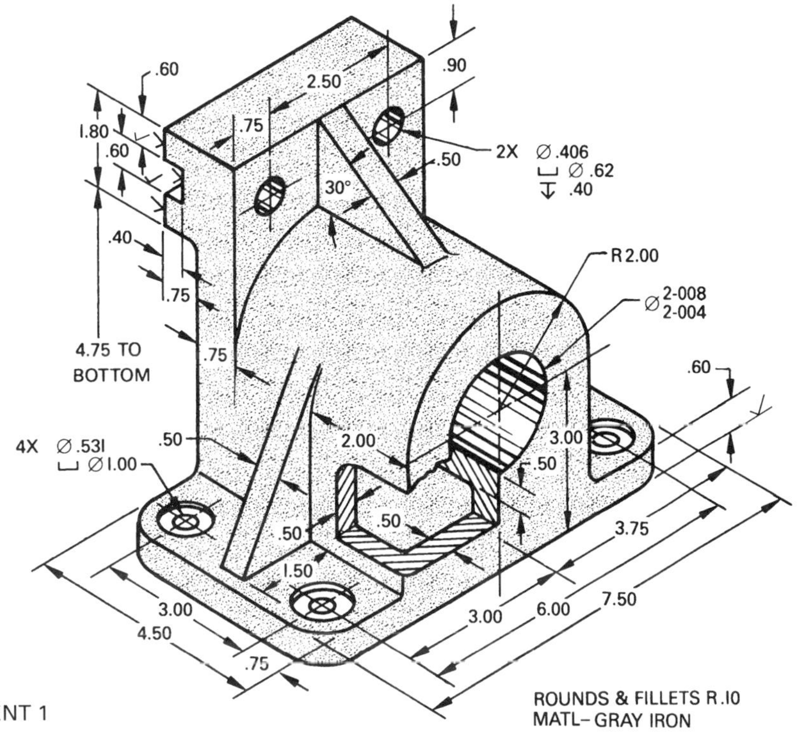

ASSIGNMENT 1

ROUNDS & FILLETS R.10
MATL- GRAY IRON

SLIDE SUPPORT

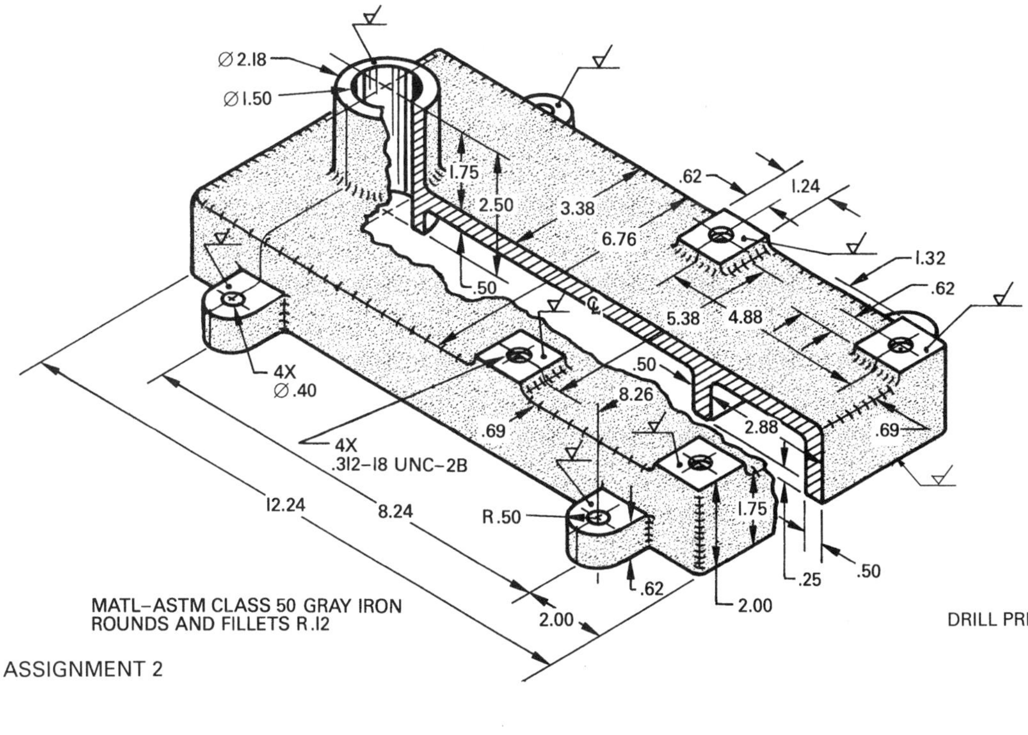

MATL–ASTM CLASS 50 GRAY IRON
ROUNDS AND FILLETS R.12

ASSIGNMENT 2

DRILL PRESS BASE

On a blank sheet of drawing paper from the back of this workbook, make a working drawing of one of the parts shown. Select the number of views, scale, and appropriate sectional views that will improve the clarity of the drawing.

SECTION REVIEW— SLIDE SUPPORT & DRILL PRESS BASE	REFERENCE UNIT **9–12**	NAME COURSE	DATE	DRAWING **G–22**

Make a detail drawing of the guide block. Draw the top, front, and left-side views. The scale is 1:1. Use symbolic dimensioning to describe the holes.

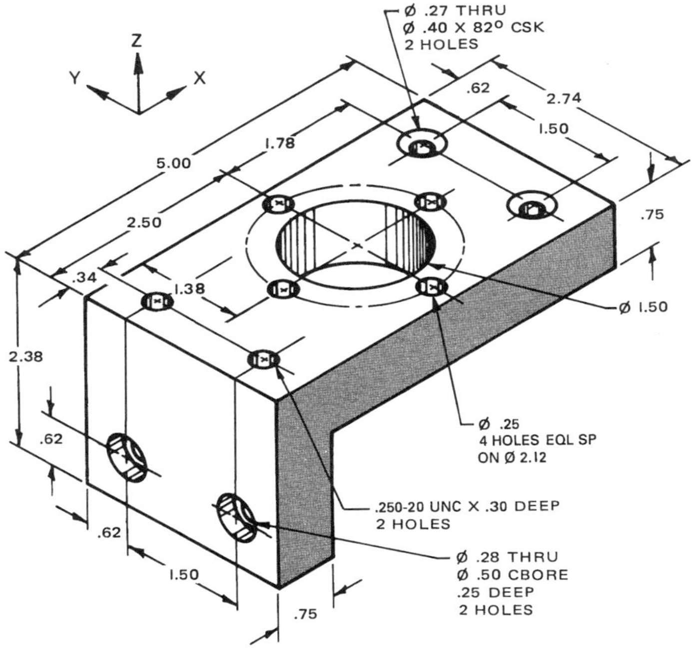

SIMPLIFIED THREAD REPRESENTATION—

GUIDE BLOCK

REFERENCE UNIT **10–1**

NAME

COURSE

DATE

DRAWING **H–1**

1. Make detail drawings of the turnbuckle parts. Show only one drawing for the two eye bolts. (*Hint*: See Fig. 6-9-5 for one method of accomplishing this.) The scale is 1:1.
2. On one of the blank drawing sheets from the back of this workbook, make a one-view assembly drawing of the turnbuckle. Show the assembly in its shortest length and also indicate the maximum position shown in phantom lines. (At least four threads must be engaged for each eye bolt.) The scale is 1:1. Include an item list and identify the parts on the assembly.

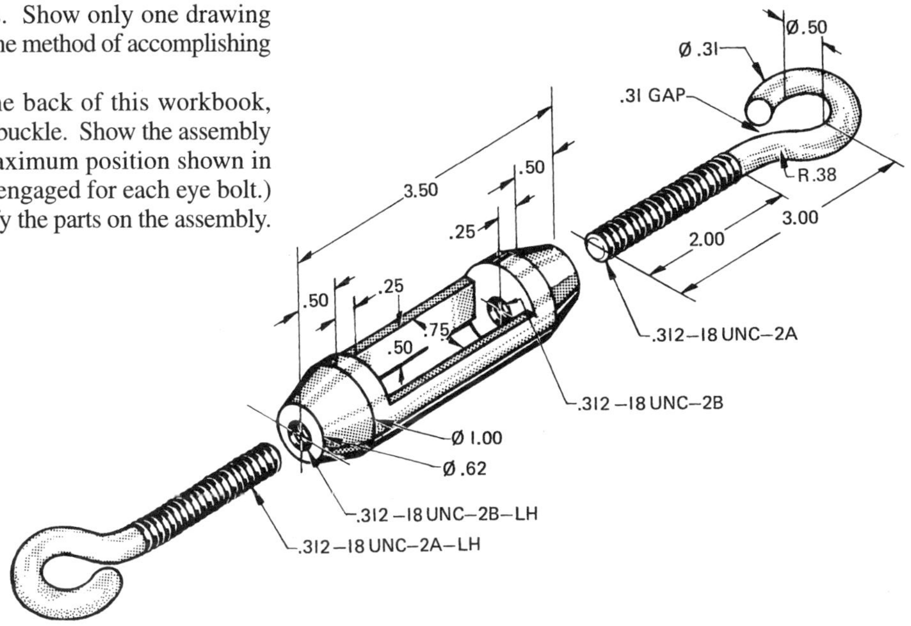

SIMPLIFIED THREAD REPRESENTATION —TURNBUCKLE DETAILS	REFERENCE UNIT 10–1	NAME COURSE DATE	DRAWING H–2

1. Make detail drawings of Parts 1 through 4 of the parallel clamps. For parts 1 and 2, draw two views for each. For parts 3 and 4, make one-view drawings and use a conventional break to shorten their lengths. The scale is 1:1.
2. On one of the blank drawing sheets from the back of this book, make a one-view assembly drawing showing the clamps open 50 mm. The scale is 1:1. Add part numbers and include a materials list.

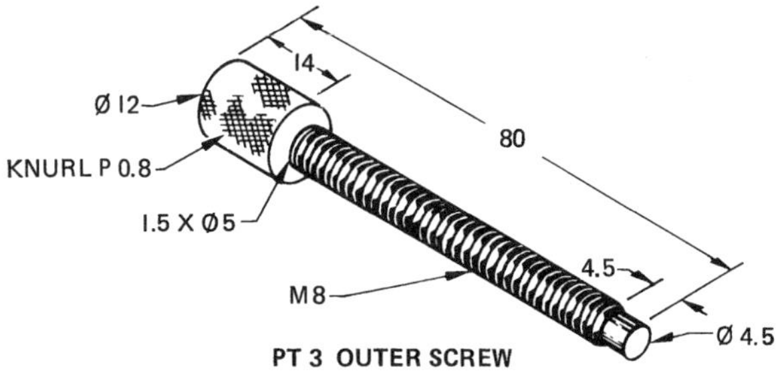

PT 3 OUTER SCREW
I REQD MATL–SAE 1112

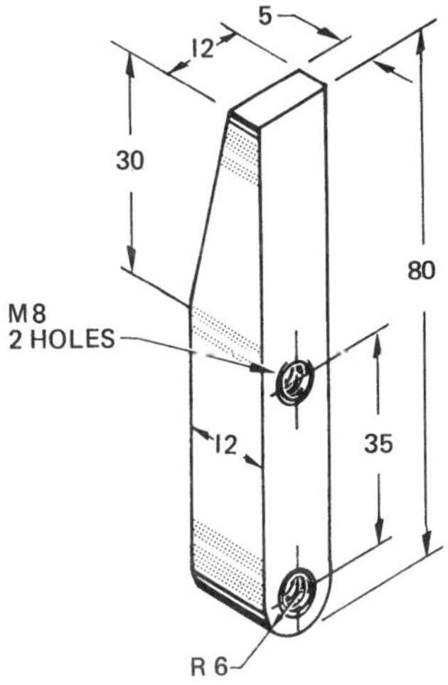

PT 1 MOVABLE JAW
I REQD
MATL–SAE 1020

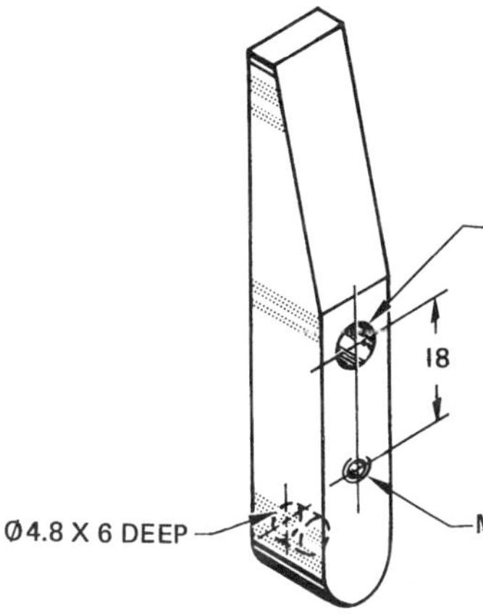

PT 2 STATIONARY JAW
I REQD MATL–SAE 1020
AS SHOWN OTHERWISE
SAME AS PT I

METRIC

PT 4 INNER SCREW
I REQD MATL–SAE 1112

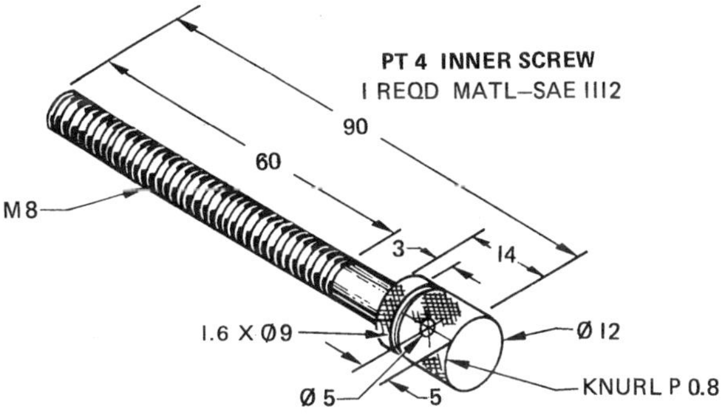

PT 6 MACHINE SCREW RD HD
M3 X 10 LG – I REQD

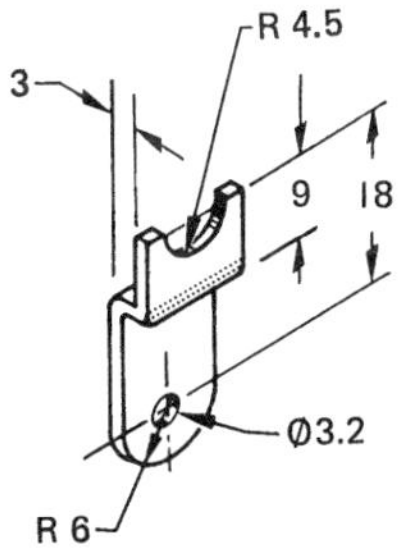

PT 5 CLIP
MATL 1.52 (16 USS)

SIMPLIFIED THREAD REPRESENTATION —PARALLEL CLAMPS DETAILS	REFERENCE UNIT **10–1**	NAME ______ COURSE ______ DATE ______	DRAWING **H–3**

Make a three-view working drawing of the terminal block. The scale is 1:2. The top view is located on the drawing surface.

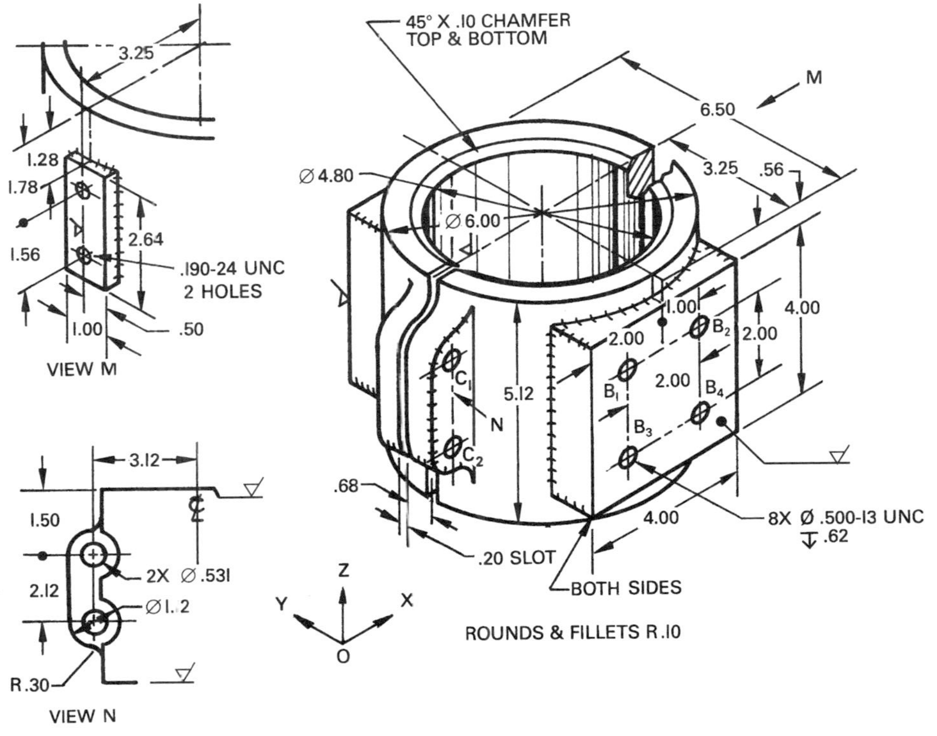

SIMPLIFIED THREAD REPRESENTATION— TERMINAL BLOCK	REFERENCE UNIT 10–1	NAME ___ COURSE ___ DATE ___	DRAWING H–4

Lay out the connector and end rods using detailed thread representation. Show approximately 30mm of each end rod in full section. The scale is 1:1.

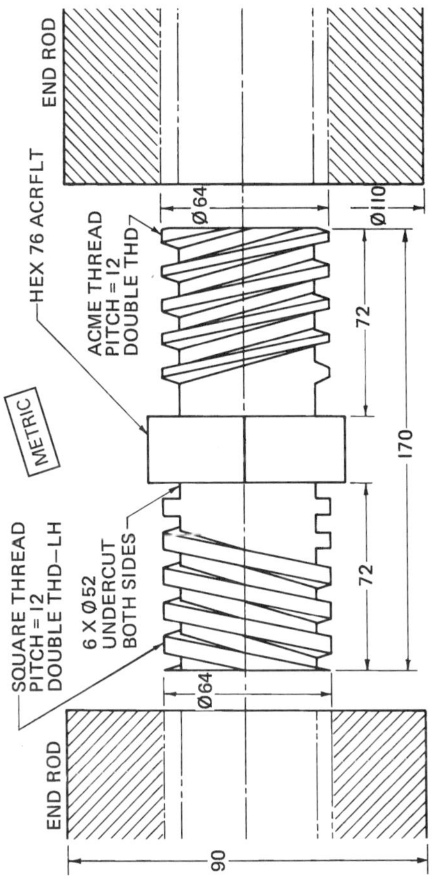

DETAILED THREAD REPRESENTATION— CONNECTOR AND SUPPORTS	REFERENCE UNIT 10–2	NAME ___ COURSE ___ DATE ___	DRAWING H–5

Draw a full-section assembly drawing of the threaded fasteners shown. Dimension both the clearance and the threaded holes. Use symbolic dimensioning wherever possible. A top view can be shown if required. The scale is 1:1.

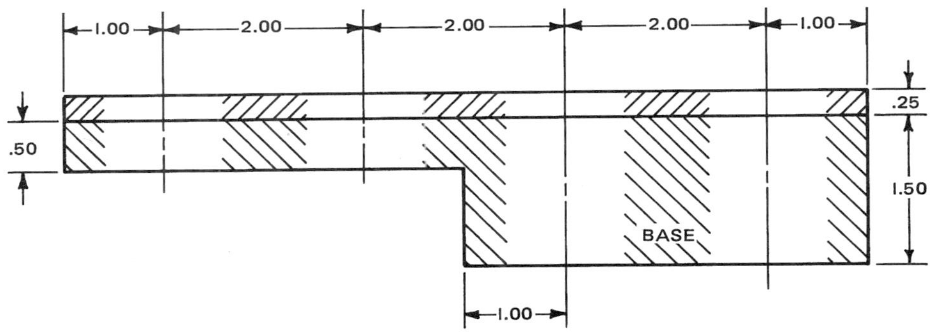

COMMON THREADED FASTENERS	REFERENCE UNIT 10–3	NAME ___ COURSE ___ DATE ___	DRAWING H–6

Make a working drawing of the shaft intermediate support. Show the top, front, and right-side views. Surfaces with ▽ should have a maximum roughness of 250 µin. and a machining allowance of .04 in. Show the limits of size for the Ø.375 hole. The scale is 1:1. The bottom left corner of the front view is located on the drawing area.

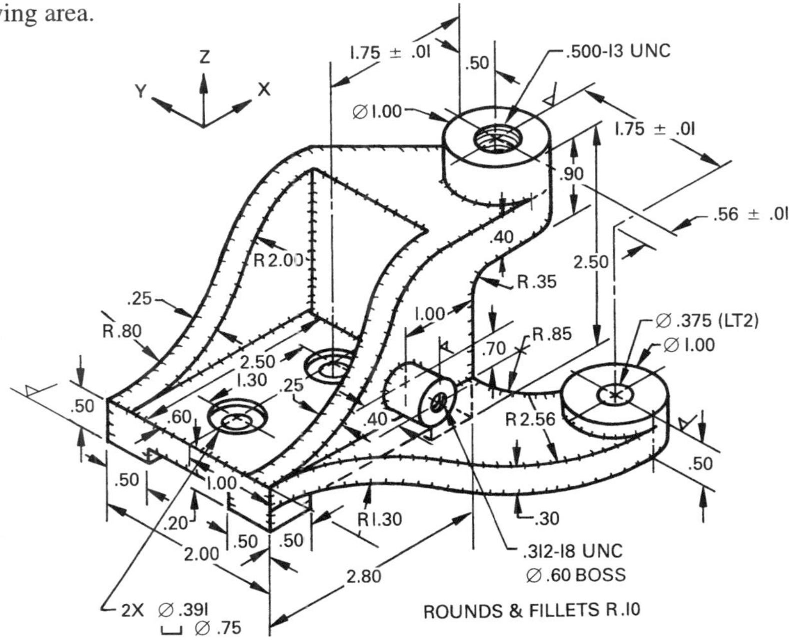

COMMON THREADED FASTENERS— SHAFT INTERMEDIATE SUPPORT	REFERENCE UNIT 10–3	NAME COURSE DATE	DRAWING H–7

1. Make detail drawings of the wheel puller parts in the space below. The scale is 1:2. Use simplified thread representation.
2. On one of the blank drawing sheets from the back of this book, make a one-view assembly drawing of the wheel puller. The scale is 1:1. Lower the center bolt or use a conventional break to shorten the assembly height and to accommodate an item list. Add part numbers to the assembly.

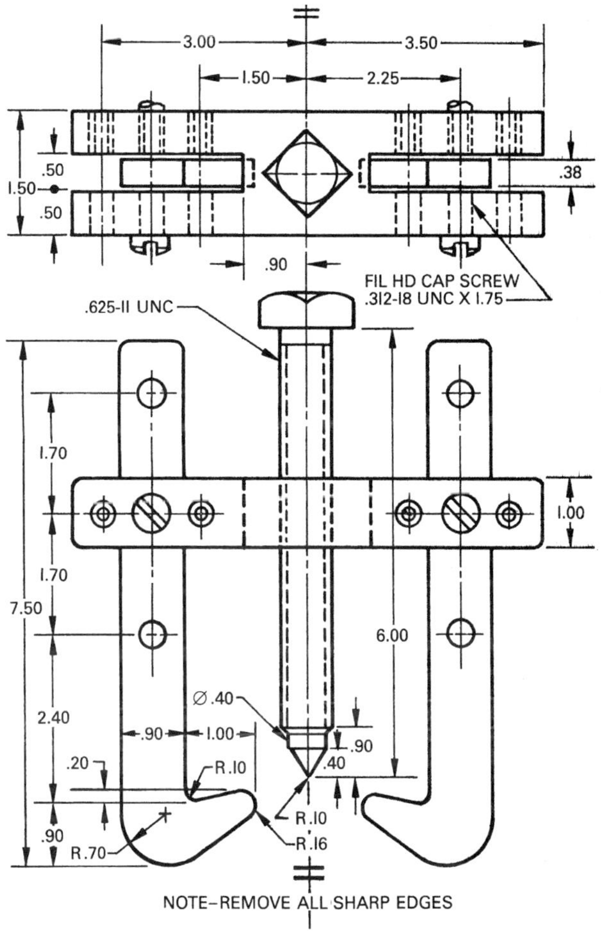

COMMON THREADED FASTENERS— WHEEL PULLER DETAILS	REFERENCE UNIT **10–3**	NAME COURSE DATE	DRAWING **H–8**

1. Make a one-view assembly drawing of the adjustable shaft support. Show the base in full-section. A broken-out section is recommended to show clearly the setscrews in the yoke. Add part numbers to the assembly. Do not dimension. The scale is 1:1.
2. On one of the blank drawing sheets from the back of this book, make working drawings of Parts 1, 3, and 4. The scale is 1:2.

PT 3 YOKE
MATL—CI I REQD
ROUNDS AND FILLETS R 3

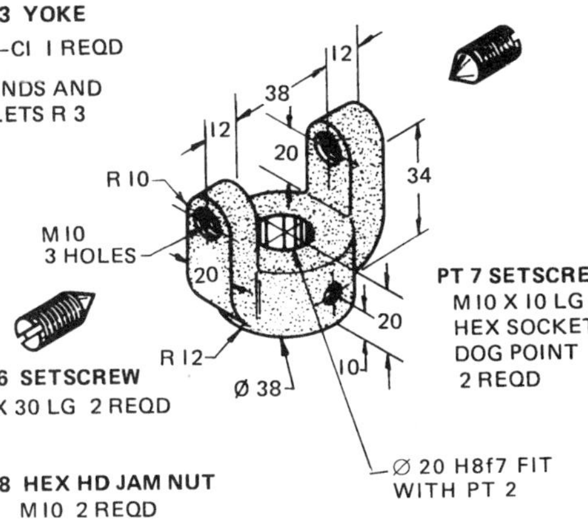

PT 7 SETSCREW
M 10 X 10 LG
HEX SOCKET
DOG POINT
2 REQD

PT 6 SETSCREW
M 10 X 30 LG 2 REQD

PT 8 HEX HD JAM NUT
M 10 2 REQD

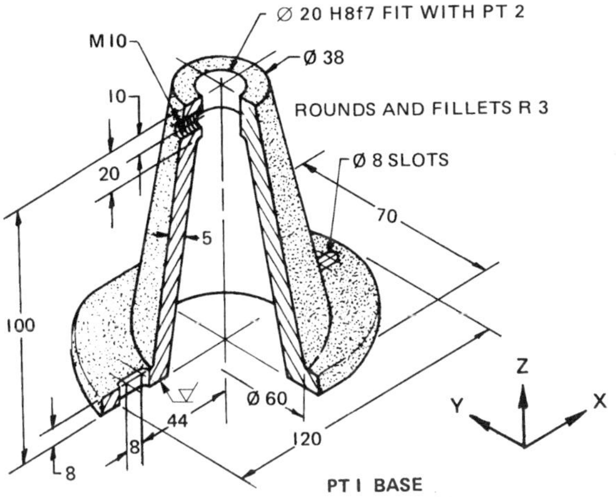

PT I BASE
MATL—CI I REQD

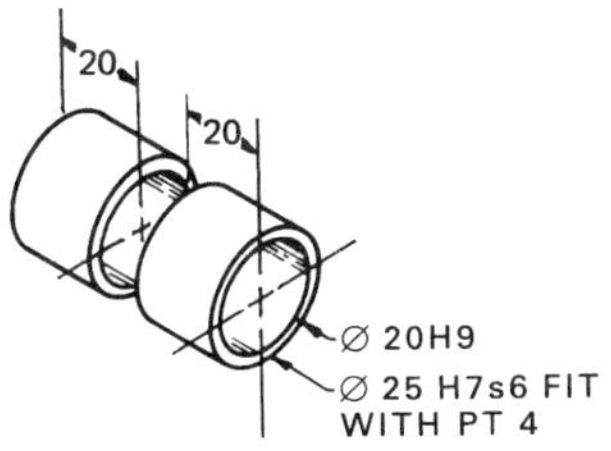

PT 5 BEARINGS
MATL—BRONZE 2 REQD

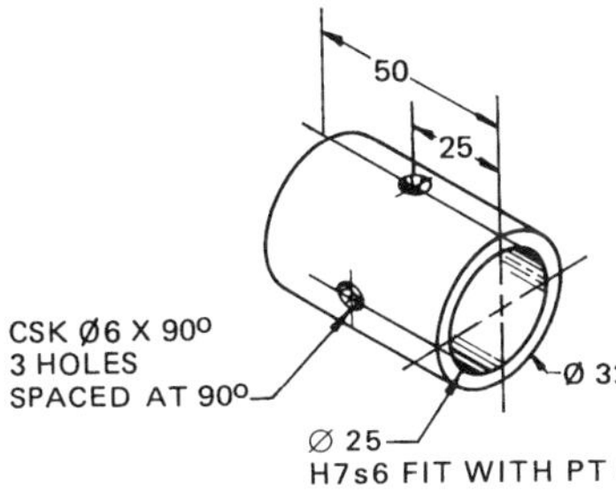

PT 4 BEARING HOUSING
MATL—STEEL I REQD

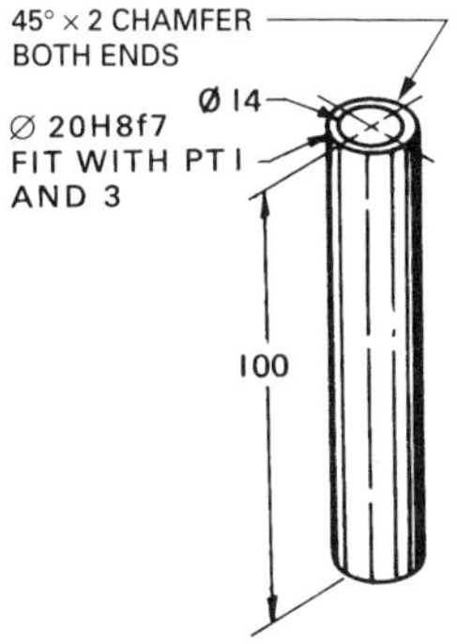

PT 2 VERTICAL SHAFT
MATL—STEEL I REQD

SPECIAL FASTENERS— ADJUSTABLE SHAFT SUPPORT ASSEMBLY	REFERENCE UNIT **10–4**	NAME_ COURSE_ DATE_	DRAWING **H–9** Solution A

1. Make a detail drawing of the front jaw in the space below. The scale is 1:1.

2. On one of the blank drawing sheets from the back of this book, make a two-view assembly drawing of the woodworking vise. Include an item list calling for all the parts. The scale is 1:2.

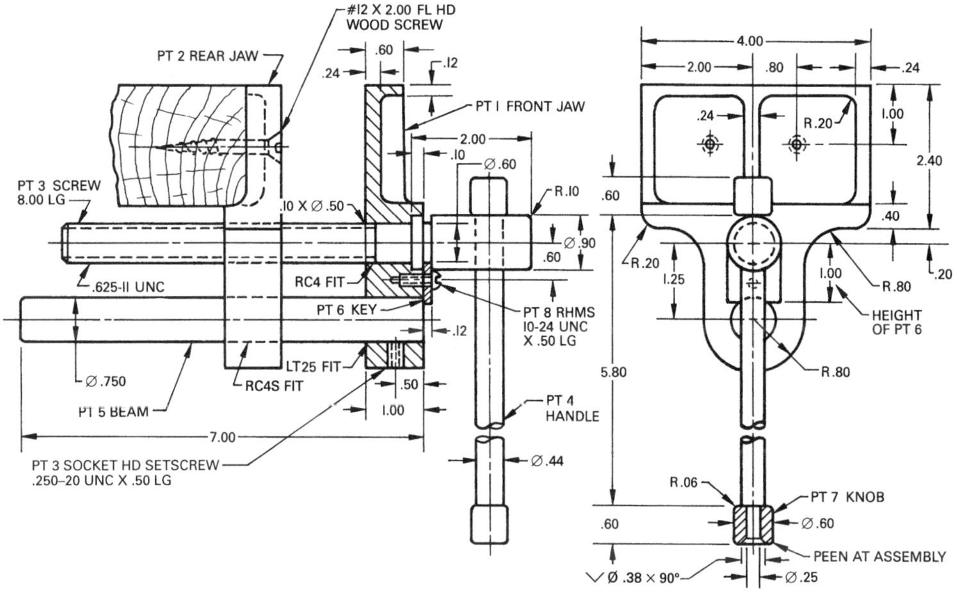

FASTENERS FOR WOOD— WOODWORKING VISE DETAILS	REFERENCE UNIT **10–5**	NAME COURSE DATE	DRAWING **H–10**

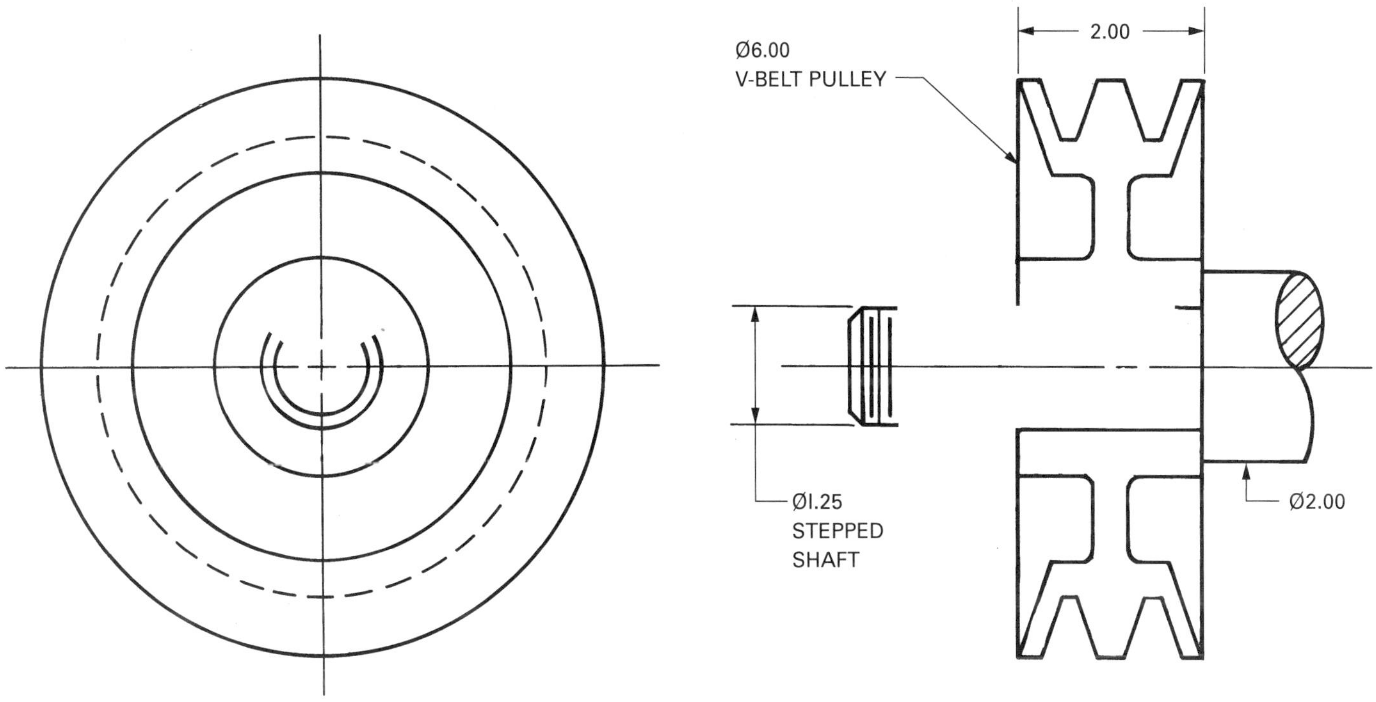

SQUARE KEY

Complete the assembly drawing by adding a square key and hex nut to lock the pulley to the shaft. The scale is 1:2.

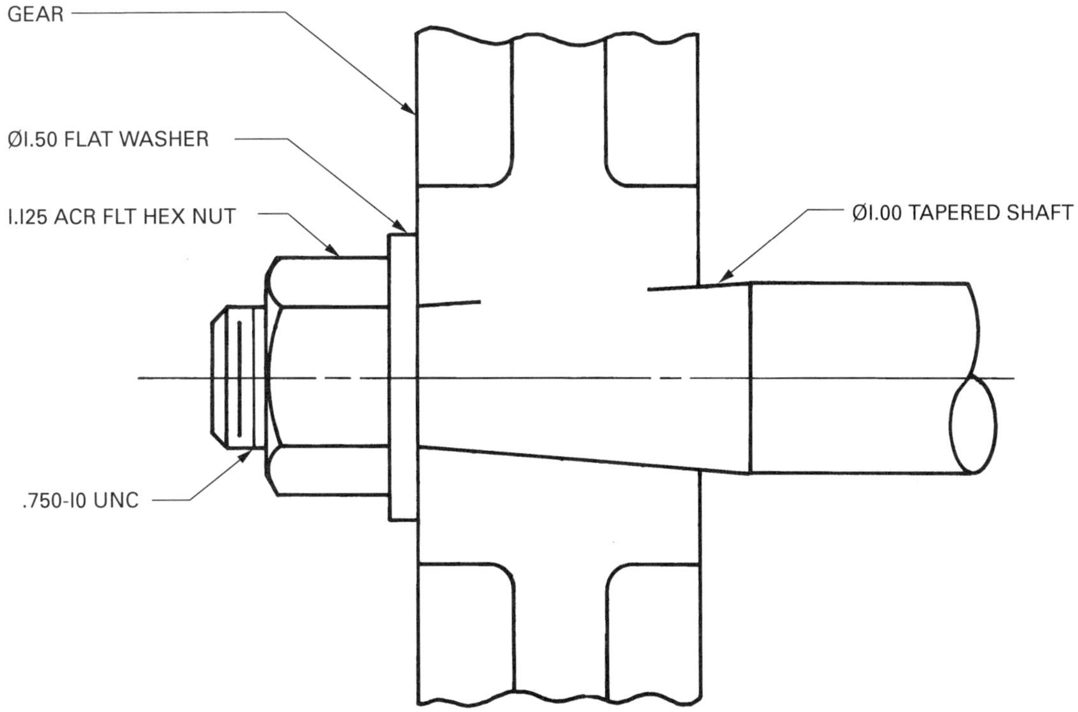

WOODRUFF KEY

Complete the assembly drawing by adding a Woodruff key to lock the gear to the shaft. The scale is 1:1.

KEYS, SPLINES, AND SERRATIONS— KEY FASTENERS	REFERENCE UNIT **11–1**	NAME COURSE DATE	DRAWING **J–1**

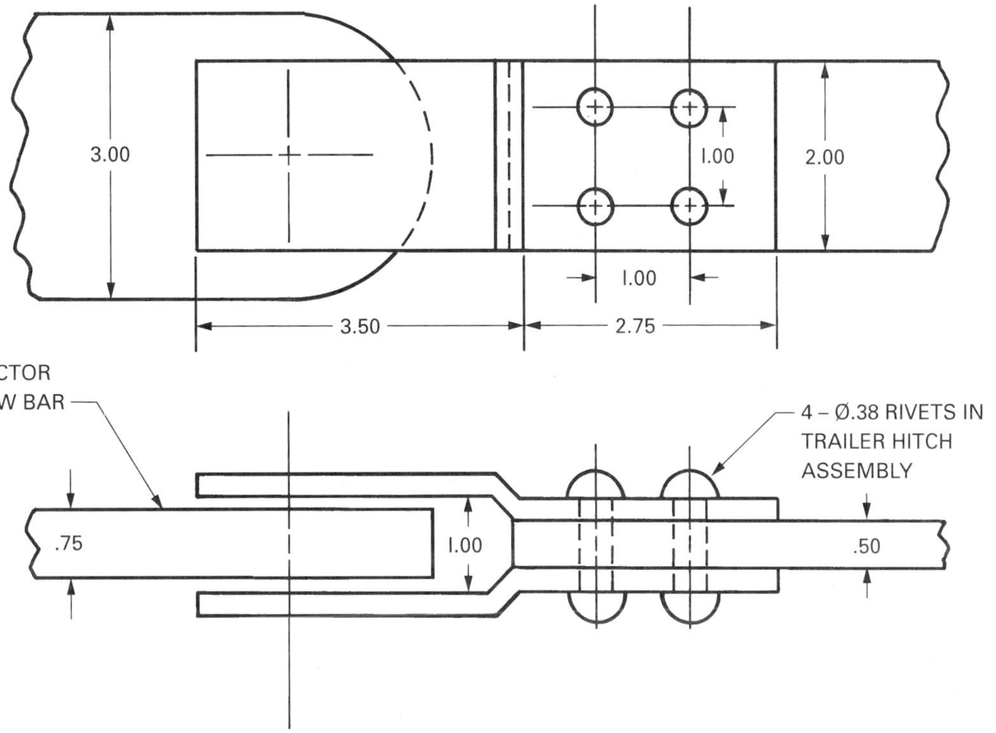

DRAW-BAR HITCH ASSEMBLY

Complete the assembly. A clevis pin with an area equal to the four rivets is used to fasten the trailer hitch to the tractor draw bar. The scale is 1:2.

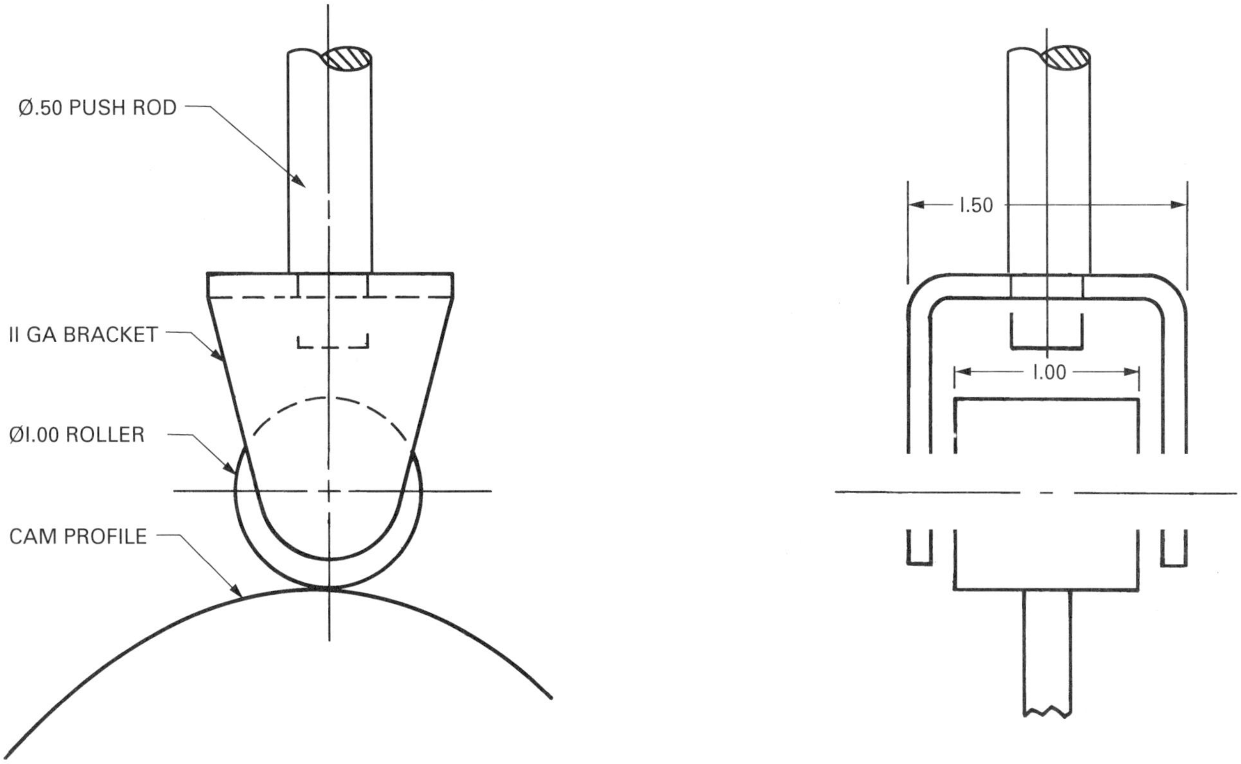

CAM FOLLOWER

Complete the assembly. A type E grooved pin holds the roller to the bracket. A washer and cotter pin are used to fasten the bracket to the push rod. The scale is 1:1.

PIN FASTENERS— DRAW-BAR HITCH AND CAM FOLLOWER	REFERENCE UNIT 11–2	NAME COURSE DATE	DRAWING J–2

Redraw the side view of the crane hook assembly. The hook is to be held to the U-frame with a slotted locknut. A cotter pin is inserted through the locknut slots to prevent the nut from turning. A clevis pin with washer and cotter pin holds the pulley to the frame. Include on the drawing an item list. Call out the fasteners used on the item list and on the assembly drawing. Show all of the part numbers. The scale is 1:1.

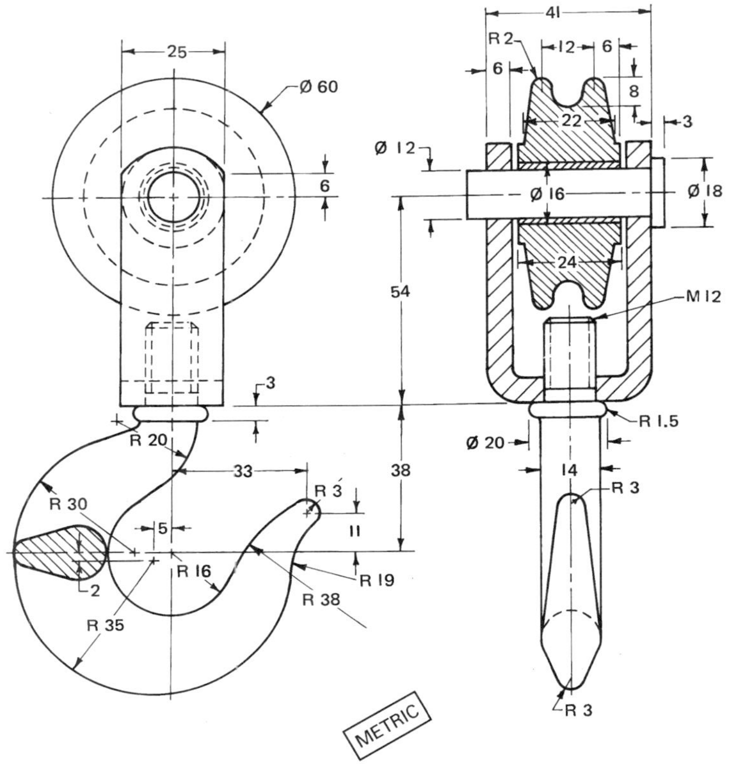

QTY	ITEM	MATL	DESCRIPTION	PT NO.

PIN FASTENERS—

CRANE HOOK

REFERENCE UNIT **11–2**

NAME

COURSE

DATE

DRAWING **J–3**

1. Complete the power drive assembly given the following information. The shaft is positioned in the housing by an SKF #6005 bearing. The end cap and a retaining ring hold the bearing in place, and a retaining ring holds the cap in the housing. Two retaining rings position the bearing on the shaft. The gear is positioned on the clutch by a retaining ring and a square key. The clutch is locked to the shaft by a square key held in position by a setscrew. The pulley drive is positioned and held to the shaft by a square key and two retaining rings. Show in the item list all of the purchased items and identify them on the drawing.
2. On one of the blank drawing sheets from the back of this workbook, make detail drawings of the end cup and the end of the shaft showing the retaining ring grooves and the key ways with their dimensions.

QTY	ITEM	MATL	DESCRIPTION	PT NO.

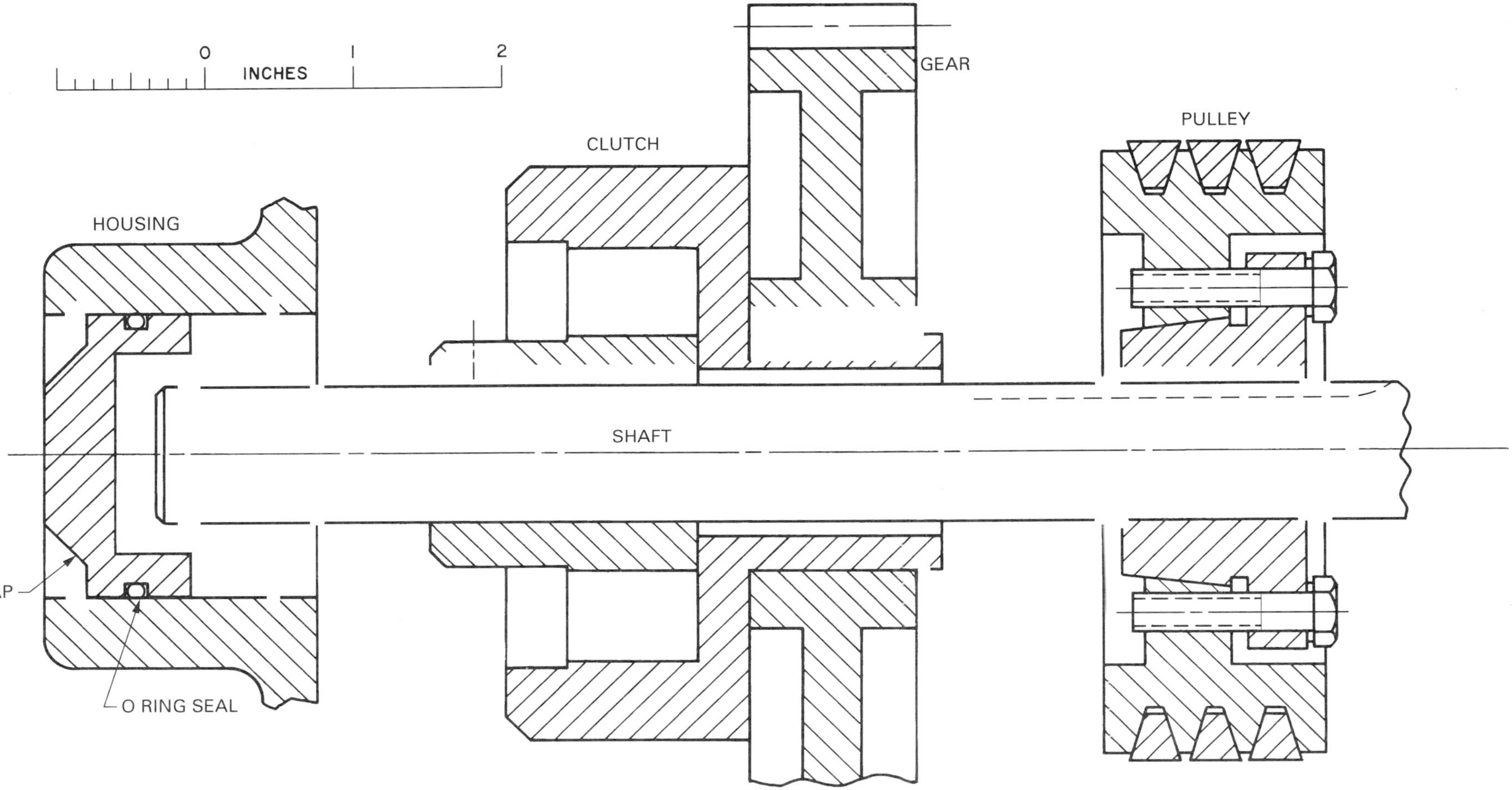

SCALE 1:1

RETAINING RINGS—

POWER DRIVE ASSEMBLY

REFERENCE UNIT	NAME	DRAWING
11-3	COURSE / DATE	J-4

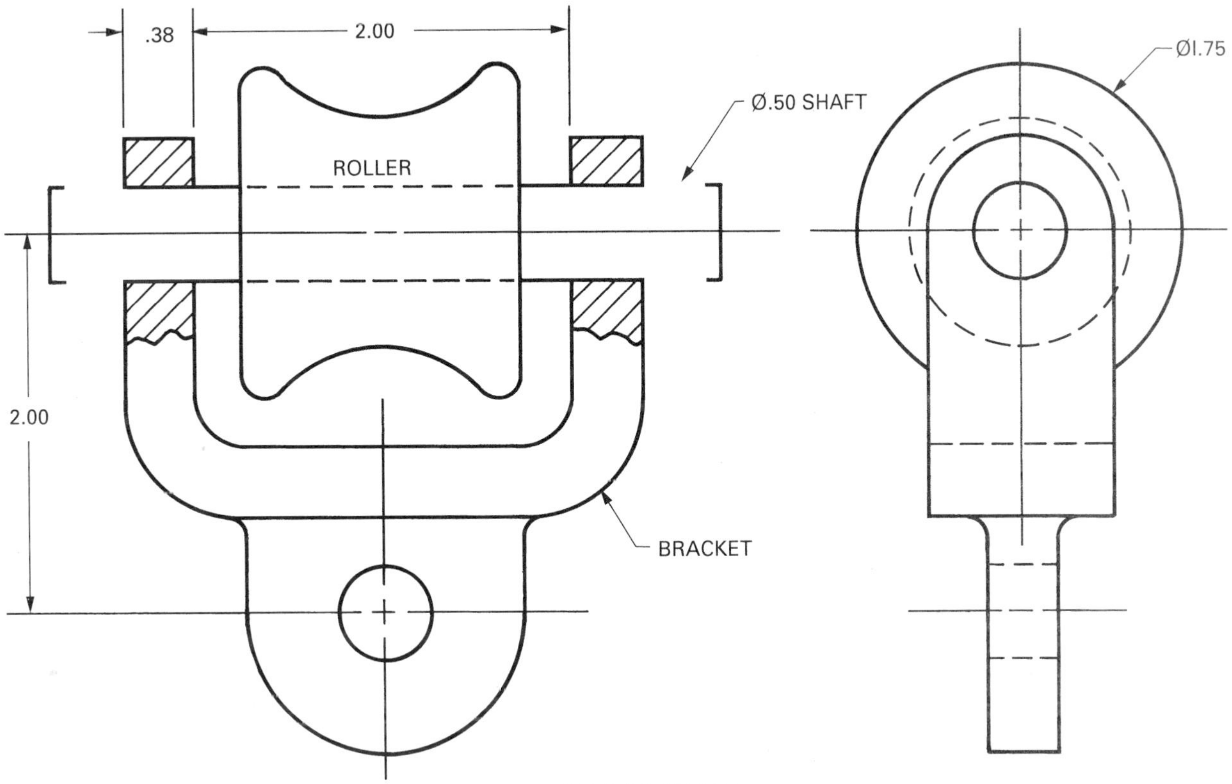

ROLLER ASSEMBLY

Complete the assembly. External self-locking retaining ring hold the roller shaft in position on the bracket. Plate washers between the retaining ring and the bracket are added to reduce friction. Add groove sizes and identify the retaining rings. The scale is 1:1.

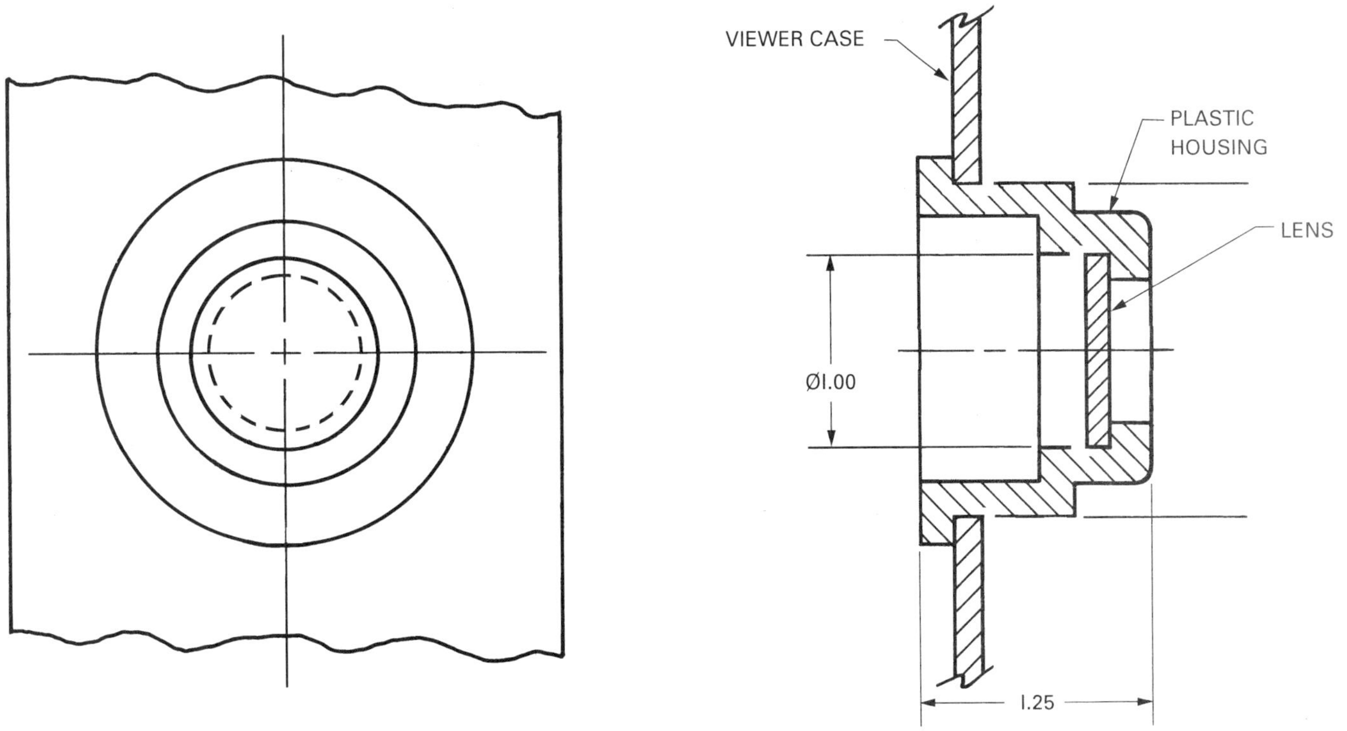

VIEWER CASE

Complete the assembly. An external self-locking retaining ring holds the plastic housing to the viewer case. An internal self-locking ring holds the lens in position. Add groove sizes and identify the retaining rings. The scale is 1:1.

RETAINING RINGS— ROLLER ASSEMBLY AND VIEWER CASE	REFERENCE UNIT 11–4	NAME _ COURSE _ DATE _	DRAWING J–5

1. Complete the punch-holder assembly. The two helical springs have plain closed ends and are Ø.06. They have a pitch of .10 in. The plunger and punch are held in the punch holder by retaining rings. An RC3 fit is required for the Ø.30 shaft. Show on the assembly drawing the limit dimensions for the Ø.30 hole and shaft and the dimensions of the retaining-ring grooves. The scale is 1:1.
2. In the space below the assembly, make detail drawings of the springs.

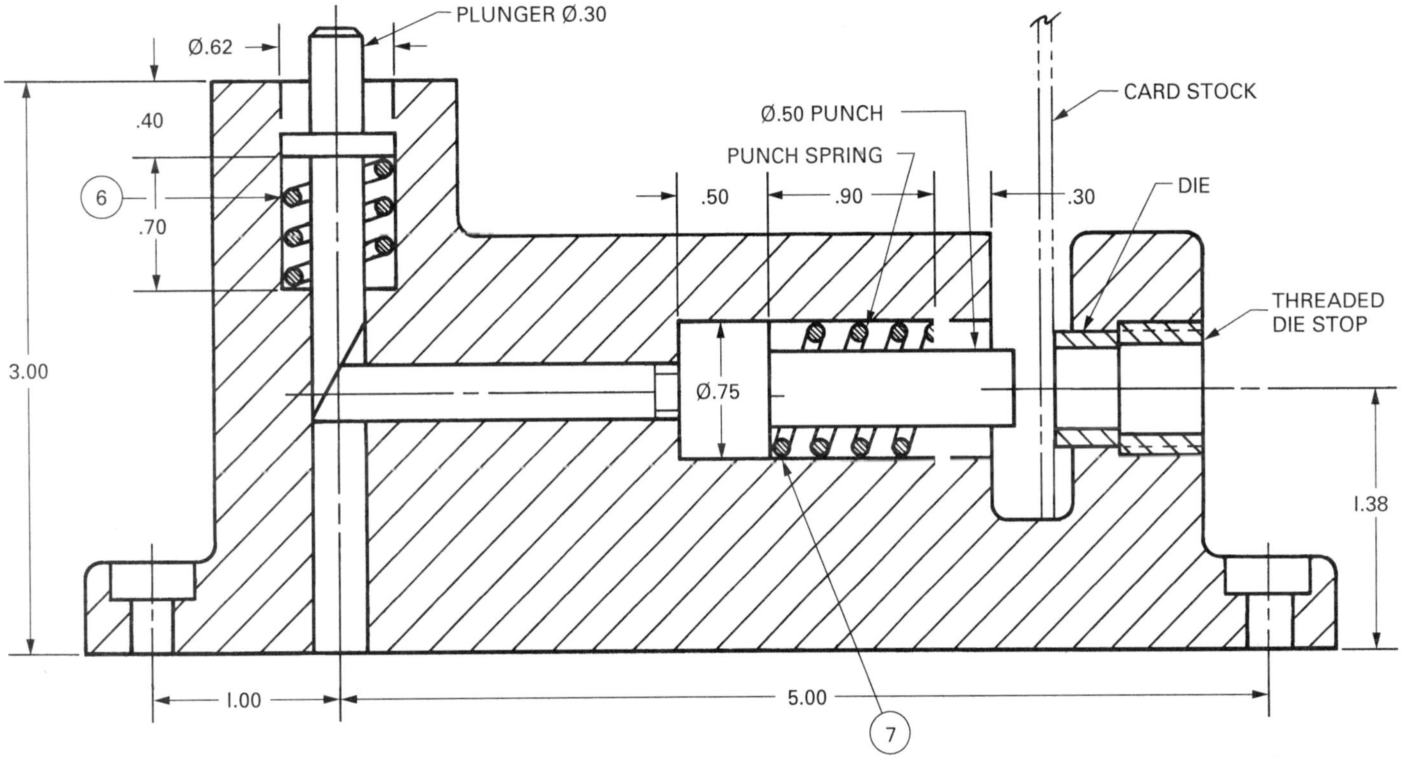

SPRINGS— PUNCH-HOLDER ASSEMBLY	REFERENCE UNIT 11–4	NAME COURSE DATE	DRAWING J–6

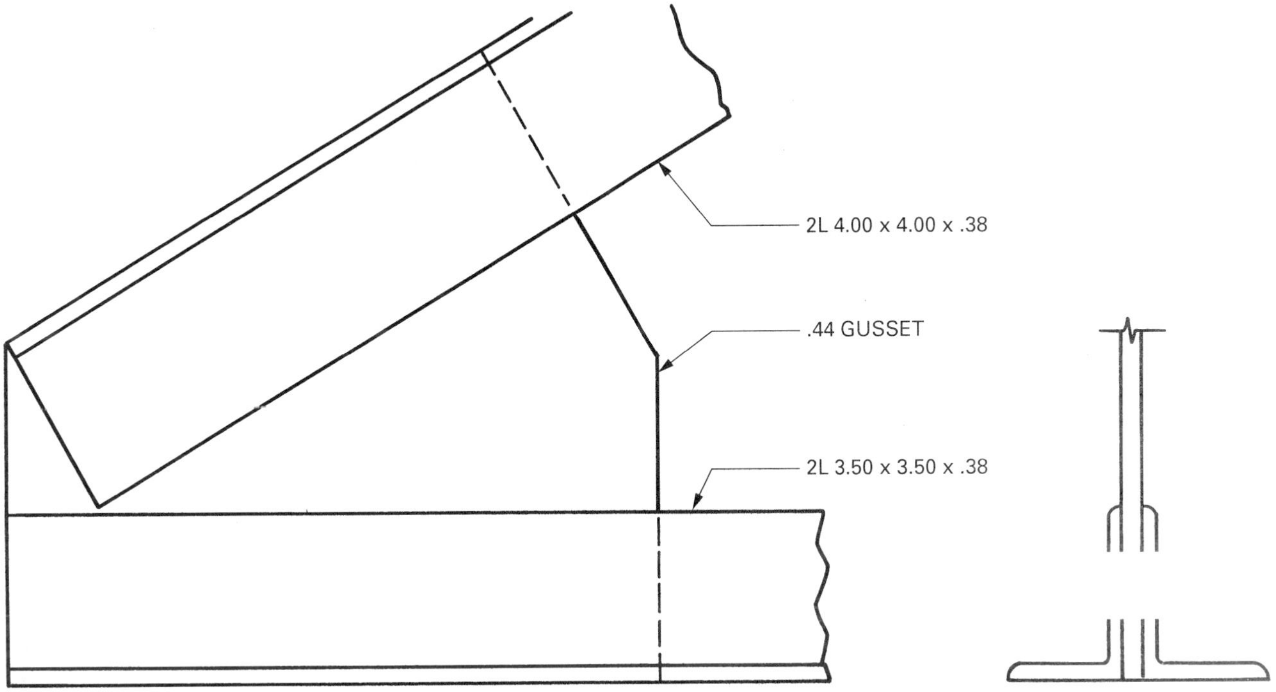

LARGE STRUCTURAL RIVETS

Complete the assembly. The roof truss is assembled in the shop with five evenly spaced Ø.50-in. rivets in each angle. The scale is 1:4.

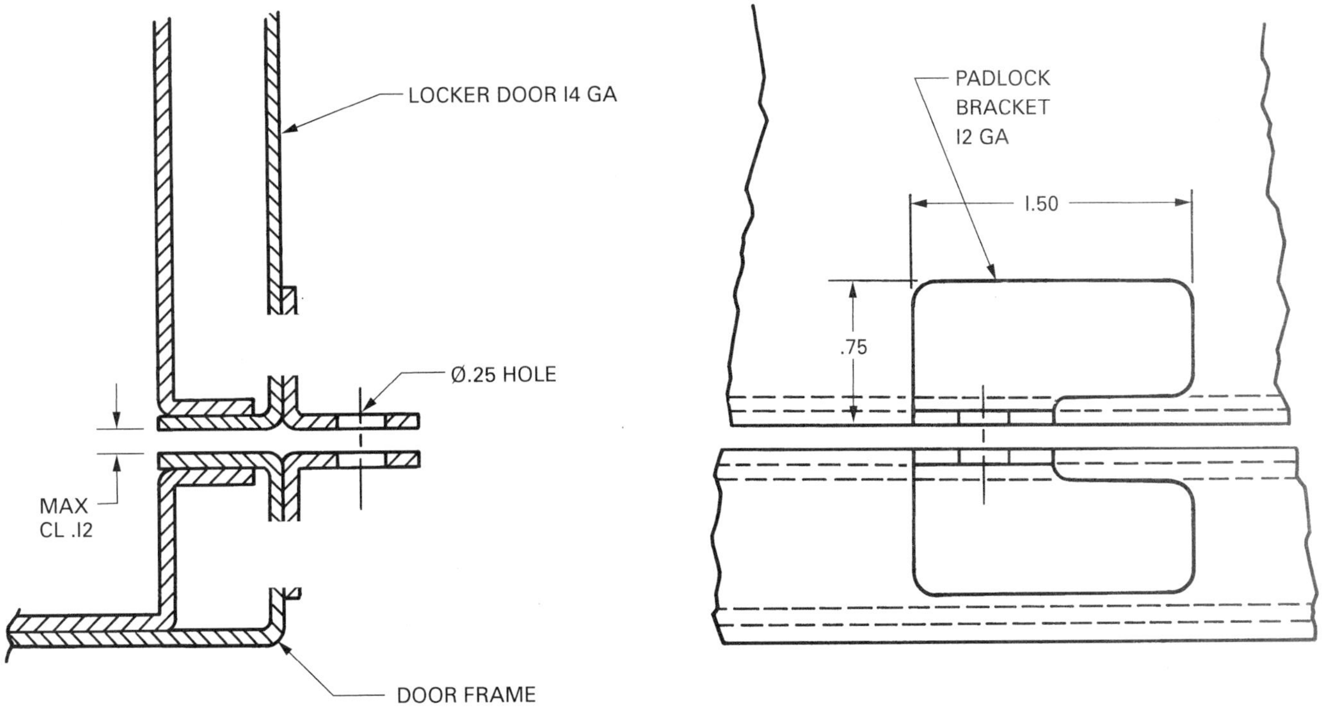

BLIND RIVETS

Complete the assembly. The padlock brackets are riveted to the locker door and door frame with two blind rivets in each bracket. The scale is 1:1.

RIVETS— STRUCTURAL AND BLIND RIVETS	REFERENCE UNIT 11–5	NAME COURSE	DATE	DRAWING J–7

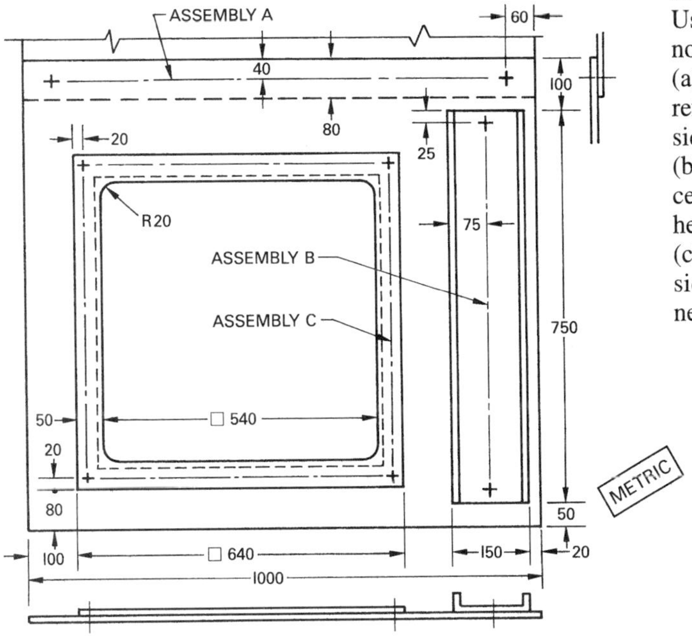

Redraw at the scale of 1:10 and complete the assembly shown. Use the graphical symbols of rivets for aerospace equipment and note the following information:

(a) *Assembly A.* Ø8 rivets equally spaced at 55 on center; item references 22; 100° countersunk both sides; preformed head near side.

(b) *Assembly B.* Ø6 combined rivets equally spaced at 50 on center; item reference 19; sleeve item reference 21; preformed head far side.

(c) *Assembly C.* Ø4 rivets equally spaced at 40 on center (4 sides); item reference 16; preformed head far side; 82° dimple near side.

RIVETS— RIVETS FOR AEROSPACE EQUIPMENT	REFERENCE UNIT 11–5	NAME COURSE DATE	DRAWING J–8

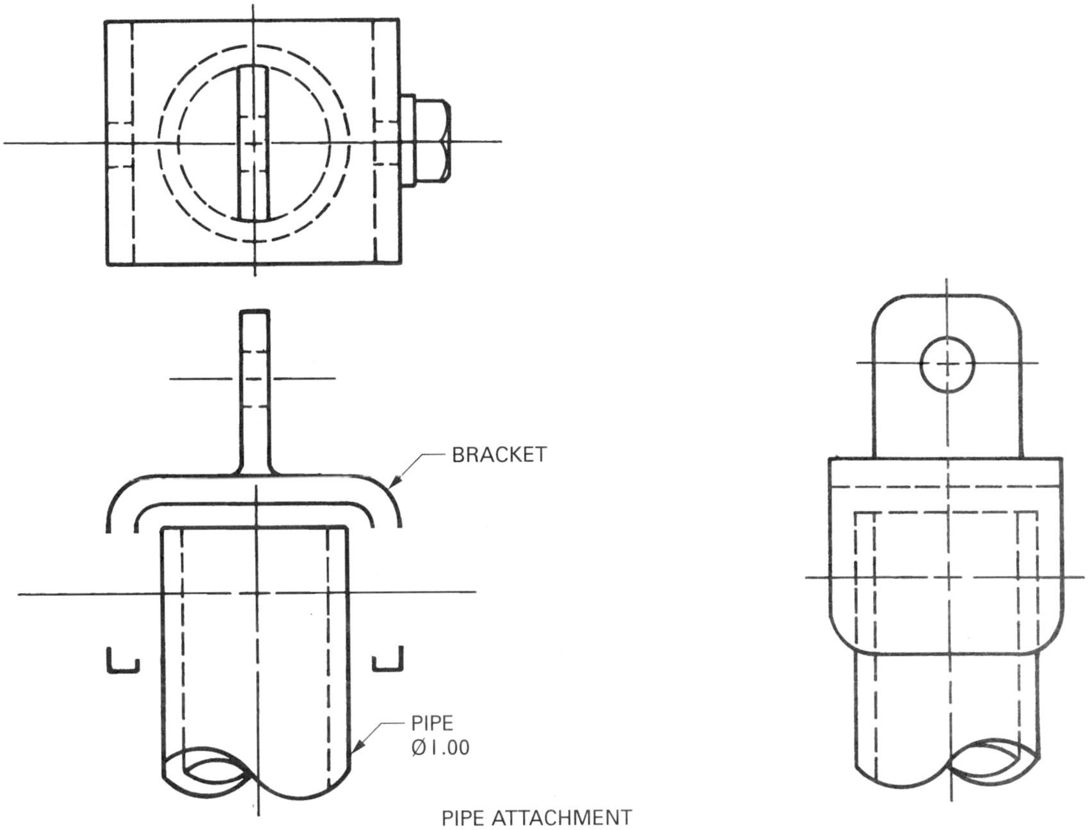

PIPE ATTACHMENT

Complete the assembly. Two resistance-welded threaded fasteners, one on each side of the pipe, are required. The bracket drops over the fasteners, and lockwashers and nuts secure the bracket to the pipe. The scale is 1:1.

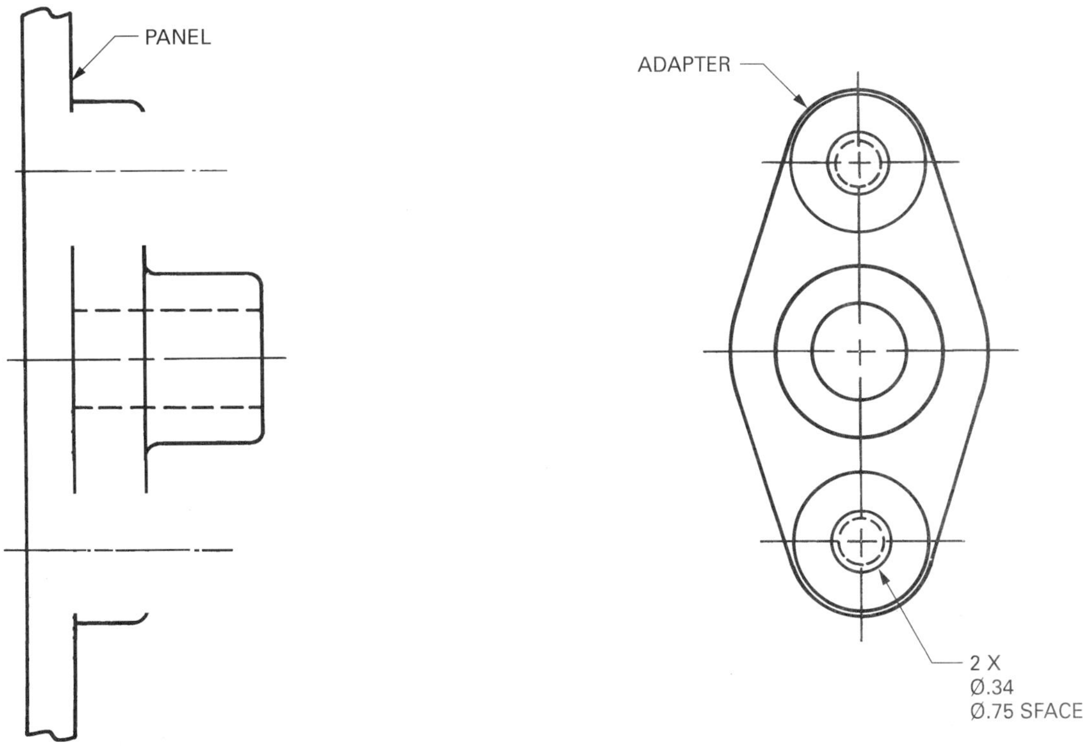

LEAKPROOF ATTACHMENT

Complete the assembly. A leakproof attachment method (stud welding) is recommended to hold the adapter to the panel. The scale is 1:1.

WELDED FASTENERS— PIPE AND LEAKPROOF ATTACHMENTS	REFERENCE UNIT 11–6	NAME COURSE DATE	DRAWING J–9

1

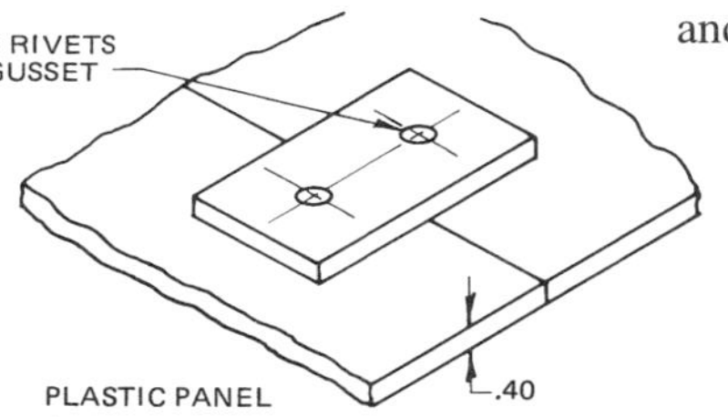

The riveted joint is to be replaced by a tongue and groove butt joint. It must be fast drying.

2

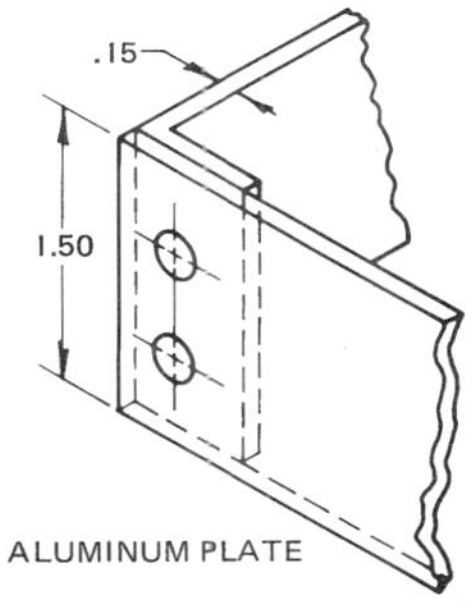

The sheet metal corner joint is to be replaced by a slip joint. It must be water-resistant.

3

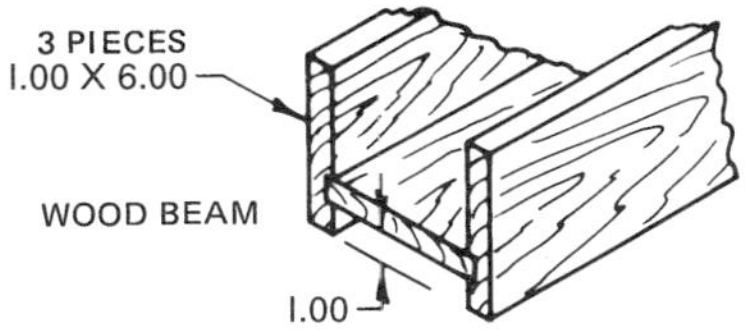

The three pieces of wood are to be assembled into the shape shown. Joint design has not been shown.

4

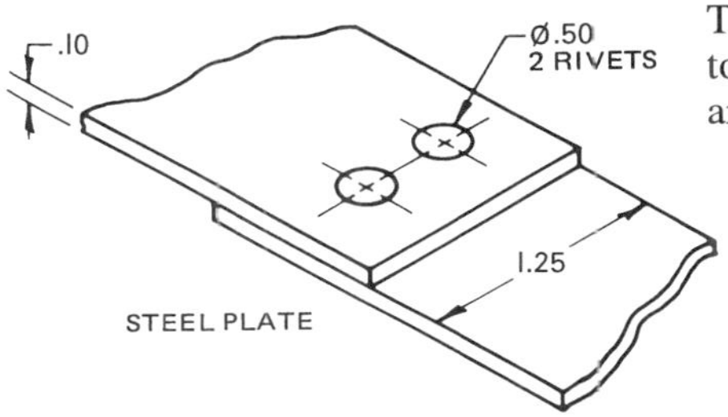

The riveted joint is to be replaced by a toggle lap joint. Military specifications are required.

With the information accompanying each drawing and the adhesive table in the appendix, list the adhesive product number and state the method of application you would recommend. Use your judgment for sizes not shown and dimension the joint. Scale is to suit.

ADHESIVE FASTENERS— JOINT DESIGN	REFERENCE UNIT **11–7**	NAME COURSE DATE	DRAWING **J–10**

1. Complete the item list for this assembly.
2. On the blank sheets from the back of this book, prepare detail drawings of the wheel assembly. The shaft should have an RC4 fit with the bushing, and the bushing an LN3 fit in the body. Use your judgment for the selection and the number of views for each part.

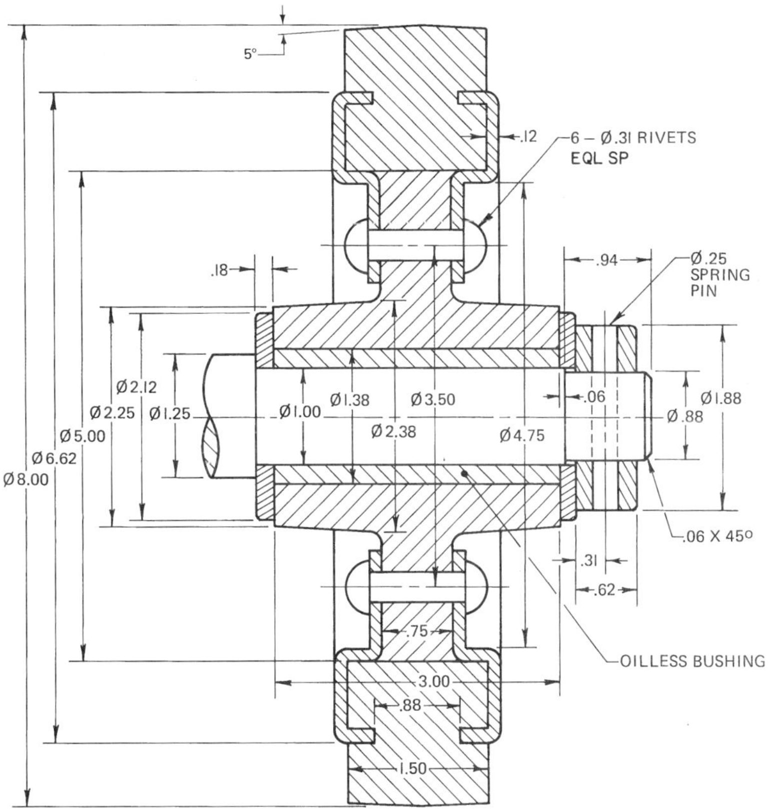

QTY	ITEM	MATL	DESCRIPTION	PT NO.

FASTENER REVIEW— WHEEL ASSEMBLY	REFERENCE UNIT 11–8	NAME COURSE DATE	DRAWING J–11

Make a one-view assembly drawing of the universal joint. The scale is 2:1. Include an item list.

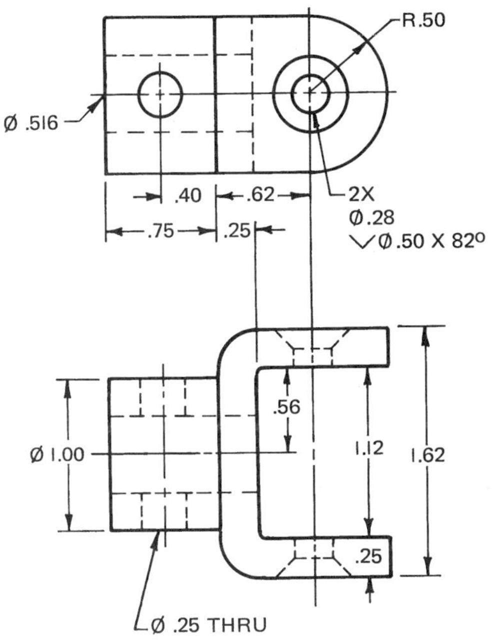

PT I - FORK - 2 REQD

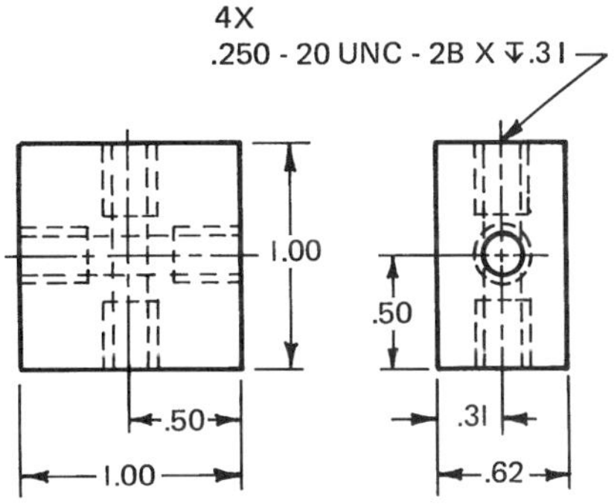

PT 2 - RING - I REQD
PT 3 - Ø .25 SPRING PIN - 2 REQD
PT 4 - .250 - 20 FHMS - .62 LG - 4 REQD

QTY	ITEM	MATL	DESCRIPTION	PT NO.

FASTENER REVIEW— UNIVERSAL JOINT	REFERENCE UNIT 11–8	NAME COURSE DATE	DRAWING J–12

Make a two-view working drawing of the door closing arm. Use a revolved section to show the center section of the arm. Select a suitable iron for the part. The scale is 1:1.

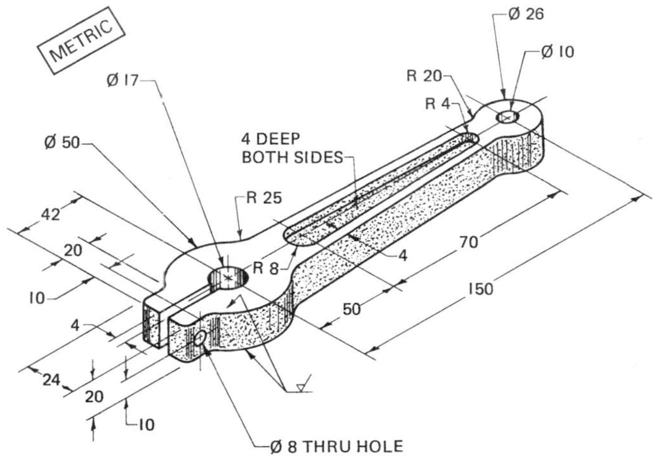

CAST IRONS—
DOOR CLOSING ARM

REFERENCE UNIT **12–1**

NAME

COURSE

DATE

DRAWING **K–1**

Make a working drawing of the raising bar. Show the acme thread using detailed representation. Select a suitable steel for the part. The scale is 1:1.

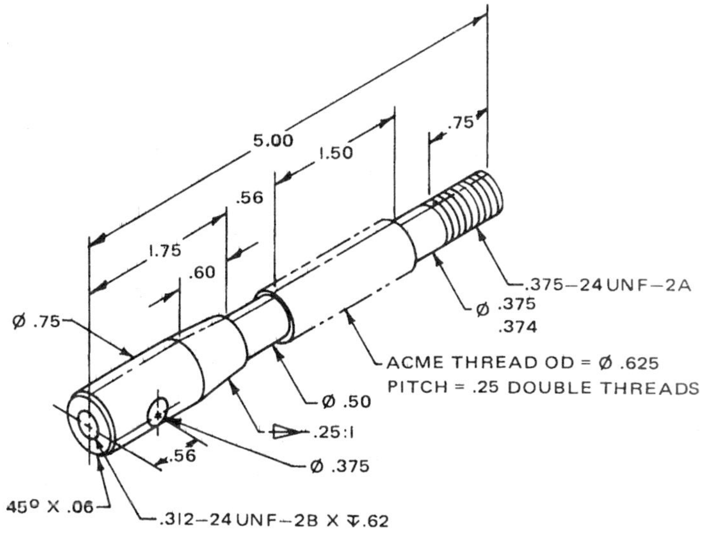

CARBON STEEL— RAISING BAR	REFERENCE UNIT 12–2	NAME COURSE DATE	DRAWING K–2

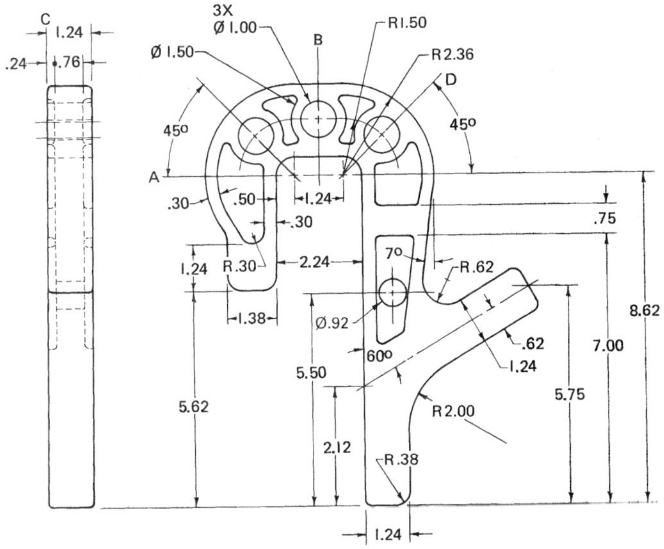

ROUNDS AND FILLETS R.06

Make a three-view working drawing of the outboard motor clamp. Add a full-section top view with the cutting plane located at line AD. Use lines or surfaces marked A, B, and C as the zero lines and use arrowless dimensioning. The scale is 1:2. Select a suitable material, noting that the part must be water-resistant, have a painted finish, have moderate strength, and have a light weight or mass.

NONFERROUS METALS— OUTBOARD MOTOR CLAMP	REFERENCE UNIT **12–3**	NAME COURSE DATE	DRAWING **K–3**

Make a one-view exploded orthographic assembly drawing of the connecting link. Complete the material list. The student is to select the materials. The scale is 1:1.

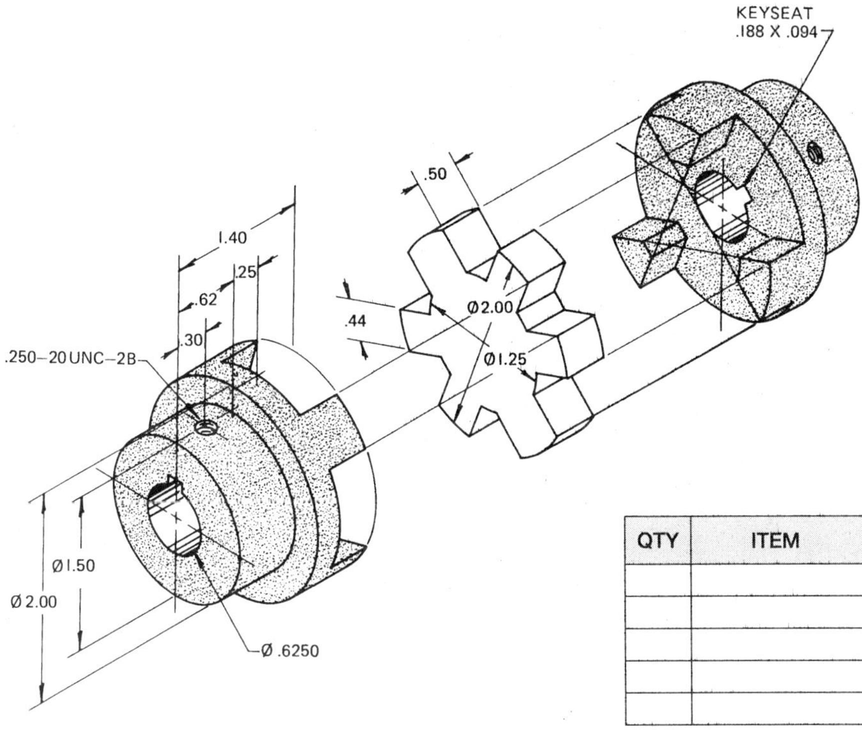

QTY	ITEM	MATL	DESCRIPTION	PT NO.

PLASTICS— CONNECTING LINK	REFERENCE UNIT 12–4	NAME COURSE DATE	DRAWING K–4

Make a one-view full-section assembly drawing of the caster assembly. Complete the item list and show the overall assembly sizes. Select a suitable material for Part 1. The scale is 1:1.

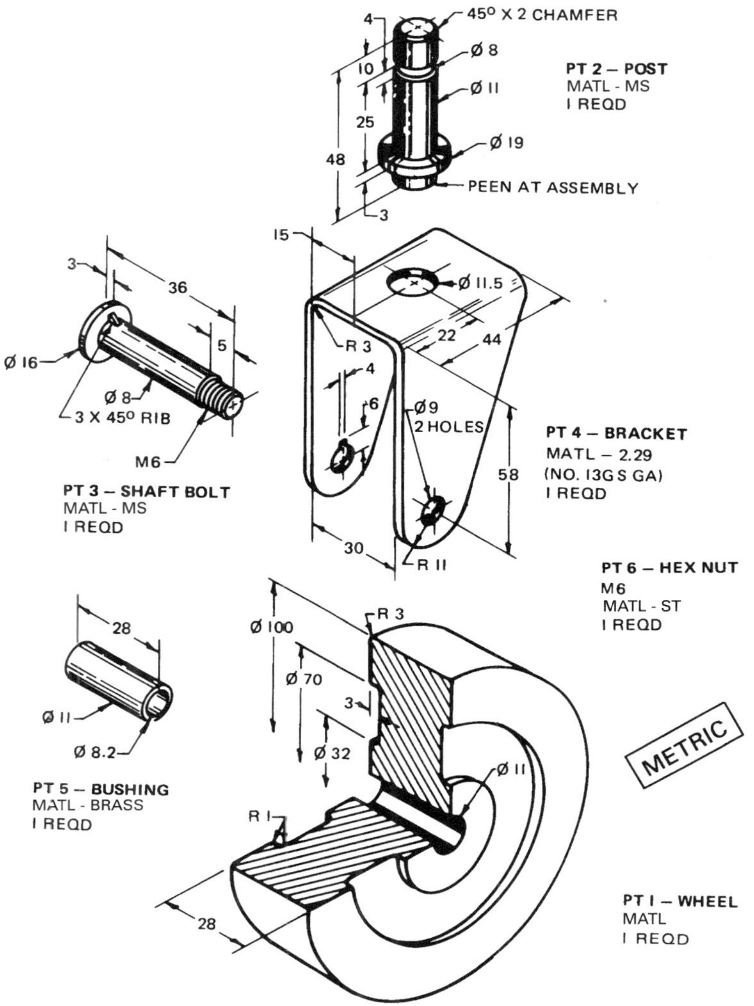

QTY	ITEM	MATL	DESCRIPTION	PT NO.

RUBBER—
CASTER ASSEMBLY

REFERENCE UNIT
12–5

NAME

COURSE

DATE

DRAWING
K–5

Prepare both the casting and the machining drawings for the connector shown. Draw a one-view full-section, complete with the necessary dimensions for each drawing. The scale is 1:1. The location of the drawings is shown on the drawing area.

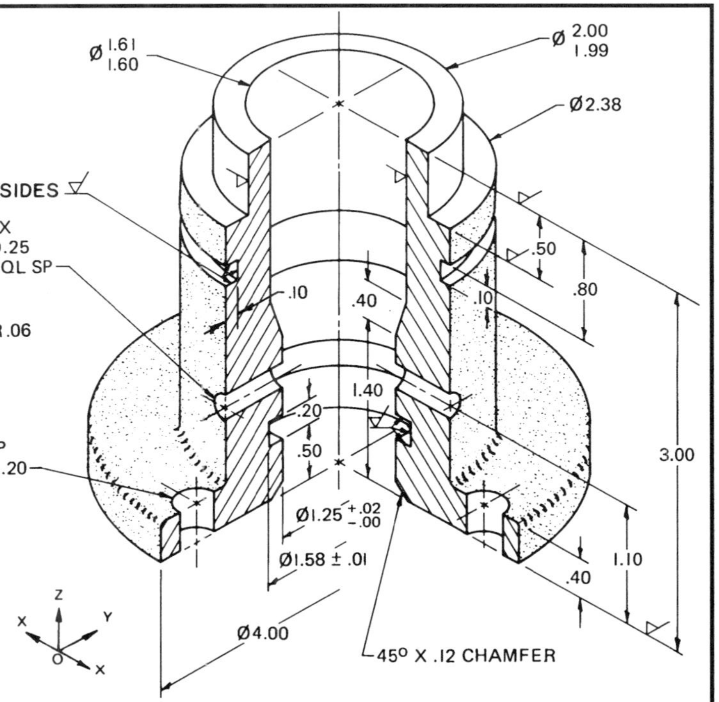

CASTINGS— CONNECTOR	REFERENCE UNIT 13–1	NAME COURSE	DATE	DRAWING L–1

Redesign the fabricated swing bracket into a cast part. Draw the front and side views. Show the limit sizes where fits are indicated. Surfaces shown by the letter A are to have a maximum roughness of 1.6 μm and a machining allowance of 2 mm. The scale is 1:1. The center of the Ø34 is located on the drawing area.

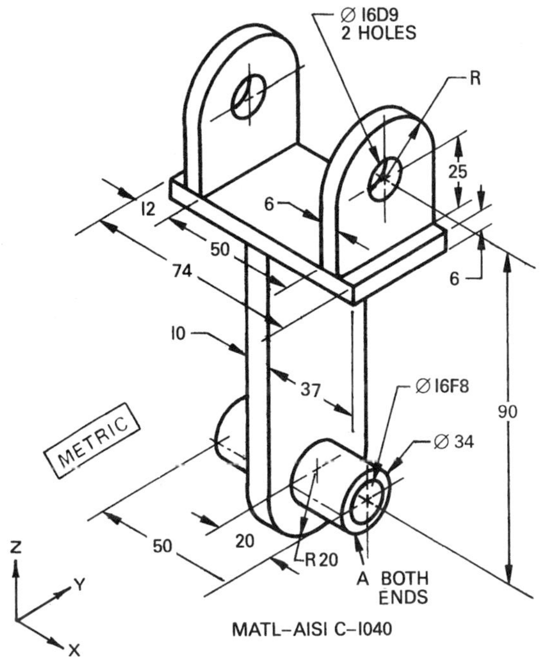

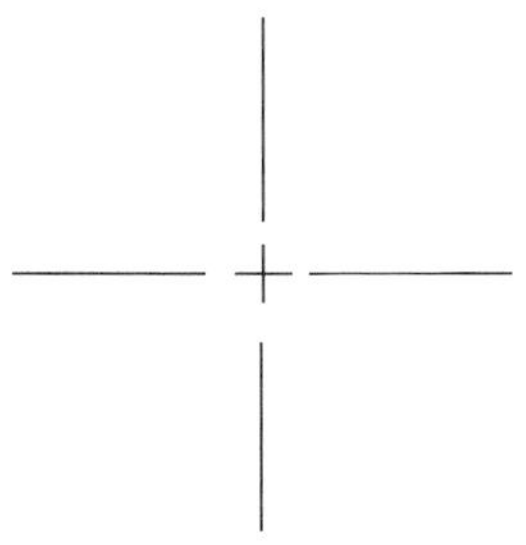

CASTINGS— SWING BRACKET	REFERENCE UNIT 13–1	NAME COURSE DATE	DRAWING L–2

Make a two-view forging drawing of the bracket shown. The scale is 1:1.

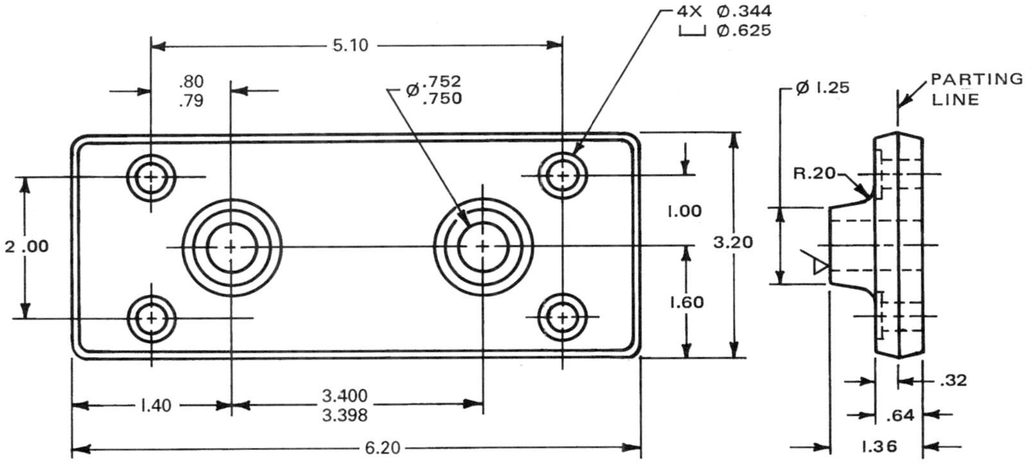

FORGINGS—
BRACKET

REFERENCE UNIT **13–2**

NAME
COURSE
DATE

DRAWING **L–3**

Prepare two drawings, one for machining the part and the second for the making of the briquet (powder metallurgy) of the bracket shown. The locations of the drawings are shown on the drawing area. The scale is 1:1.

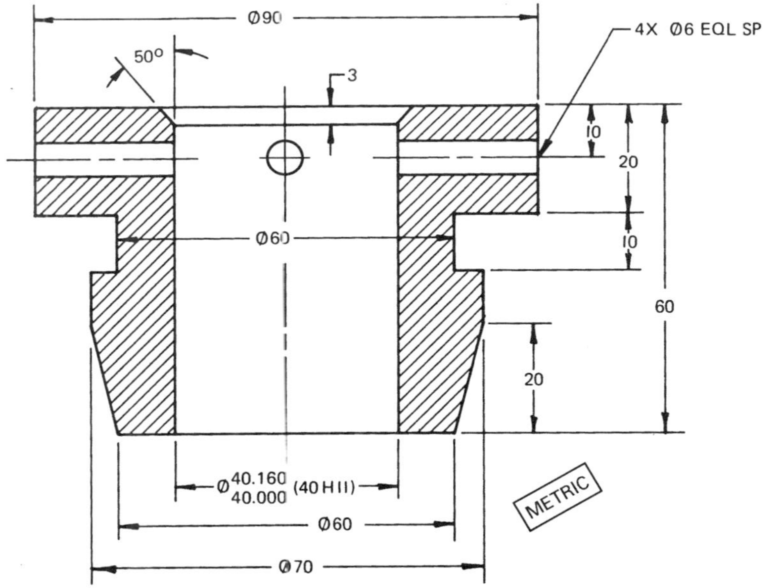

POWDER METALLURGY— BRACKET	REFERENCE UNIT 13–3	NAME COURSE DATE	DRAWING L–4

Using a plastic molding design, add threaded inserts to the pivot arm shown. Use your judgment for dimensions not shown and the type and number of views required. The scale is 1:1.

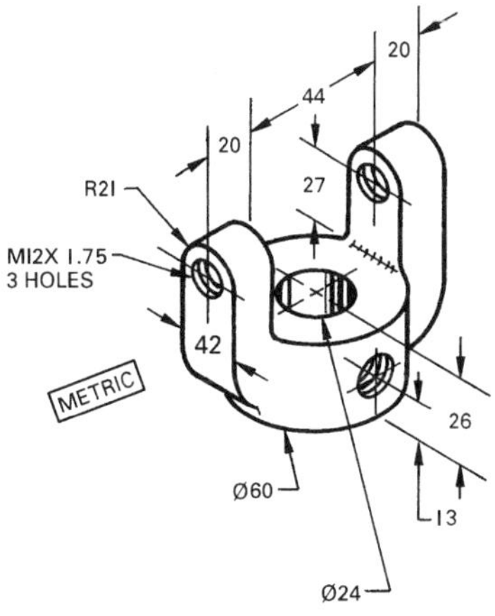

ROUNDS AND FILLETS R5

PLASTIC MOLDED PARTS— PIVOT ARM	REFERENCE UNIT 13–4	NAME COURSE	DATE	DRAWING L–5

After the number of drawings made over the last six months was reviewed, it was discovered that many cable straps were similar in design. Prepare a standard tabulated drawing similar to Fig. 14–1–1, reducing the number of standard parts to 4. The scale is 1:1.

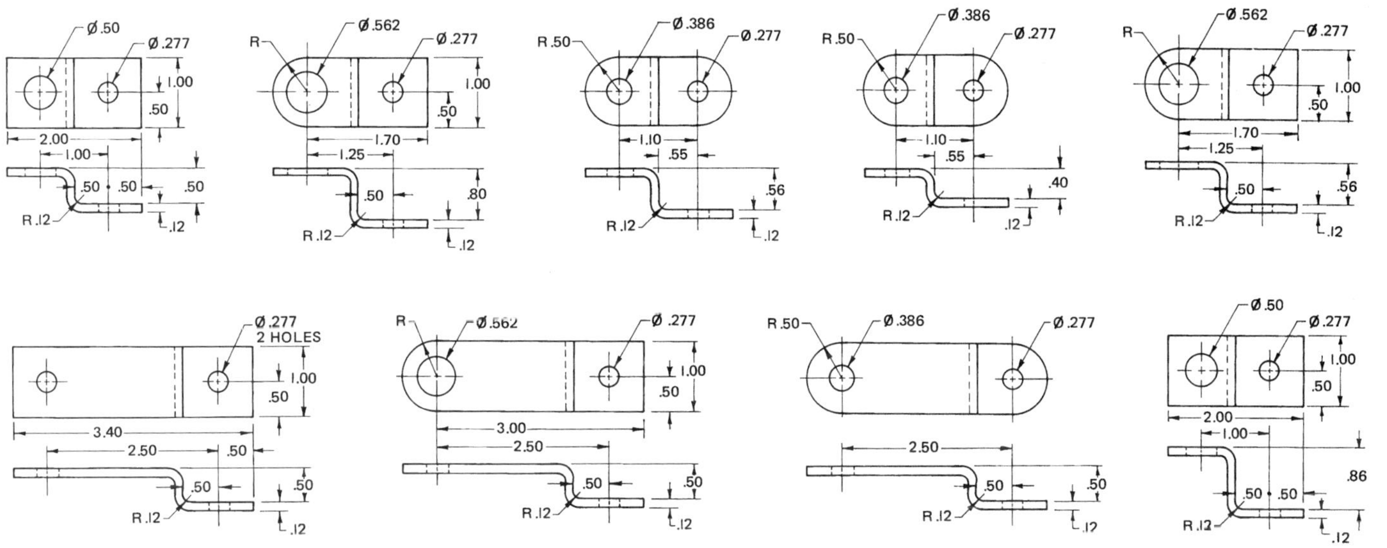

PT NO.	A	B	C	D	ROUND ENDS	SQUARE ENDS

FUNCTIONAL DRAFTING— CABLE STRAPS	REFERENCE UNIT 14–2	NAME COURSE DATE	DRAWING M–1

Redraw the cover plate shown using arrowless dimensioning and simplified drawing practices. Use the bottom left-hand corner of the top view as the datum surfaces. The scale is 1:12.

SIZE	SYMBOL	FROM X	FROM Y	FROM Z

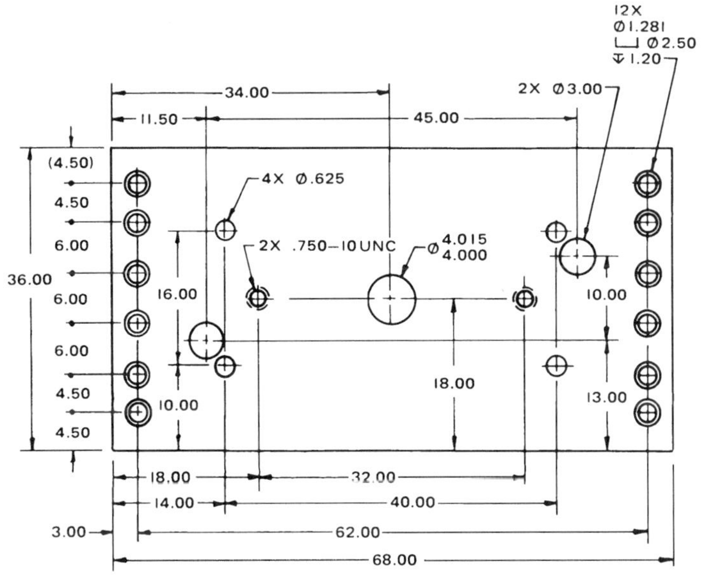

MATL – 3.00 THK ACETAL RESIN

FUNCTIONAL DRAFTING— COVER PLATE	REFERENCE UNIT 14–2	NAME ___ COURSE ___ DATE ___	DRAWING M–2

Make a detail drawing of the base plate shown.
Select the views and the scale required.

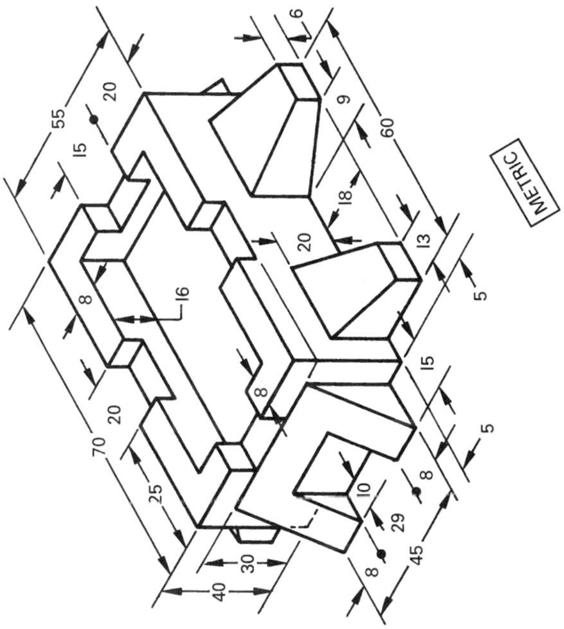

DETAIL DRAWINGS— BASE PLATE	REFERENCE UNIT 14–3	NAME COURSE DATE	DRAWING M–3

Make a three-view detail drawing of the offset bracket shown. The scale is 1:1.

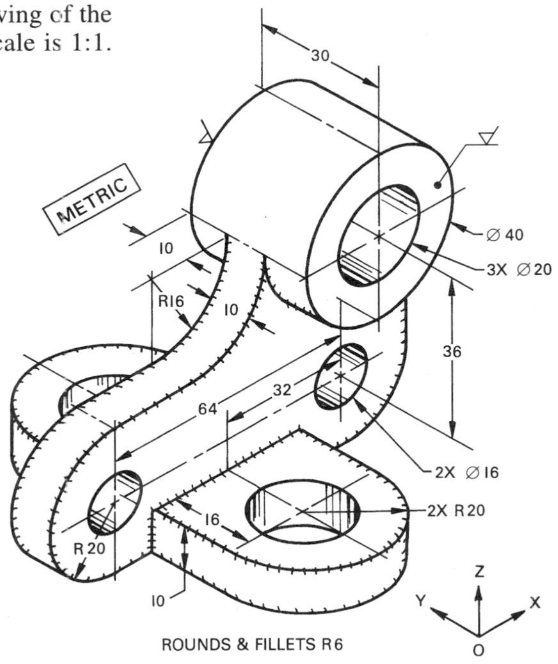

DETAIL DRAWINGS—
SWIVEL HANGER

REFERENCE UNIT **14–3**

NAME

COURSE

DATE

DRAWING **M–4**

Make a three-view working drawing of the locating stand shown. Dimensions are to be converted to millimeters. Only the dovetail and the T-slot dimensions must be taken to an accuracy of two points beyond the decimal point. The bottom left corner of the front view is located on the drawing area. The scale is 1:1.

–EXCEPT WHERE NOTED, SURFACE FINISH TO BE 63/▽

–FINISH ON SURFACES ON THE TEE AND DOVETAIL SLOTS TO BE 32/▽

–ROUNDS AND FILLETS R.12

–MATL–MALLEABLE IRON

METRIC

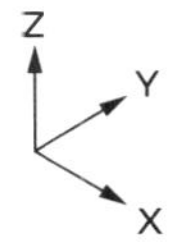

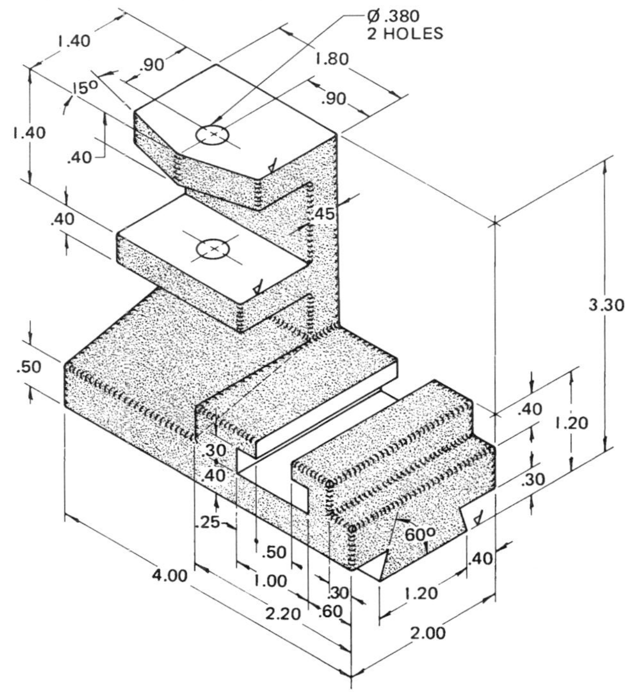

DETAIL DRAWINGS— LOCATING STAND	REFERENCE UNIT 14–3	NAME COURSE DATE	DRAWING M–5

As assigned by your instructor, on one or more of the blank drawing sheets from the back of this book, prepare detail drawings of any of the parts assigned by your instructor from the assembly drawings shown. For the pulley assembly, the following fits are to be used: Ø28H9/d9, Ø45H7/s6, and Ø35H8/f7. For the adjustable pulley, an RC4 fit is required for the Ø1.20 shaft. The scale and selection of views are to be decided by the student. Include an item list for the parts.

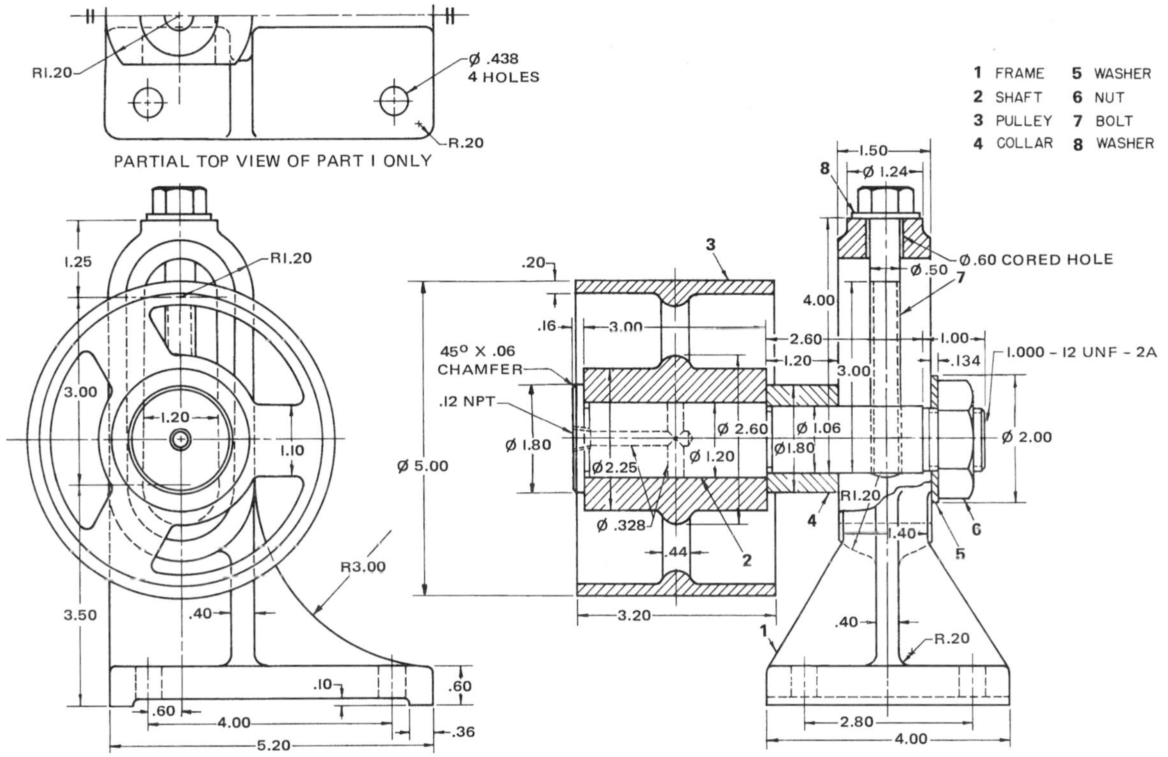

ASSIGNMENT I — ADJUSTABLE PULLEY

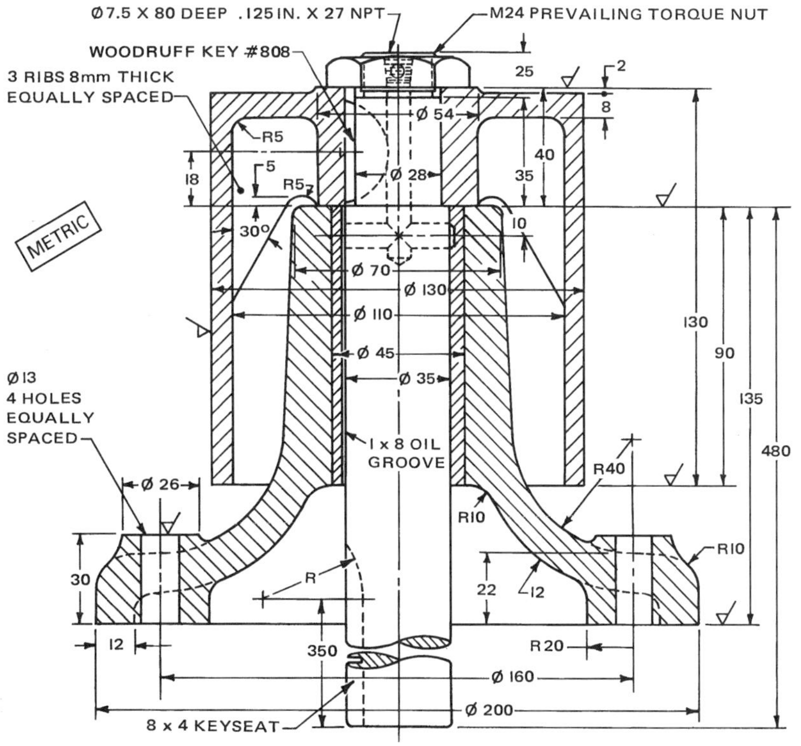

ASSIGNMENT 2 — PULLEY ASSEMBLY

MULTIPLE DETAIL DRAWINGS— ADJUSTABLE PULLEY AND PULLEY ASSEMBLY	REFERENCE UNIT **14–4**	NAME COURSE DATE	DRAWING **M–6**

Make the following drawing changes to the axle cap and record the changes in the boxes located below the drawing:
1) Ø.50 changed to Ø.53, 2) Ø5.75 changed to Ø6.00, 3) 2.25 changed to 2.30, 4) Ø2.88 changed to 2.75.

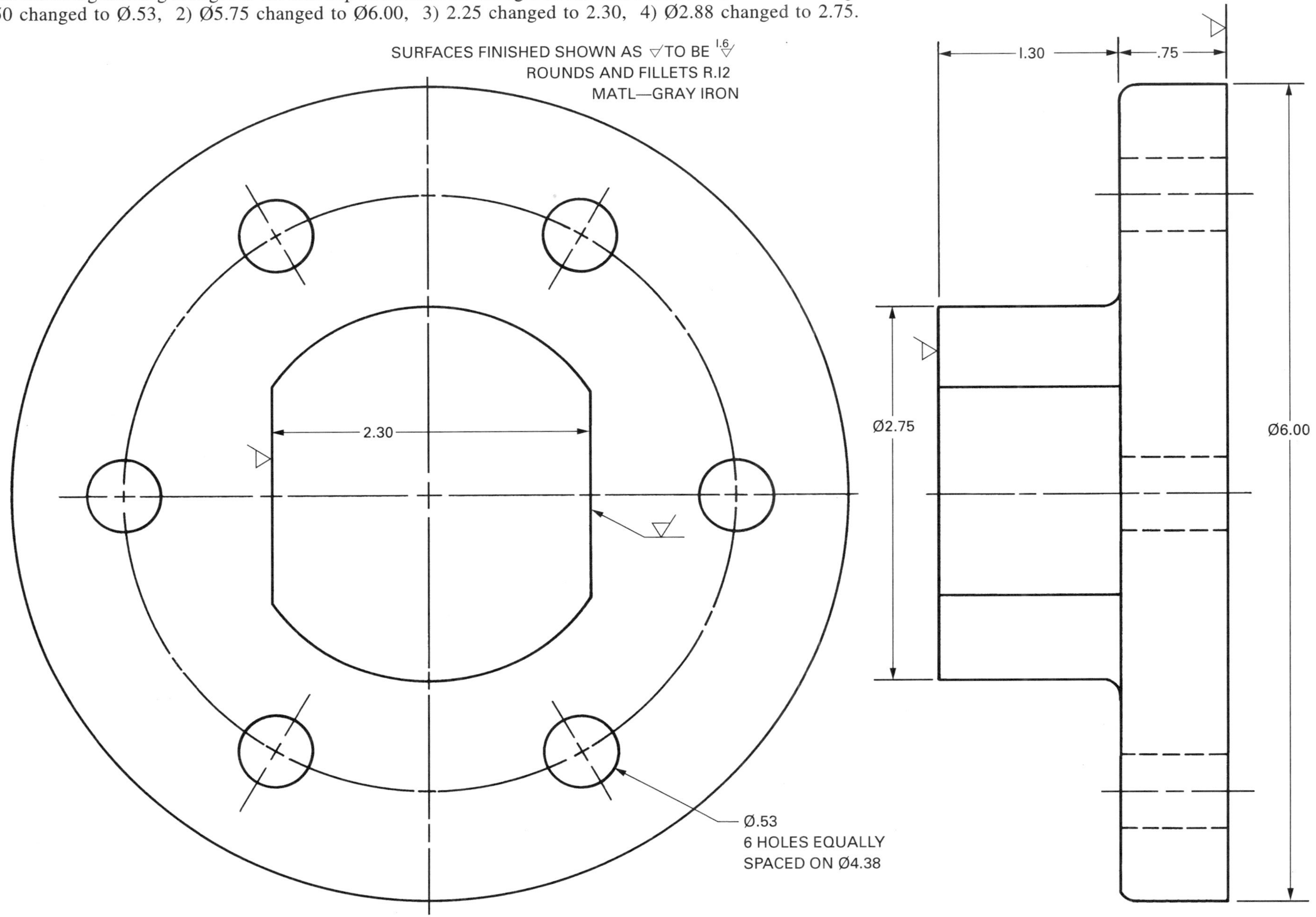

DRAWING REVISIONS—
AXLE CAP

REFERENCE UNIT **14–5**

NAME

COURSE

DATE

DRAWING **M–7**

Make a one-view assembly drawing of the tool post holder shown. Identify the parts on the assembly and complete the item list. Sectioning can be added to improve the clarity of the drawing. The scale is 1:1.

QTY	ITEM	MATL	DESCRIPTION	PT

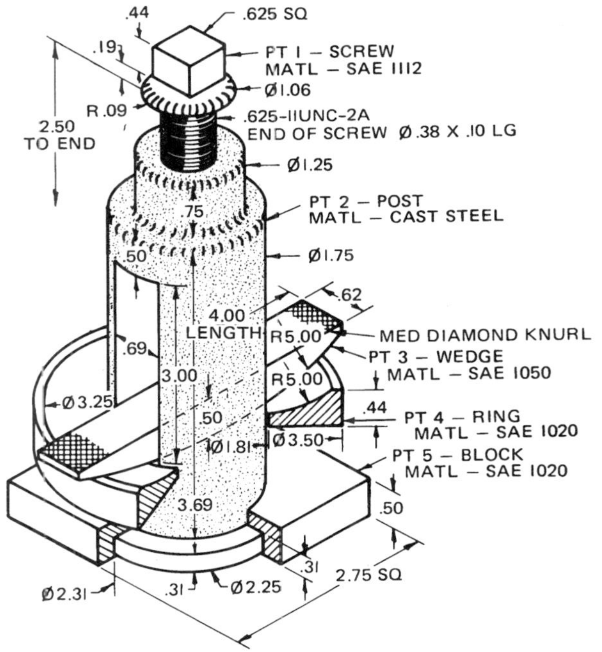

ASSEMBLY DRAWINGS— TOOL POST HOLDER	REFERENCE UNIT **14–6**	NAME ___ COURSE ___ DATE ___	DRAWING **M–8**

Make a one-view assembly drawing of the V-block clamp assembly shown. Depict a Ø24-mm bar in phantom lines held in position. Identify the parts on the assembly and complete the item list. The scale is 1:1.

QTY	ITEM	MATL	DESCRIPTION	PT NO.

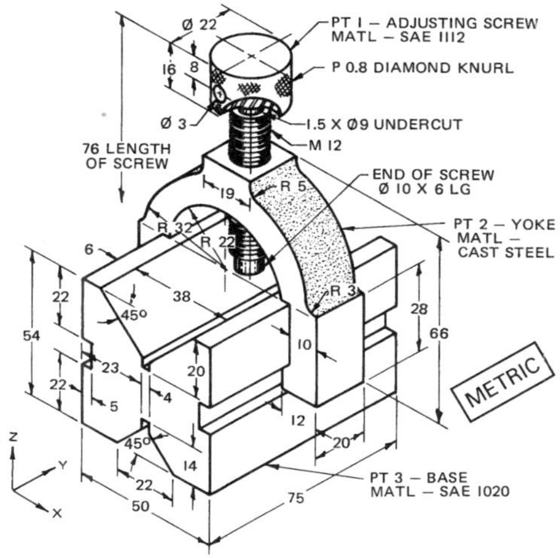

ASSEMBLY DRAWINGS—
V-BLOCK CLAMP

REFERENCE UNIT 14–6

NAME
COURSE
DATE

DRAWING M–9

QTY	ITEM	MATL	DESCRIPTION	PT NO.

Make an exploded isometric assembly drawing of the coupling shown. Allow .70 in. space between parts and align the parts on the center line shown. Add identification part numbers and complete the item list. The scale is 1:1.

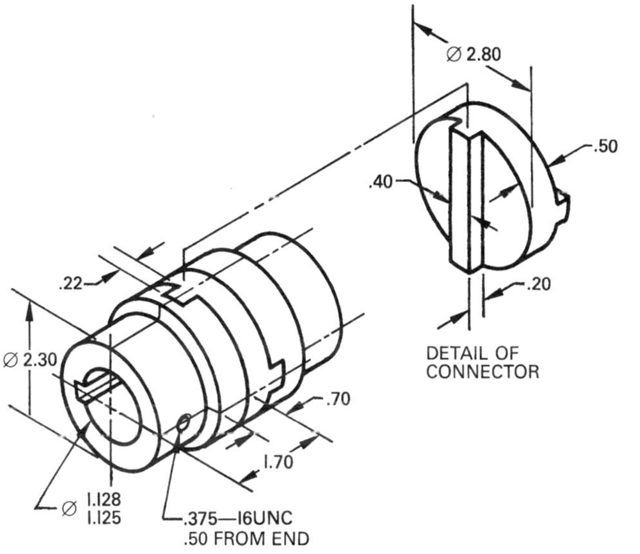

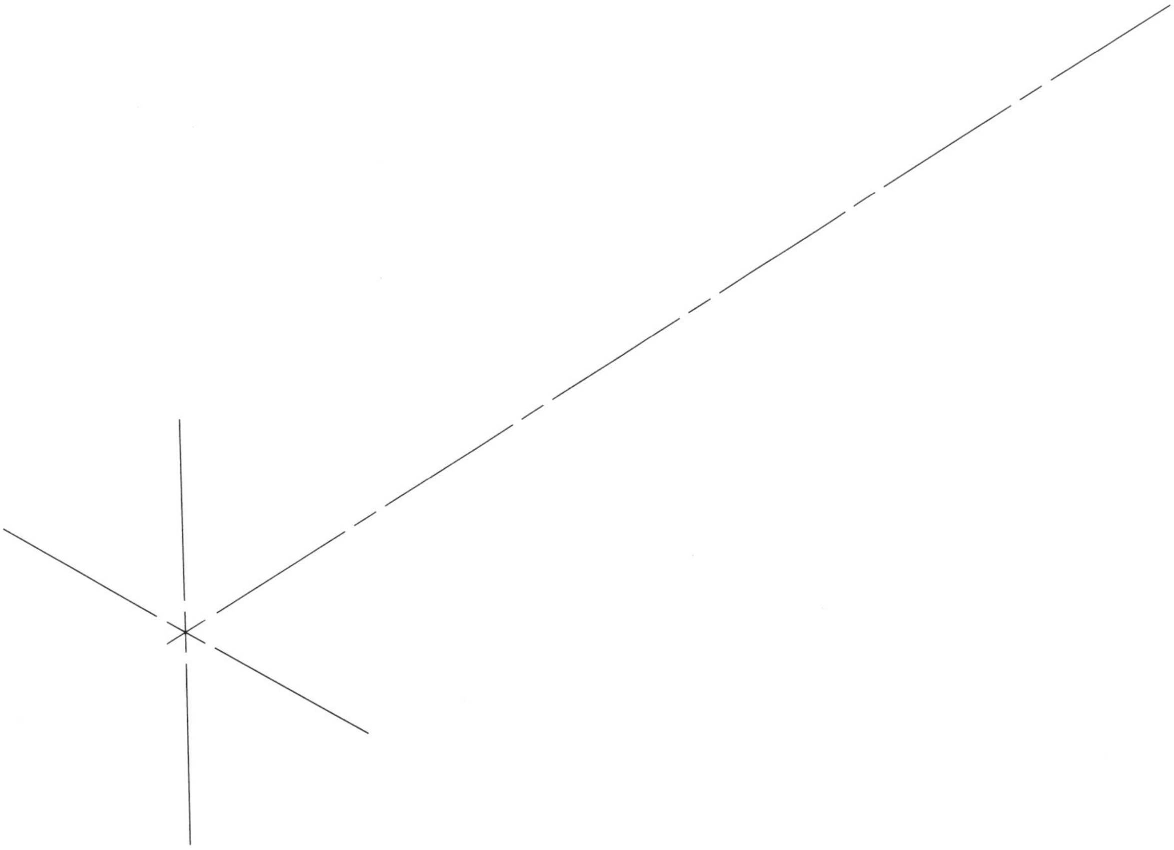

EXPLODED ASSEMBLY DRAWINGS—

COUPLING

REFERENCE UNIT **14-7**

NAME

COURSE

DATE

DRAWING **M-10**

QTY	ITEM	MATL	DESCRIPTION	PT NO.

Make a three-view detailed assembly drawing of the sawhorse to the scale of 1:8 ($1\frac{1}{2}$ in. = 1 ft). Show all parts on the item list and add identification part numbers to the assembly.

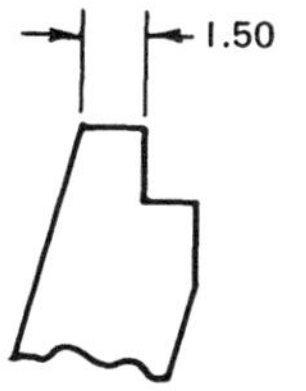

DETAIL OF LEG

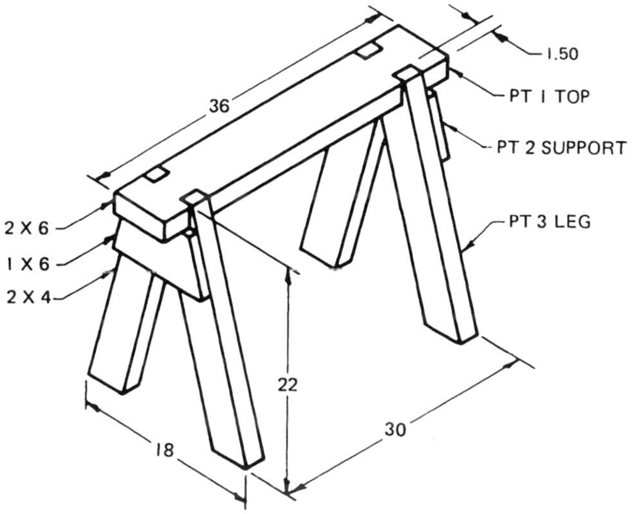

MATL – CONSTRUCTION GRADE SPRUCE

NOTE: WOOD SIZES (THICKNESS AND WIDTH) ARE NOMINAL INCH SIZES.

DETAILED ASSEMBLY DRAWINGS— SAWHORSE	REFERENCE UNIT 14–8	NAME _ COURSE _ DATE _	DRAWING M–11

Make a one-view subassembly drawing of the wheel. A broken-out or partial section view is recommended to show the interior features. Include on the drawing pertinent dimensions and identification part numbers. Four Ø10-mm bolts fasten the wheel to an 8-mm plate. Show the bolted connection on the right side only. The scale is 1:1. The center line of PT4 is located on the drawing area.

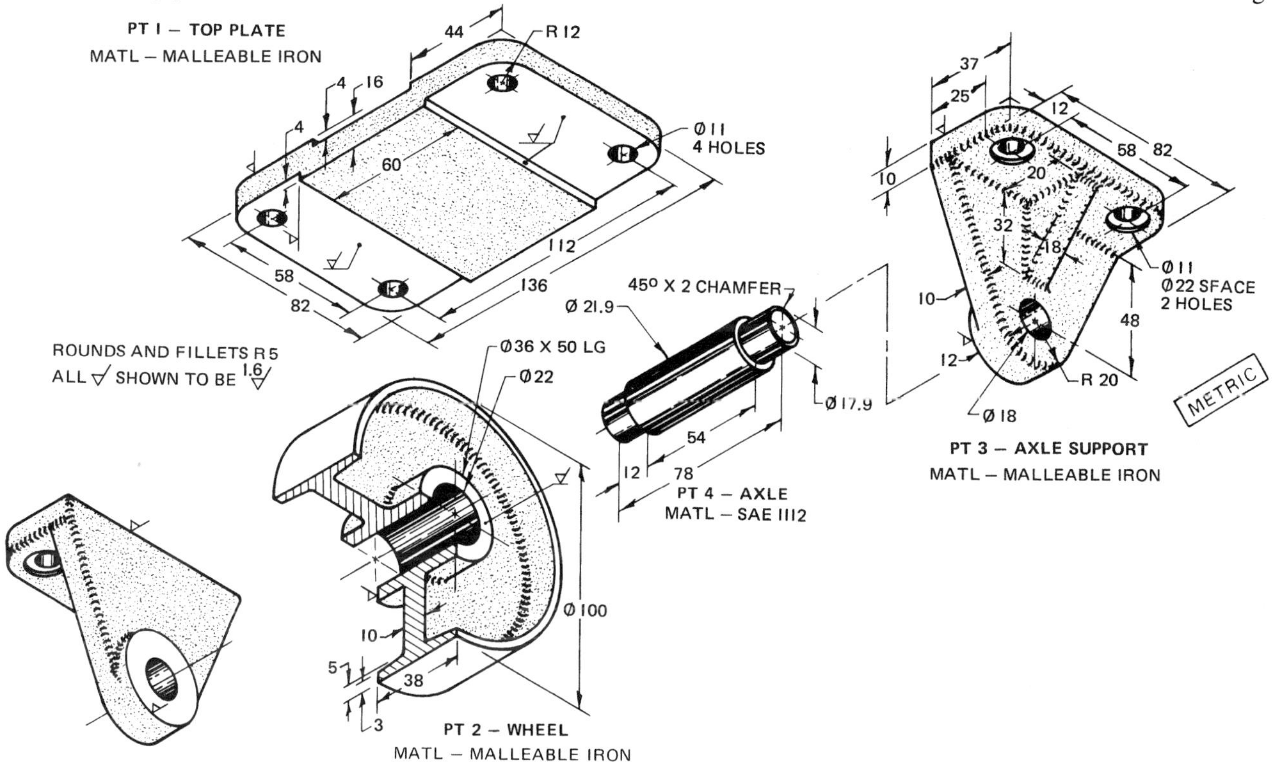

SUBASSEMBLY DRAWINGS— WHEEL ASSEMBLY	REFERENCE UNIT 14–9	NAME COURSE DATE	DRAWING M–12

1

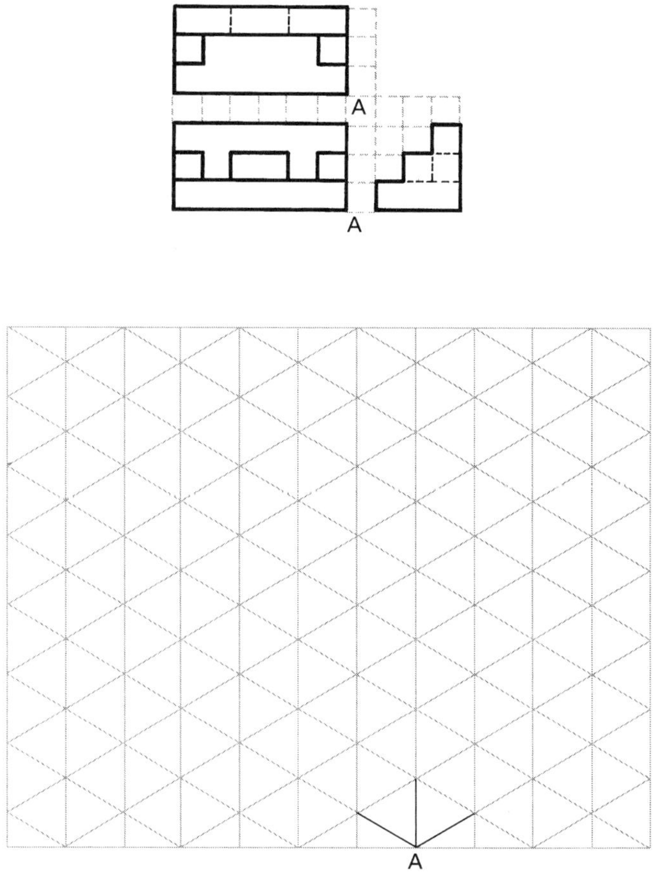

2

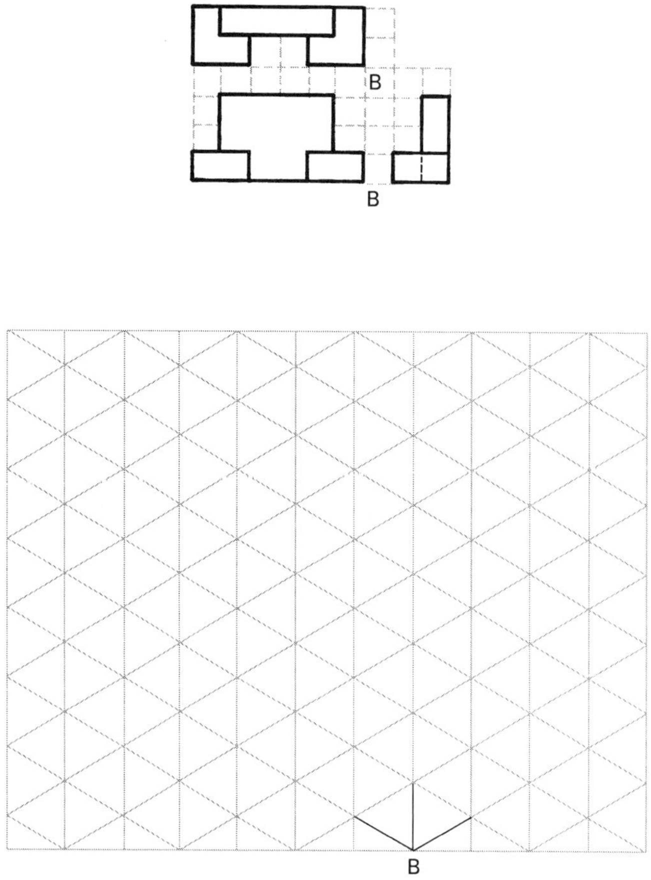

3

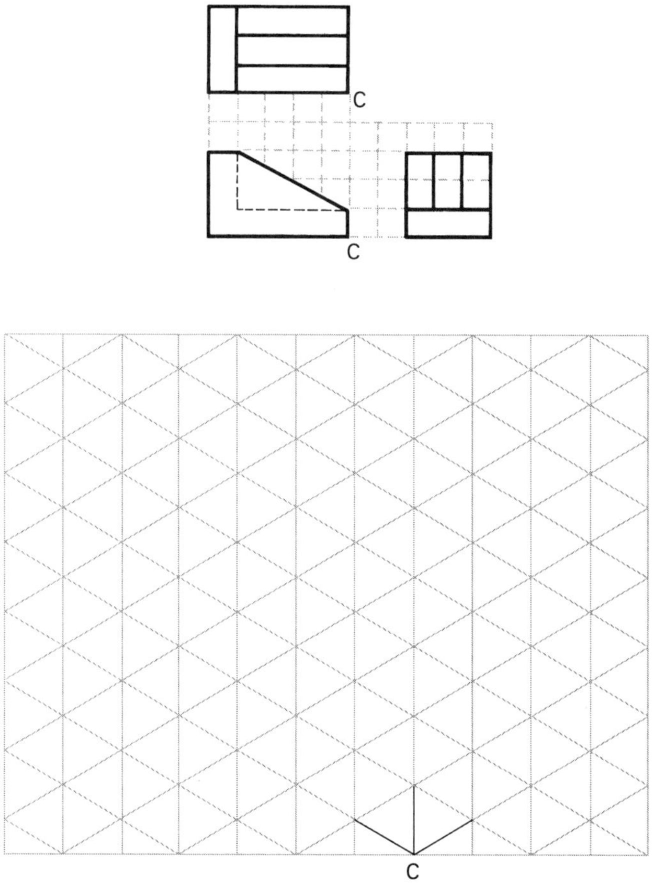

4

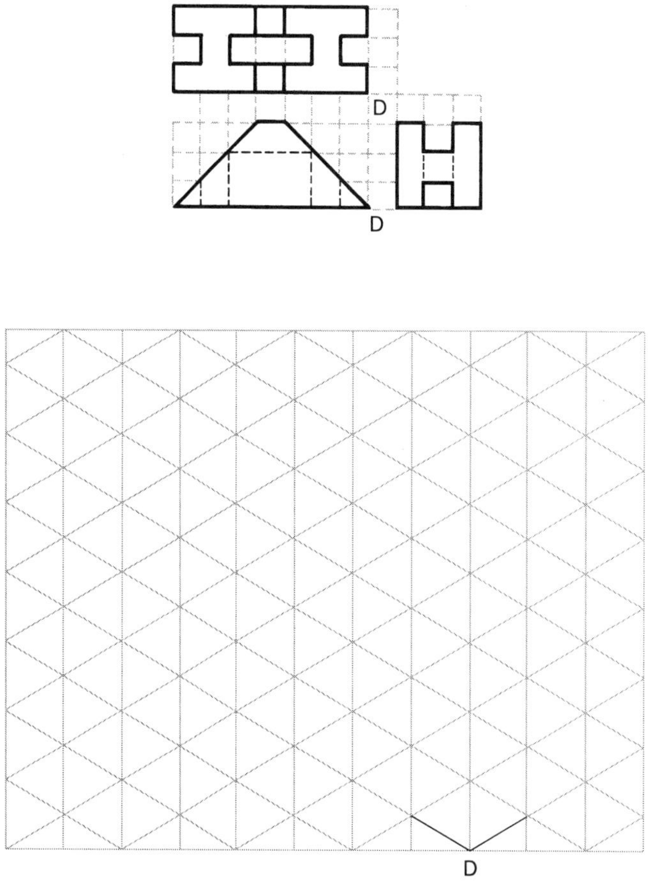

On the isometric grid provided, draw the parts shown. Do not show hidden lines. Each square shown on the drawing represents one isometric square on the grid.

ISOMETRIC DRAWINGS— FLAT SURFACES	REFERENCE UNIT 15–1	NAME COURSE / DATE	DRAWING N–1

ISOMETRIC DRAWINGS—

STEP BLOCK AND PLANTER BOX

REFERENCE UNIT 15–1

NAME

COURSE

DATE

DRAWING N–2

Make an isometric drawing, complete with dimensions, of the step block. Point A is located on the drawing area. The scale is 1:2.

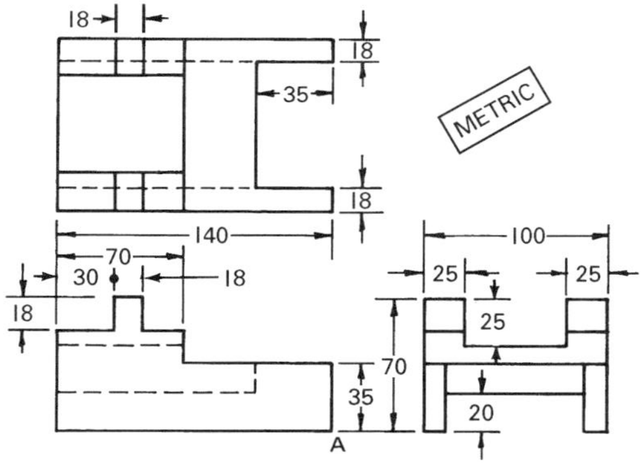

Make an isometric drawing, complete with dimensions, of the planter box. Point B is located on the drawing area. The scale is 1:12.

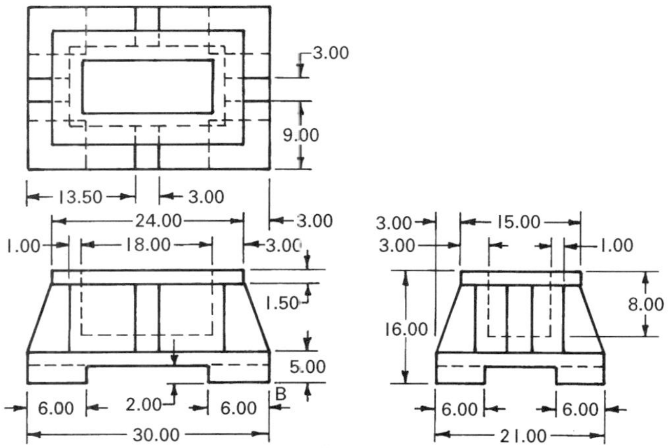

1

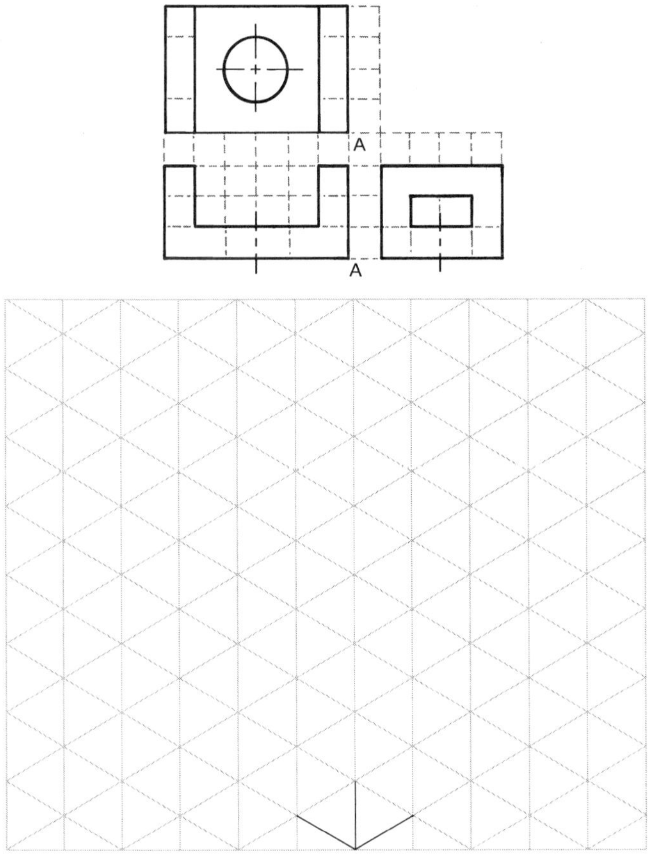

2

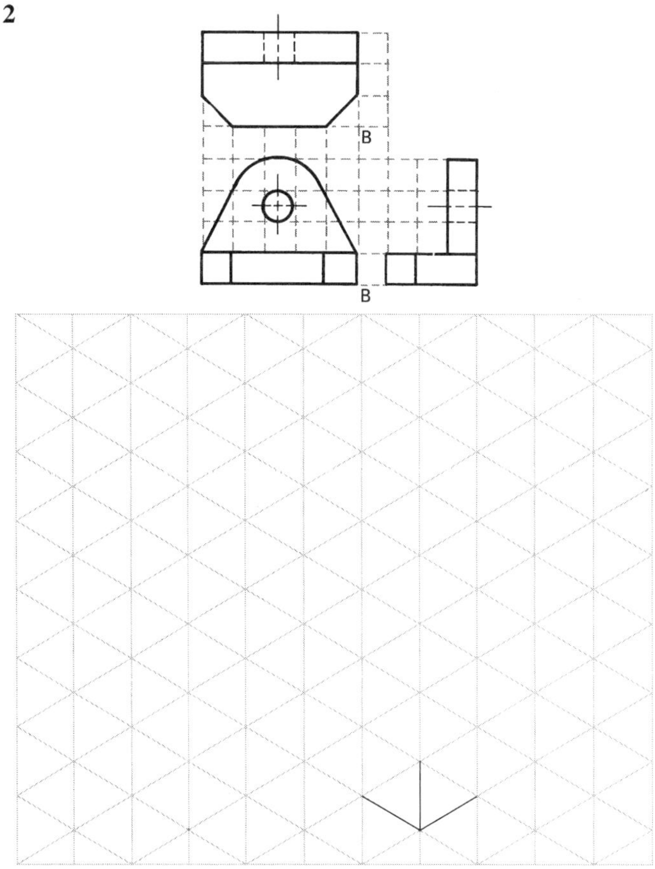

3

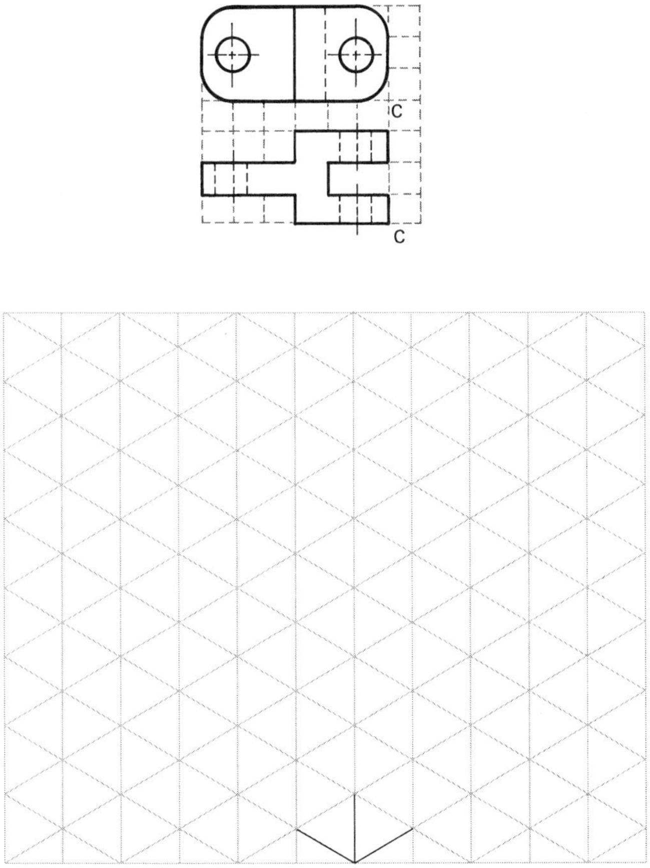

4

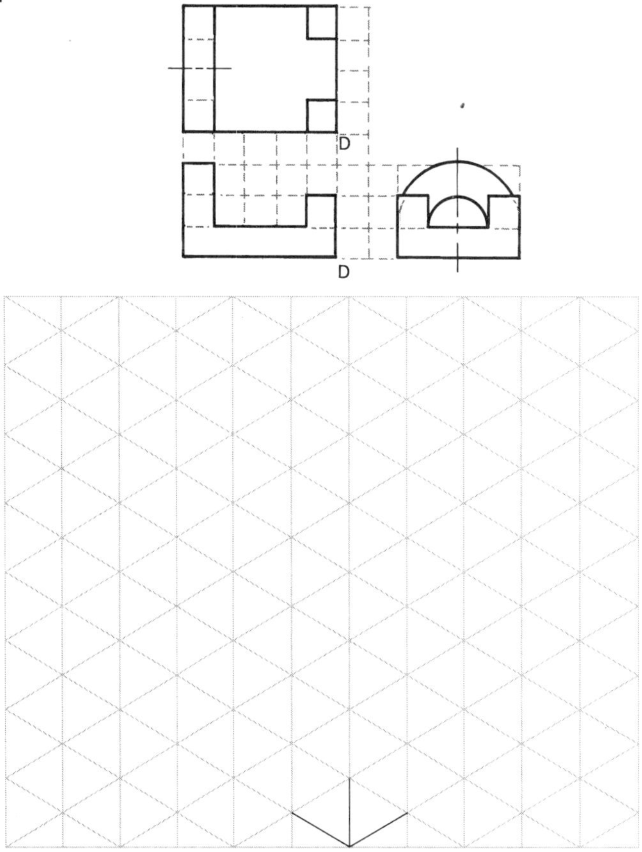

On the isometric grids provided, draw the parts shown. Do not show hidden lines. Each square shown on the drawing represents one isometric square on the grid.

ISOMETRIC DRAWINGS— CURVED SURFACES	REFERENCE UNIT 15–2	NAME COURSE DATE	DRAWING N–3

Make an isometric drawing, complete with dimensions, of the base. The location of point A is shown on the drawing area. The scale is 1:2.

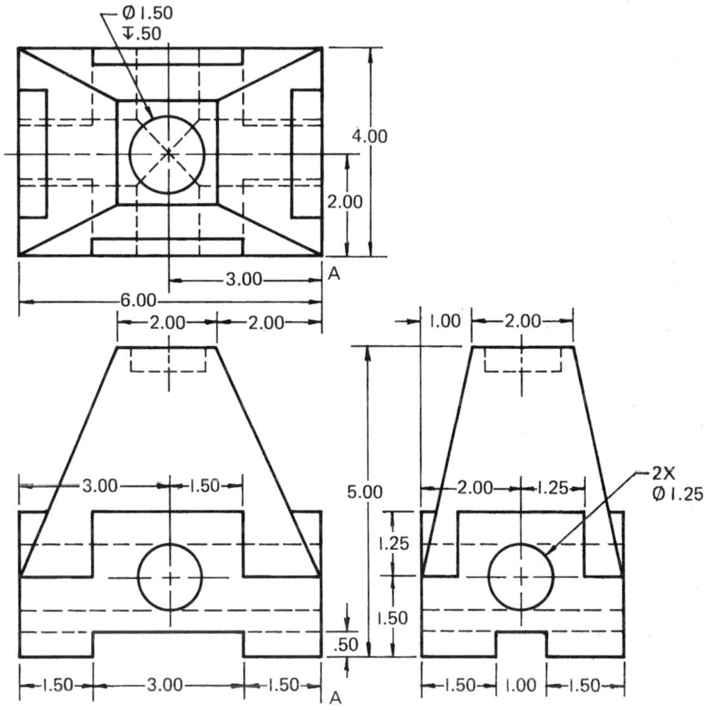

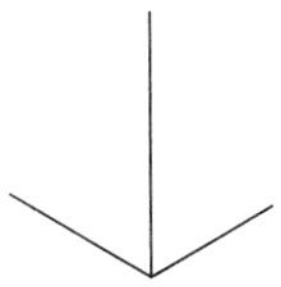

CURVED SURFACES— BASE	REFERENCE UNIT **15–2**	NAME COURSE DATE	DRAWING **N–4**

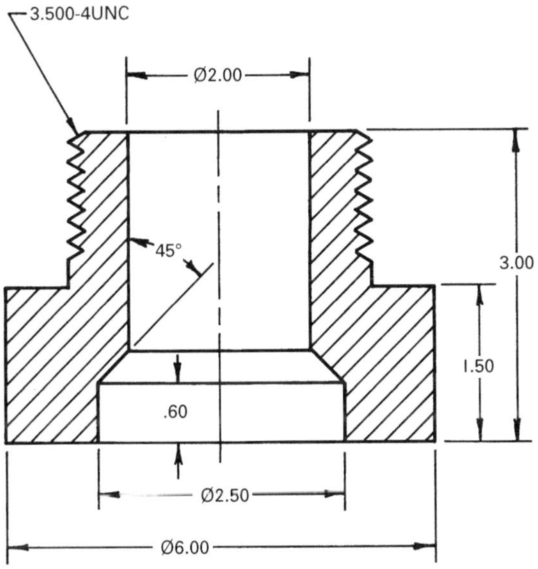

Make an isometric full-section drawing, complete with dimensions, of the adapter shown. Do not show hidden lines. The Ø2.50 is located on the drawing surface. The scale is 1:1.

COMMON FEATURES IN ISOMETRIC— ADAPTER	REFERENCE UNIT **15–3**	NAME COURSE DATE	DRAWING **N–5**

Make an isometric assembly drawing of the two-post die set, Model 304. Allow 2.00 in. between the top and base. The scale is 1:4. Do not dimension. Complete the item list. Using part numbers, identify the parts on the assembly. The bottom of the base is located on the drawing area.

SIZE	NO. 302 DIE SET	NO. 304 DIE SET
A	11.12	14.00
B	6.00	7.50
C	6.50	8.00
D	1.62	1.75
E	1.50	1.62
F	1.25	1.25
G	1.00	1.12
H	1.25	1.38
J	2.00	2.25
K	2.00	2.25
L	2.00	2.00

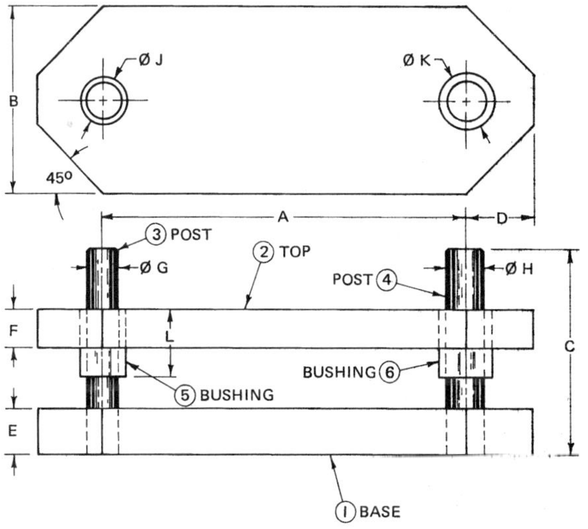

QTY	ITEM	MATL	DESCRIPTION	PT NO.

COMMON FEATURES IN ISOMETRIC— TWO-POST DIE SET	REFERENCE UNIT 15–3	NAME COURSE DATE	DRAWING N–6

1

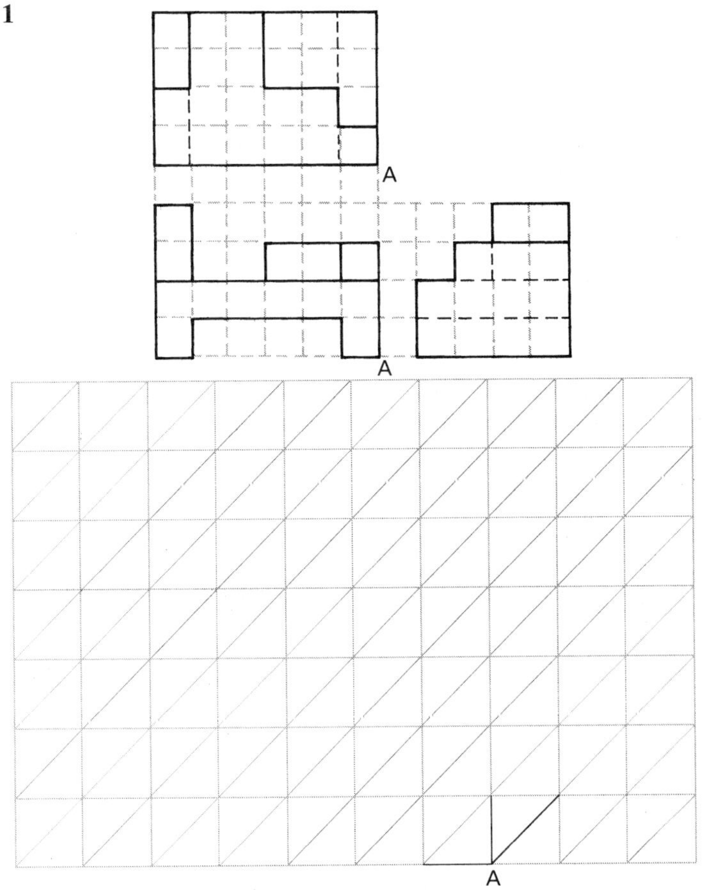

2

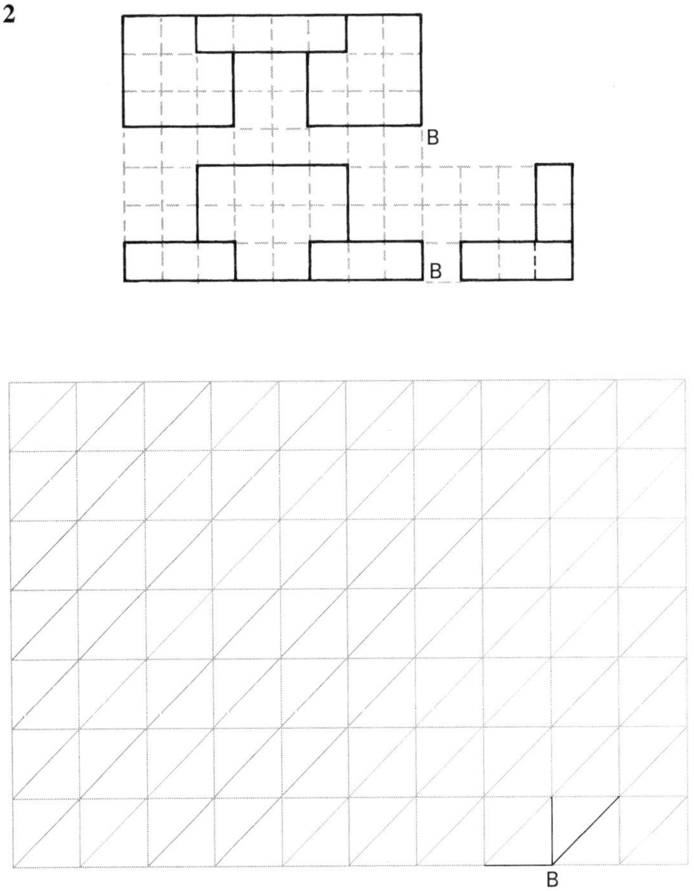

3

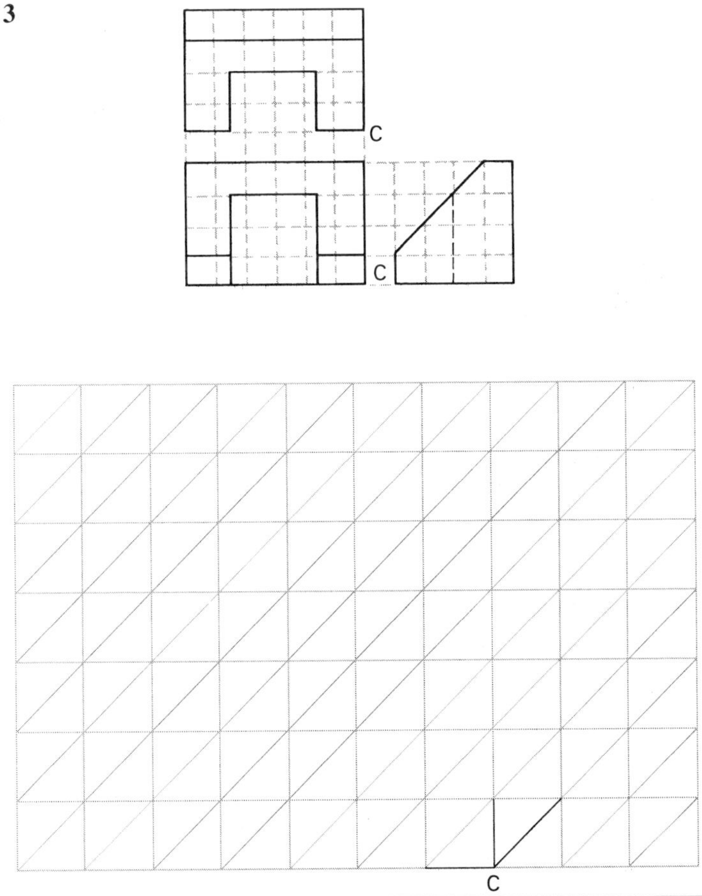

4

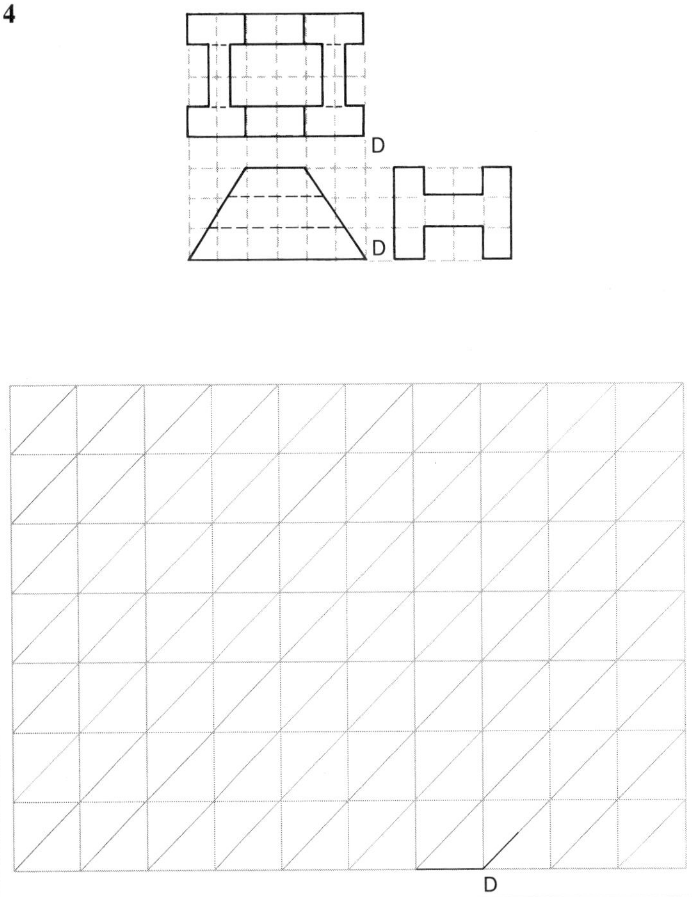

On the oblique grids provided, draw the parts shown in cabinet oblique projection. Do not show hidden lines. Remember to halve the distance for receding lines. The starting points are located on the grid.

OBLIQUE PROJECTION— FLAT SURFACES	REFERENCE UNIT **15–4**	NAME COURSE DATE	DRAWING **N–7**

Make a cabinet oblique drawing of the spacer block shown to the scale 1:1. Add dimensions. The lower left-hand corner of the spacer block is located on the drawing area.

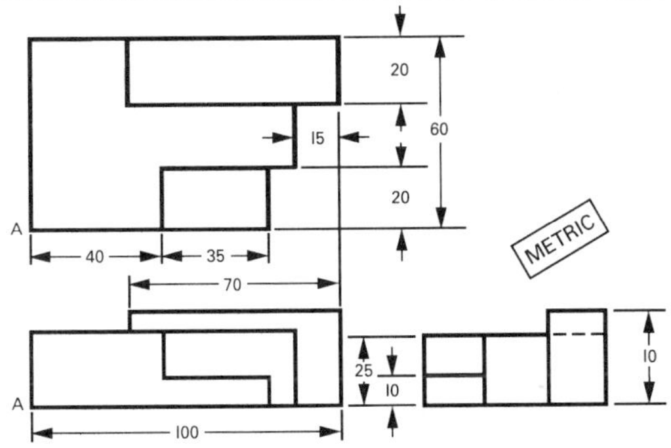

Make a cavalier oblique drawing of the dovetail guide shown to the scale of 1:2. Add dimensions. The front horizontal line of the part is located on the drawing area.

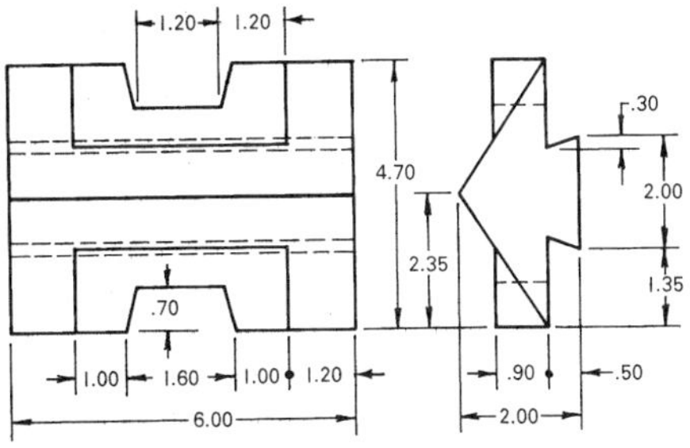

OBLIQUE PROJECTION— SPACER BLOCK AND DOVETAIL GUIDE	REFERENCE UNIT 15–4	NAME COURSE DATE	DRAWING N–8

Make a half-section cabinet oblique drawing of the step pulley to the scale 1:1. The location of the Ø1.00 hole is shown on the drawing area. Add dimensions. Use 30° as the receding axis.

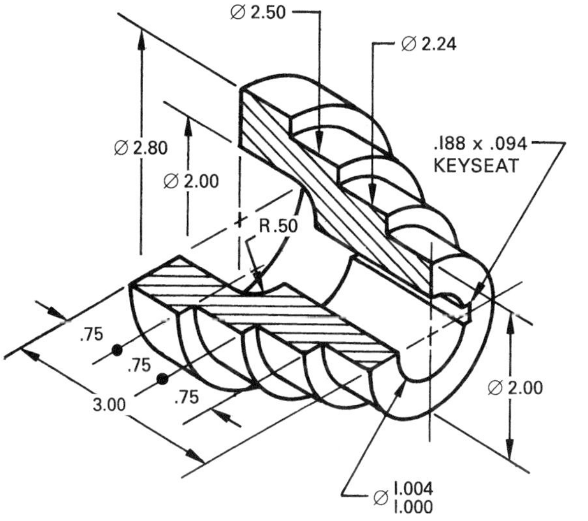

Make a cabinet oblique drawing of the connector to the scale 1:1. Add dimensions. The Ø20 hole is located on the drawing area. Use 30° as the receding axis.

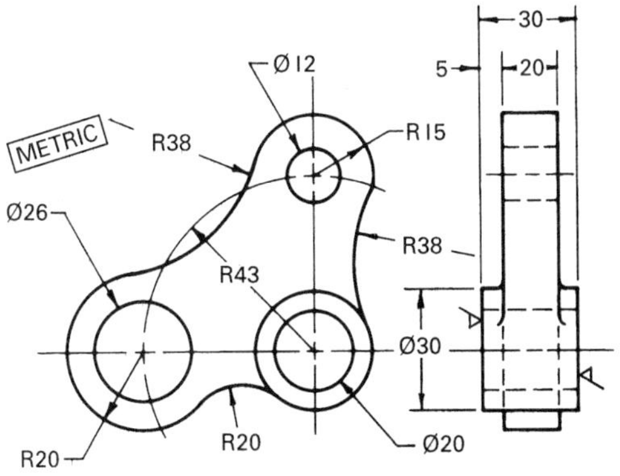

COMMON FEATURES IN OBLIQUE— STEP PULLEY AND CONNECTOR	REFERENCE UNIT **15–5**	NAME COURSE DATE	DRAWING **N–9**

Place a sheet of tracing paper located at the back of this text over the grid below and draw a one-point perspective of the vise base shown to the scale 1:1. Add dimensions.

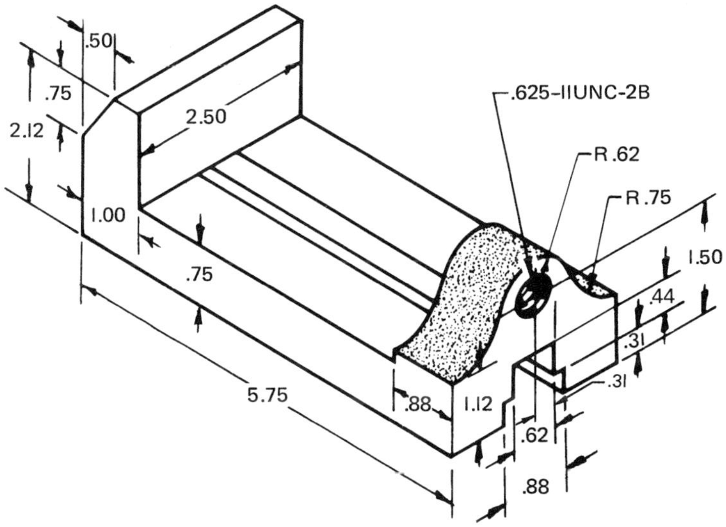

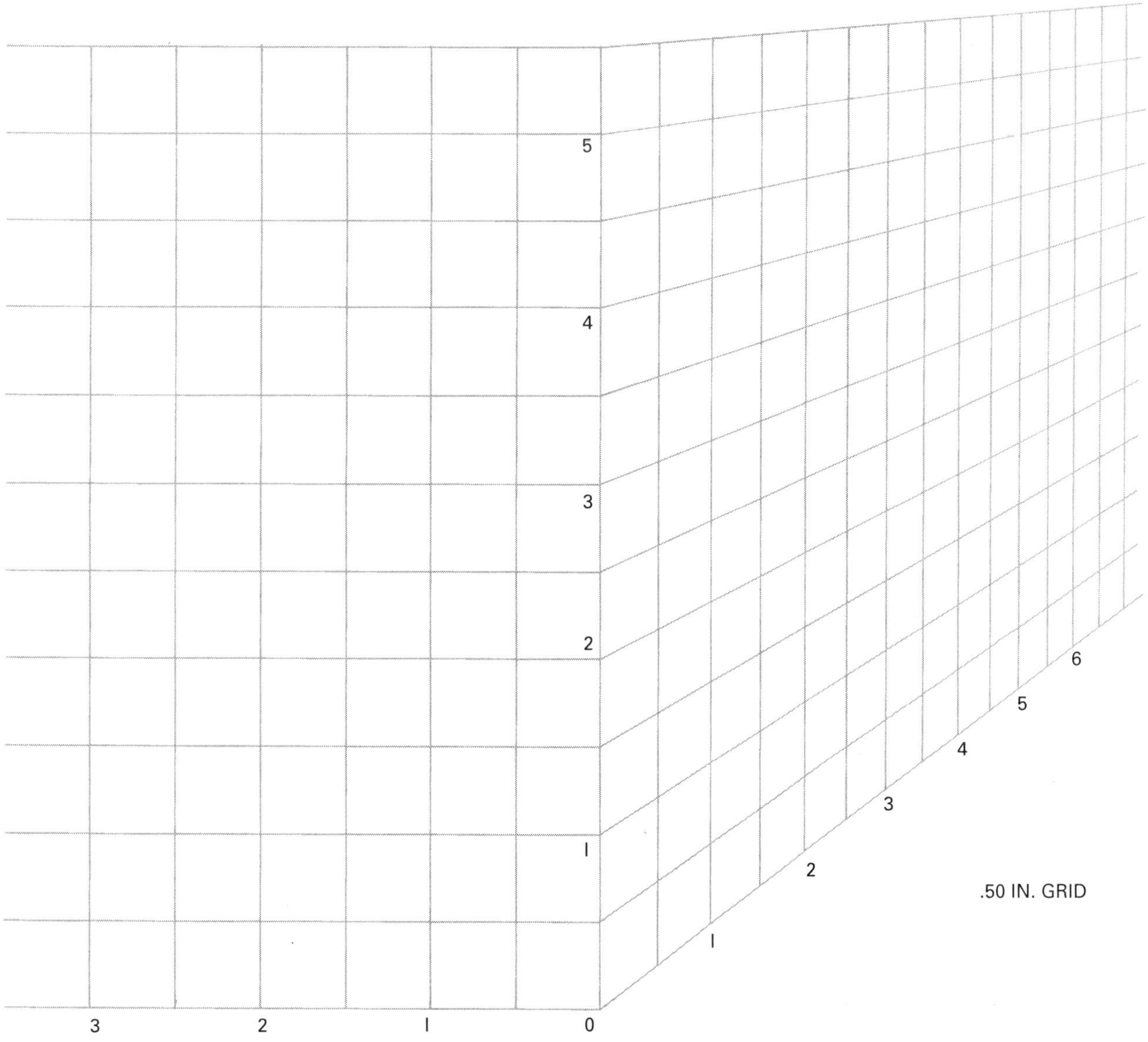

ONE-POINT PERSPECTIVE— VISE BASE	REFERENCE UNIT 15–6	NAME COURSE DATE	DRAWING N–10

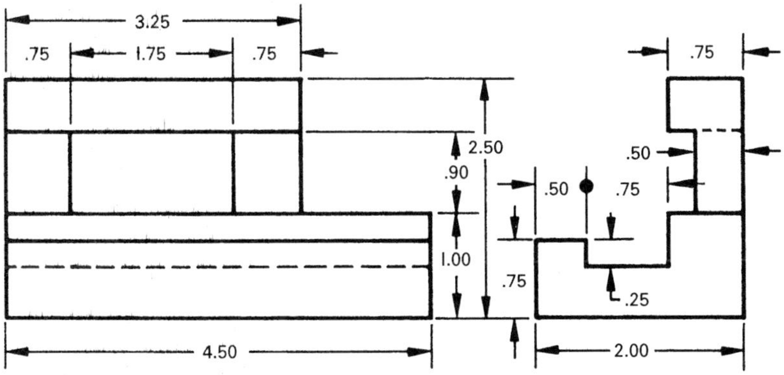

Place a sheet of tracing paper located at the back of book over the grid below and draw a two-point perspective of the support guide shown to the scale 1:1. Add dimensions.

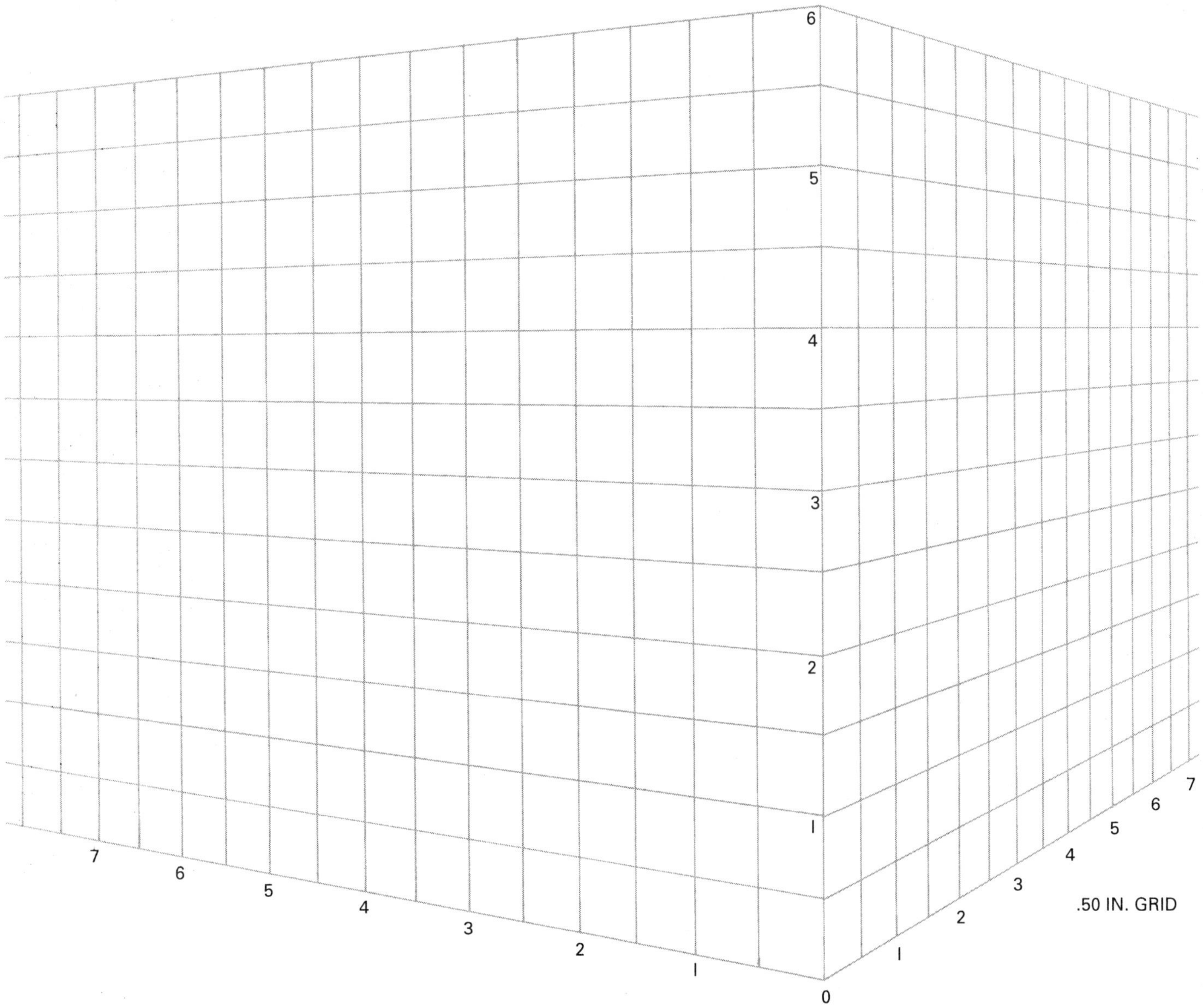

TWO-POINT PERSPECTIVE— SUPPORT GUIDE	REFERENCE UNIT **15–7**	NAME COURSE	DATE	DRAWING **N–11**

Complete the Boolean operations for each of the objects given below. Darken in the object lines for the resulting object created by the Boolean operation for each problem.

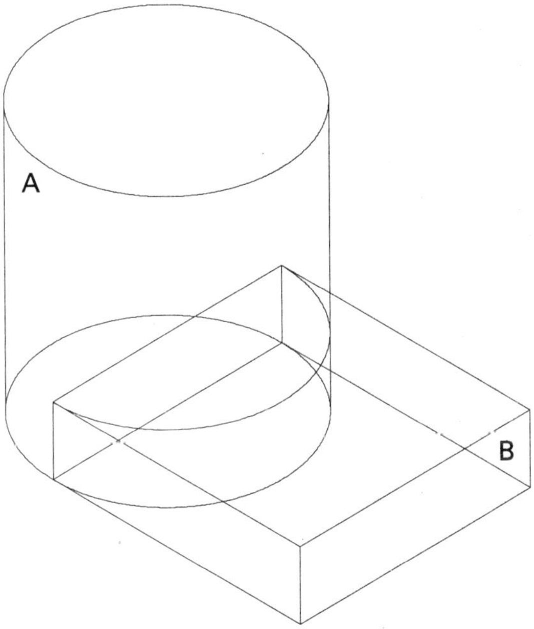

(A+B)

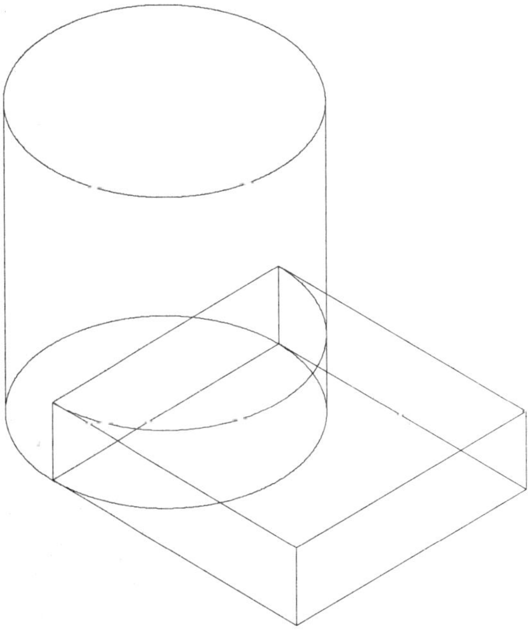

(A-B)

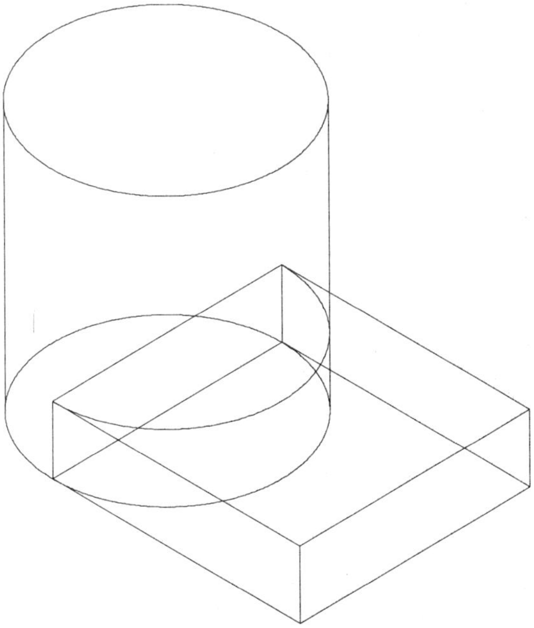

(B-A)

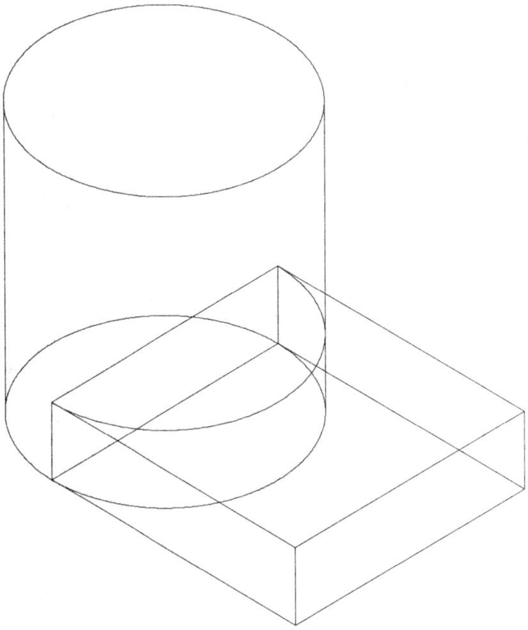

(A intersect B)

CSG MODELING BOOLEAN EXERCISE 1	REFERENCE UNIT 15-8	NAME COURSE	DATE	DRAWING N-12

Darken the outline of the object for each step of the Boolean operations indicated below. Include hidden lines if needed for clarity.

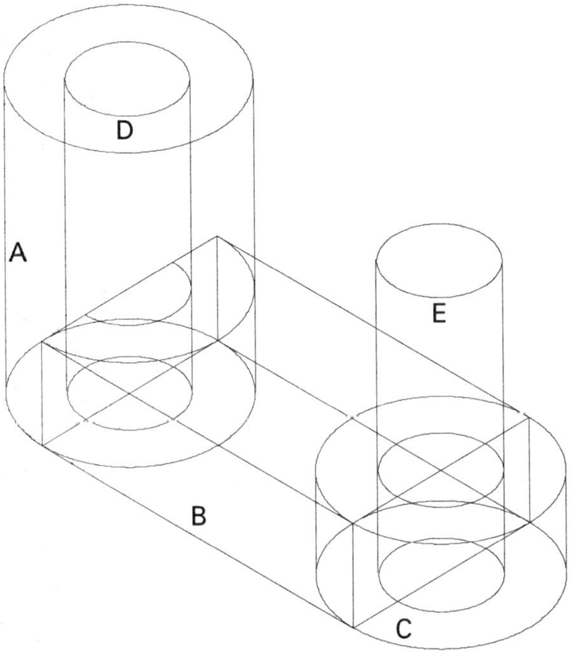

(A+B)

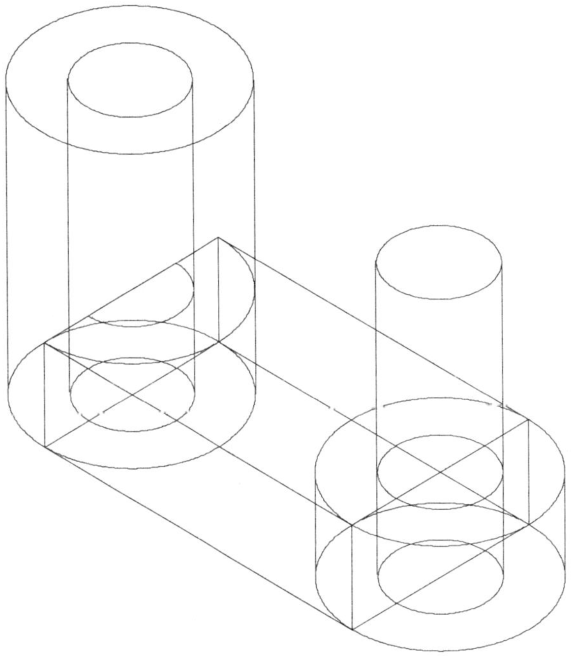

(results + C)

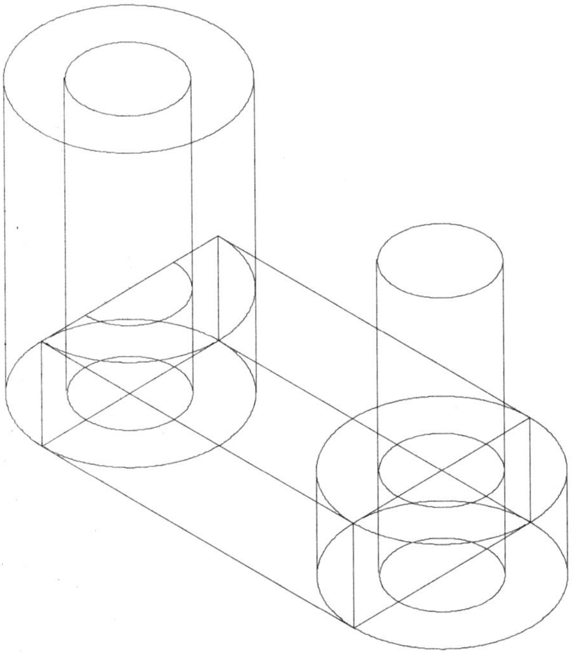

(results - D)

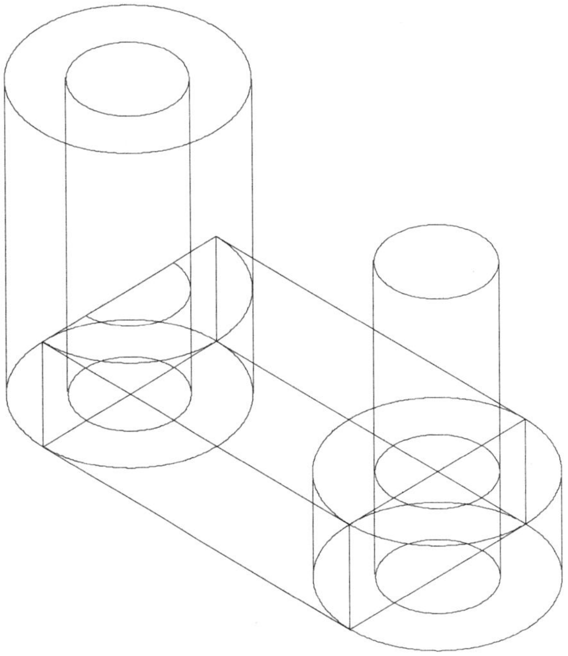

(results - E)

CSG MODELING BOOLEAN EXERCISE 2	REFERENCE UNIT **15-8**	NAME COURSE DATE	DRAWING **N-13**

Draw each of the profiles needed to produce each of the objects below by the intersection of the extruded profiles.

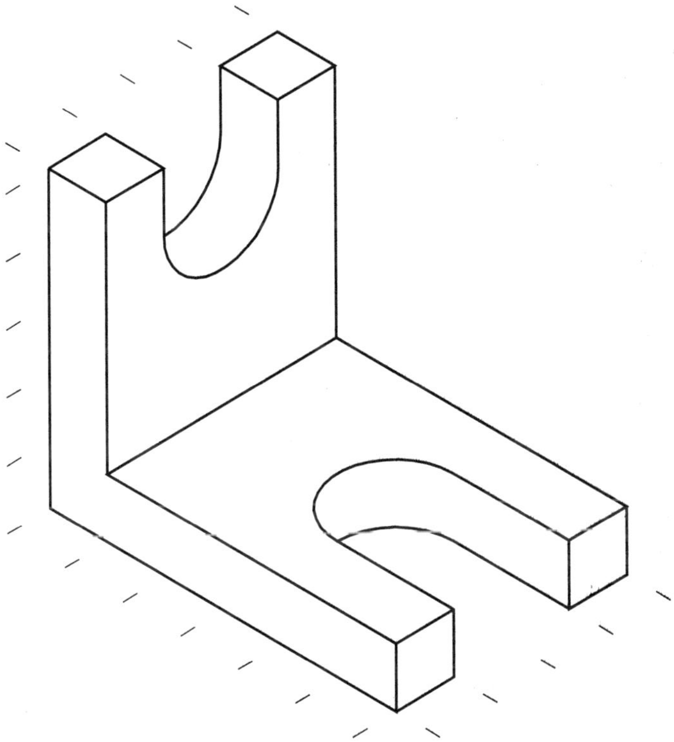

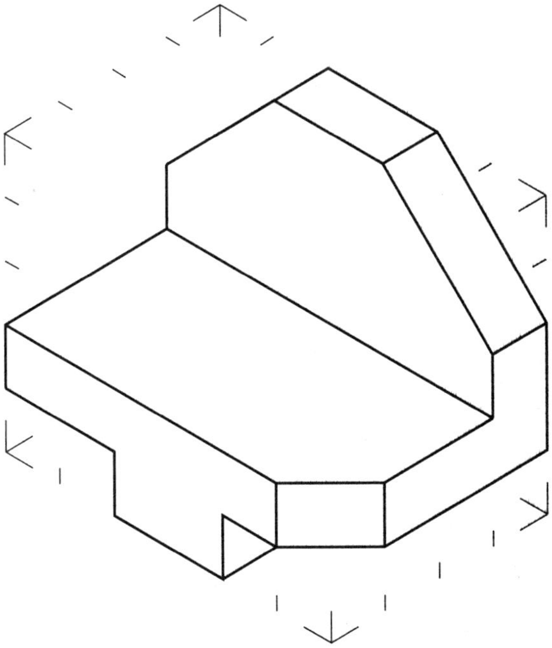

CSG MODELING	REFERENCE UNIT	NAME		DRAWING
INTERSECTION PROFILE EXERCISE	15-8	COURSE	DATE	N-14

1. On the grids to the right of Figs. 1 and 2, sketch two different acceptable form variations for each part. Add dimensions.
2. On the grid to the right of Fig. 3, make a tolerance block diagram showing the deviations and limits of size.

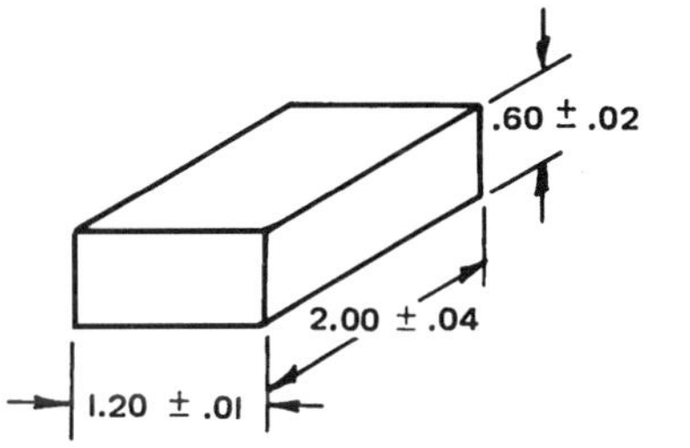

FIGURE 1

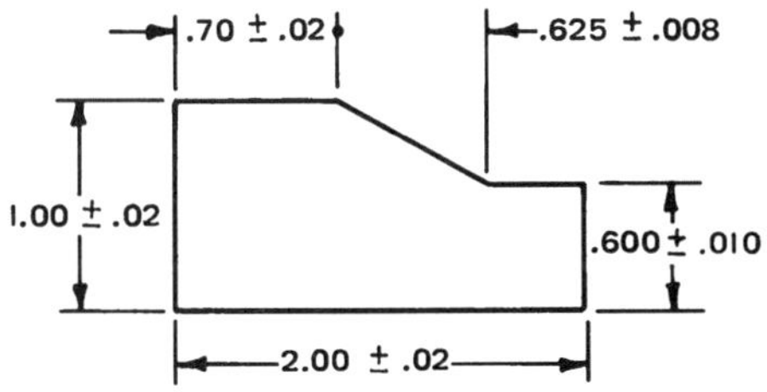

FIGURE 2

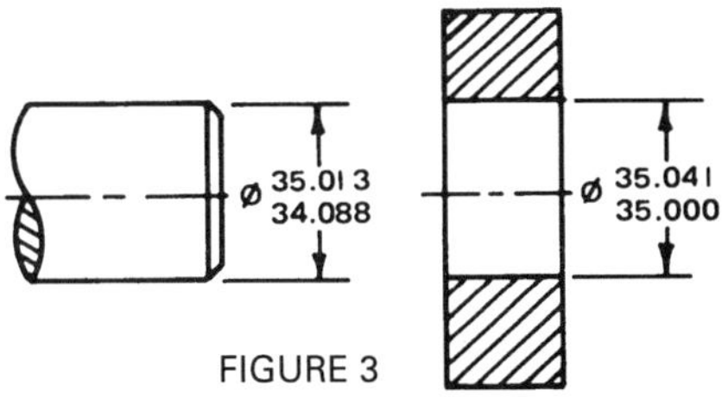

FIGURE 3

MODERN ENGINEERING TOLERANCING— DEVIATIONS AND BLOCK DIAGRAM	REFERENCE UNIT 16–1	NAME COURSE DATE	DRAWING P–1

Given the following information, add the feature control frame to the parts shown below.

Part 1. Surface A to have a straightness tolerance of .004 in.
Part 2. Surface M to have a straightness tolerance of .006 in.
Surface N to have a straightness tolerance of .008 in.
Part 3. Surface R to be straight within .006 in. for direction A and straight within .002 in. for direction B.
Part 4. With straightness specified as shown, what is the maximum permissible deviation from straightness of the line elements if the radius is (a) .496 in., (b) .501 in., (c) .504 in.?
Part 5. Eliminate the top view and place the feature control frames on the front and side views.

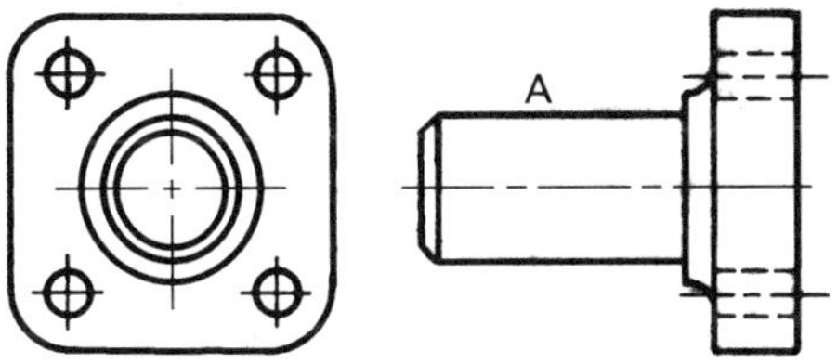

PART I

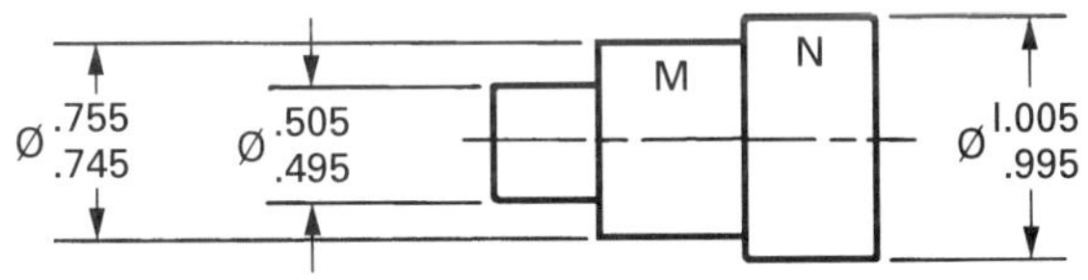

PART 2

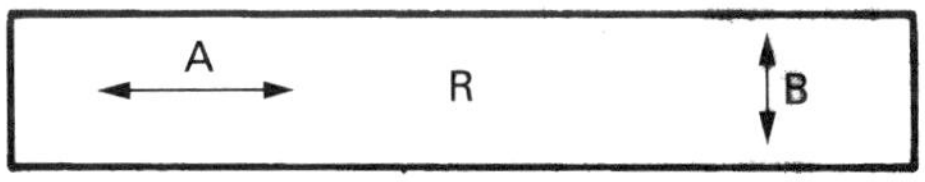

PART 3

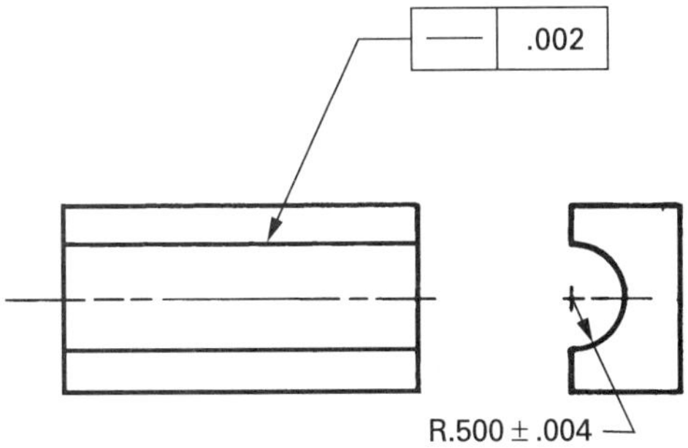

PART 4

A =

B =

C =

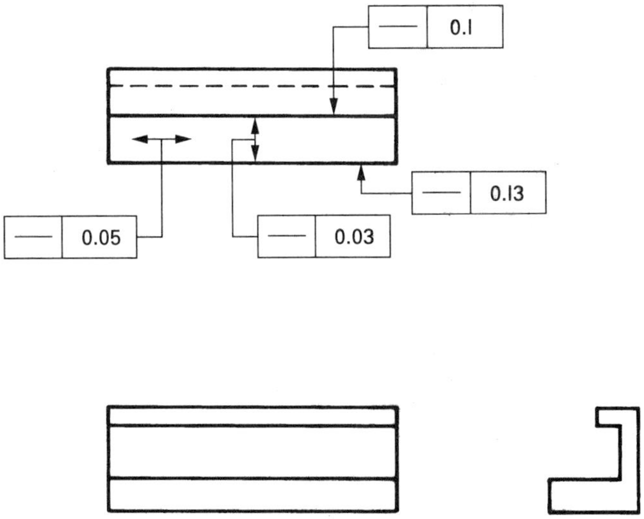

PART 5

GEOMETRIC TOLERANCING— STRAIGHTNESS TOLERANCE	REFERENCE UNIT 16–2	NAME COURSE DATE	DRAWING P–2

1. Add a flatness tolerance of 0.03 to the base of the flange shown in Part 1.
2. Add the following tolerances to surface B of the base shown in Part 2: (a) Maximum flatness tolerance of .010 in. for entire surface, (b) limited area flatness tolerance of .005 for any 2.00 × 2.00 in. area.
3. Part 3 is required to fit into Part 4 so there will be no interference and the maximum clearance will never exceed .005 in. Add the maximum limits of size to Part 4. Flatness tolerances of .001 in. are to be added to the two surfaces of each part.
4. Show the tolerance zones and limits of size dimensions for Parts 5 and 6 on the graph provided.

PART I

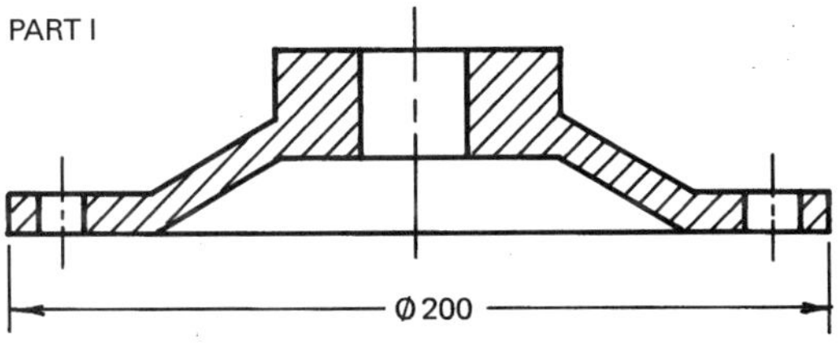

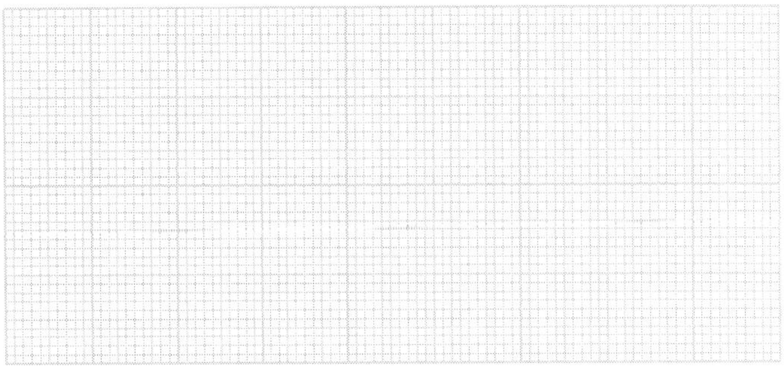

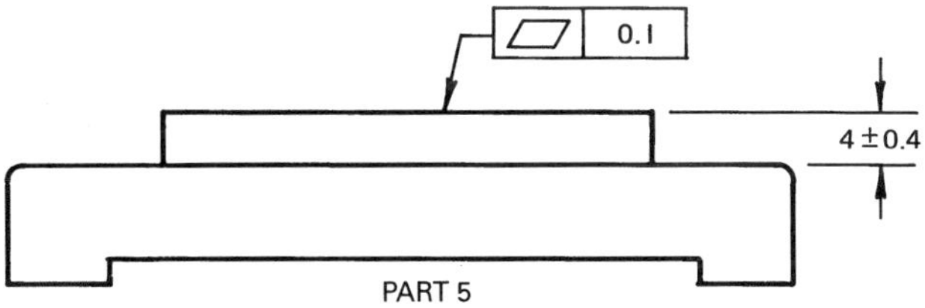

PART 5

PART 2

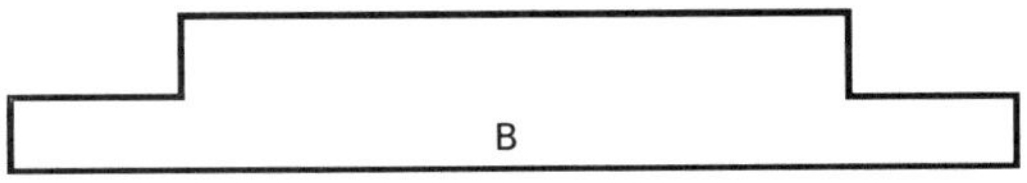

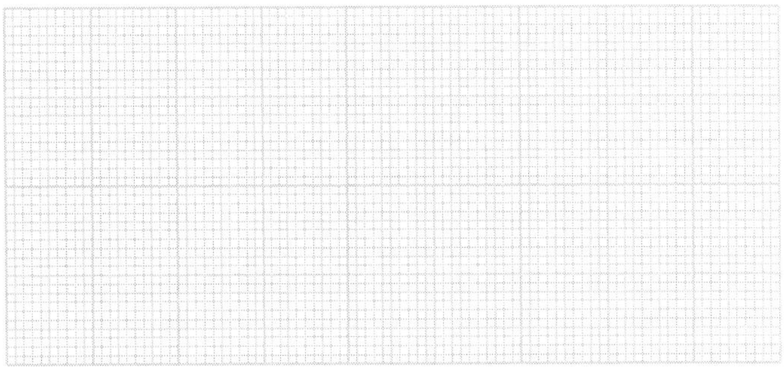

PART 3

PART 4

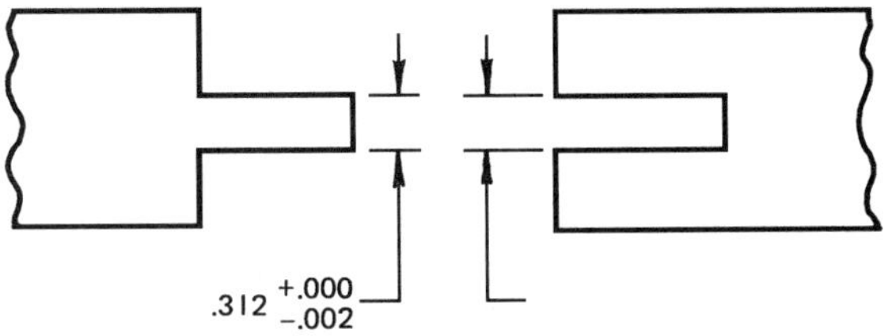

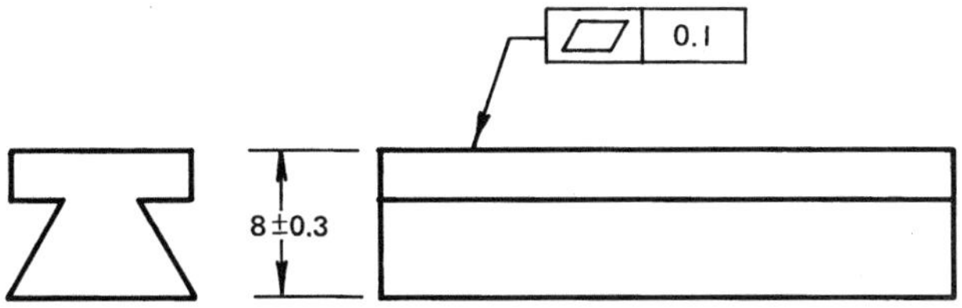

PART 6

GEOMETRIC TOLERANCING— FLATNESS TOLERANCE	REFERENCE UNIT 16–3	NAME COURSE DATE	DRAWING P–3

1. Complete the charts in Figs. 1 and 2 showing the largest permissible straightness error for the feature sizes shown.
2. If the maximum straightness was not added to the part in Fig. 3, indicate which parts would be acceptable or unacceptable.
3. Refer to Fig. 4 and complete the chart showing the maximum deviation permitted from straightness for the shaft sizes indicated.

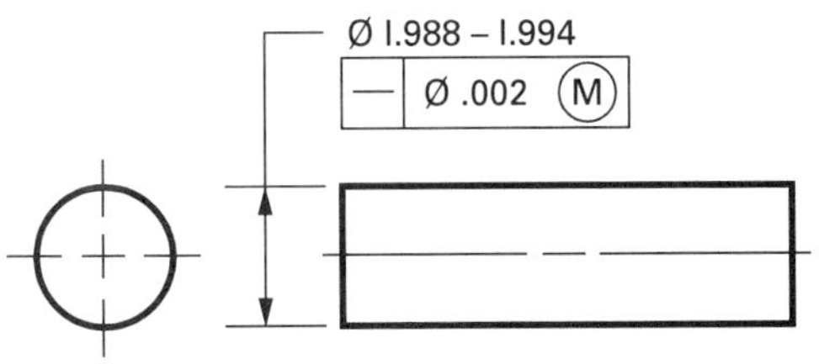

FEATURE SIZE (DIA)	PERMISSIBLE STRAIGHTNESS TOLERANCE (DIA)
1.994	
1.993	
1.992	
1.991	
1.990	
1.989	
1.988	

FIGURE 1

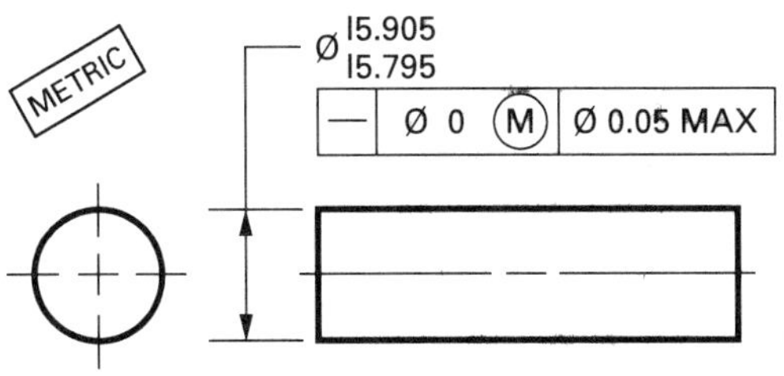

FEATURE SIZE (DIA)	PERMISSIBLE STRAIGHTNESS TOLERANCE (DIA)
15.905	
15.895	
15.885	
15.875	
15.865	
15.845	
15.795	

FIGURE 2

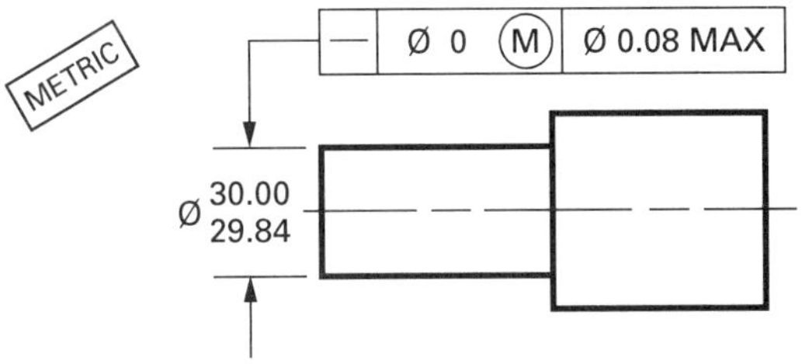

PART	FEATURE SIZE Ø	STRAIGHTNESS DEVIATION	ACCEPTABLE YES OR NO
A	29.94	0.08	
B	30.02	0.008	
C	29.80	0.06	
D	29.86	0.10	
E	29.97	0.04	

FIGURE 3

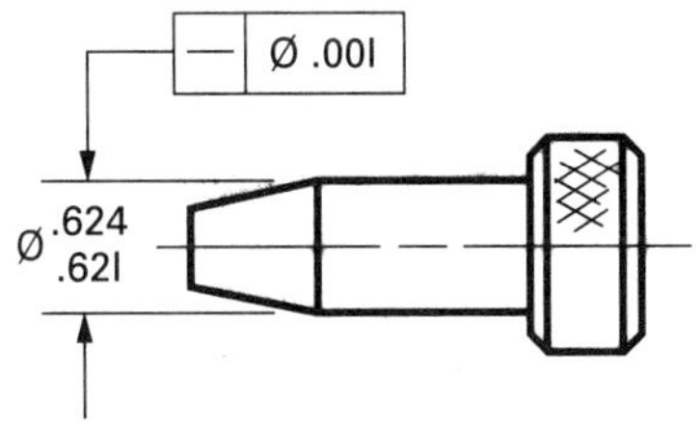

SHAFT SIZE	MAXIMUM DEVIATION PERMITTED
MMC	
LMC	
Ø .623	

FIGURE 4

STRAIGHTNESS OF A FEATURE OF SIZE	REFERENCE UNIT 16–4	NAME COURSE DATE	DRAWING P–4

Add the following geometric tolerances to the shaft support shown. The scale is 1:1 (decimal inch). The hole is Ø.400–.406 and the shaft is Ø.592–.600.

(a) All datums and tolerances should be on an MMC basis unless otherwise specified.
(b) Surfaces marked A, B, and C should be datums A, B, and C, respectively.
(c) The shaft should have a straightness tolerance of .003 at MMC and should be datum E.
(d) The bottom should be flat within .005 for the entire surface, but the flatness error must not exceed .002 for any 1.00 × 1.00 area.
(e) Both sides of the notch should be flat within .001.
(f) The hole should have a straightness tolerance of .002 at MMC and be datum D.

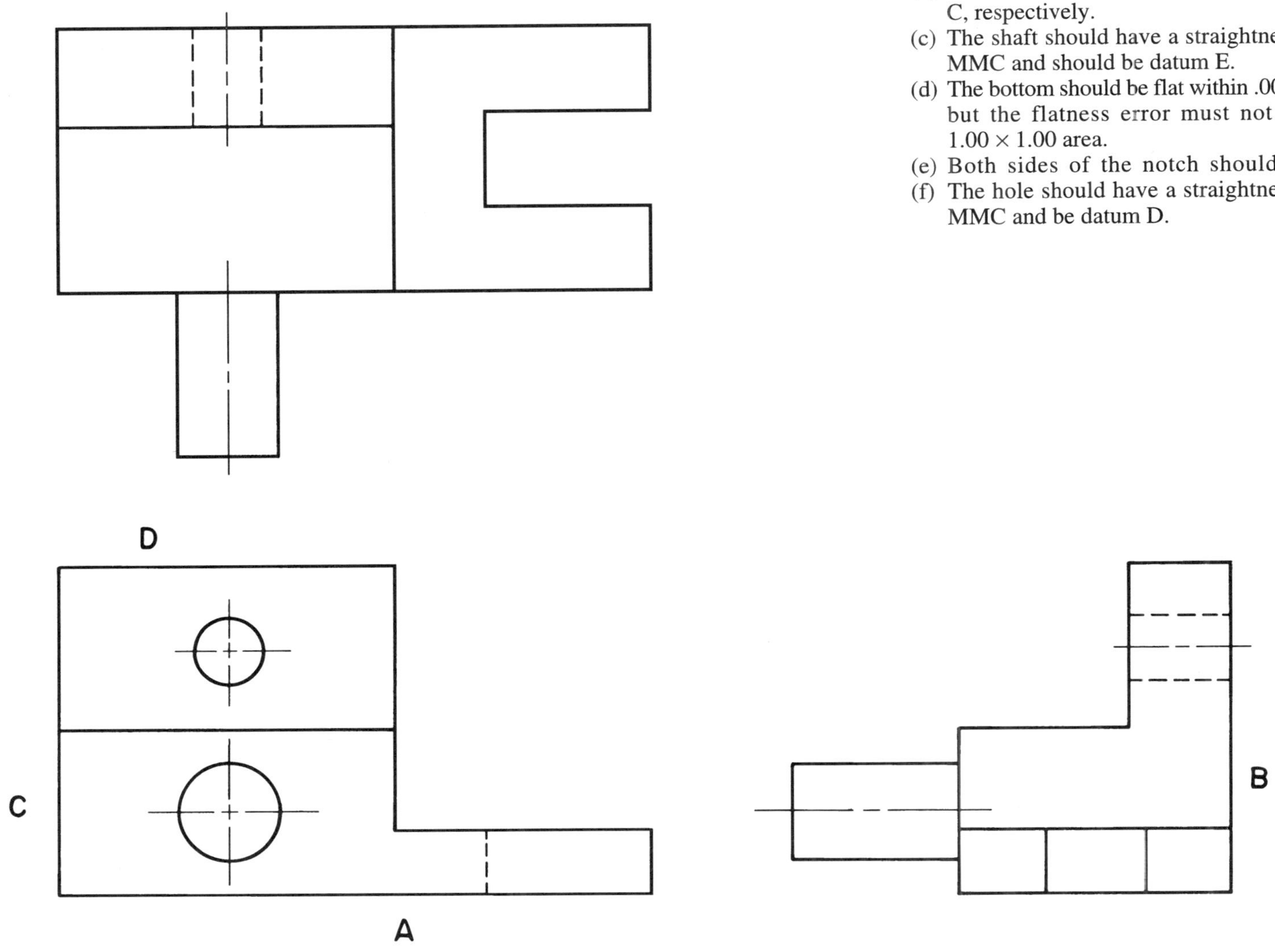

DATUMS AND THE THREE-PLANE CONCEPT	REFERENCE UNIT 16–5	NAME _ COURSE _ DATE _	DRAWING P–5

Add the following geometric tolerances to the stand shown. The scale is 1:1 (decimal inch). The hole is Ø.406 LC3 fit.

(a) Surfaces marked A, B, D should be datums A, B, and D, respectively.
(b) The back should be perpendicular to the bottom within .01 and flat within .006.
(c) The top should be parallel to the bottom within .005.
(d) The inclined surface should have an angularity tolerance of .008 with the bottom. Surface D should be the secondary datum for this requirement.
(e) The bottom should be flat within .002.
(f) The sides of the slot should be parallel to each other within .001. One side of the slot should be datum E.
(g) The slot should be perpendicular to the back of the stand within .002.

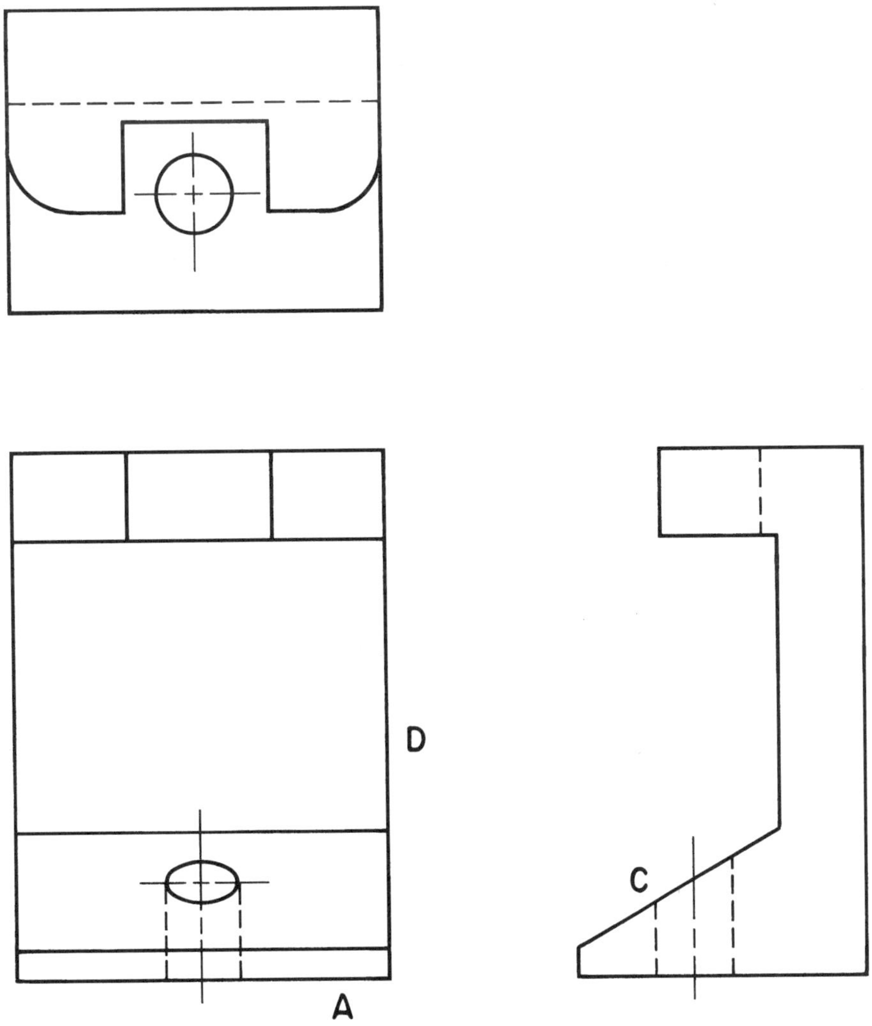

ORIENTATION TOLERANCING OF FLAT SURFACES	REFERENCE UNIT 16–6	NAME COURSE	DATE	DRAWING P–6

Determine the size of the gaging element to evaluate datum A, given the drawing callout and the measured size of the related datum features shown.

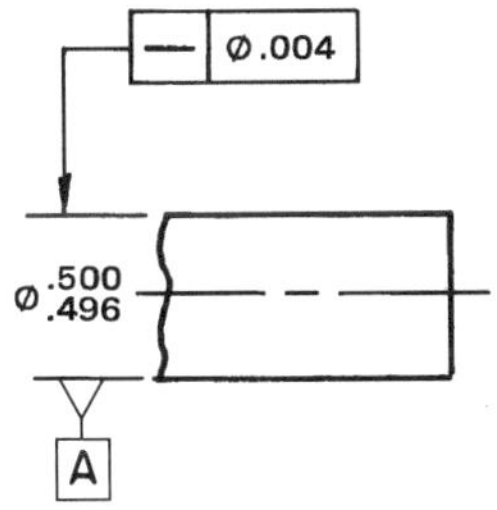

EXAMPLE 1

MEASURED SIZE OF DATUM FEATURE A = .499
GAGE SIZE = Ø.503

EXAMPLE 2

MEASURED SIZE OF DATUM FEATURE A = .496
GAGE SIZE = Ø.500

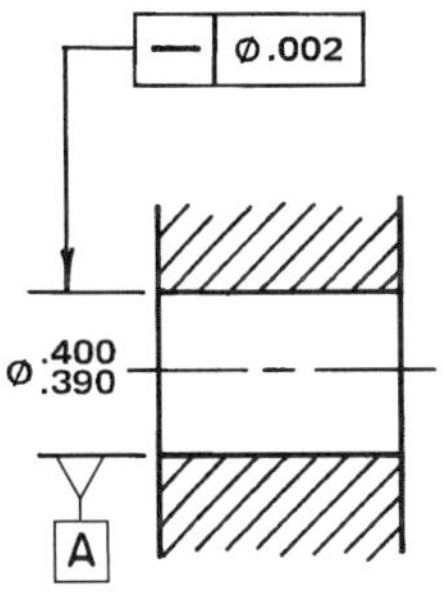

EXAMPLE 7

MEASURED SIZE OF DATUM FEATURE A = .398
GAGE SIZE = Ø.396

EXAMPLE 8

MEASURED SIZE OF DATUM FEATURE A = .390
GAGE SIZE = Ø.388

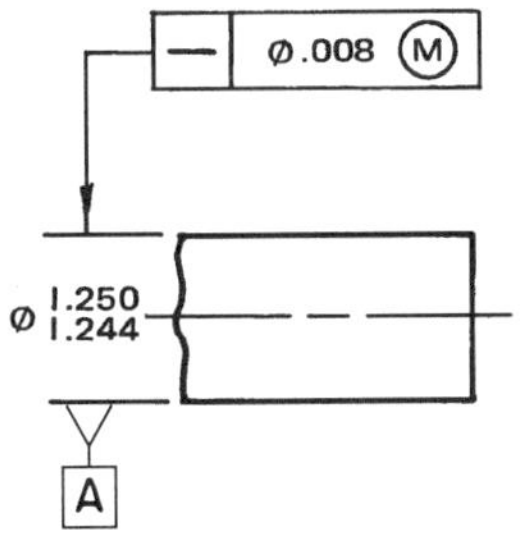

EXAMPLE 3

MEASURED SIZE OF DATUM FEATURE A = 1.248
GAGE SIZE = Ø1.258

EXAMPLE 4

MEASURED SIZE OF DATUM FEATURE A = 1.245
GAGE SIZE = Ø1.258

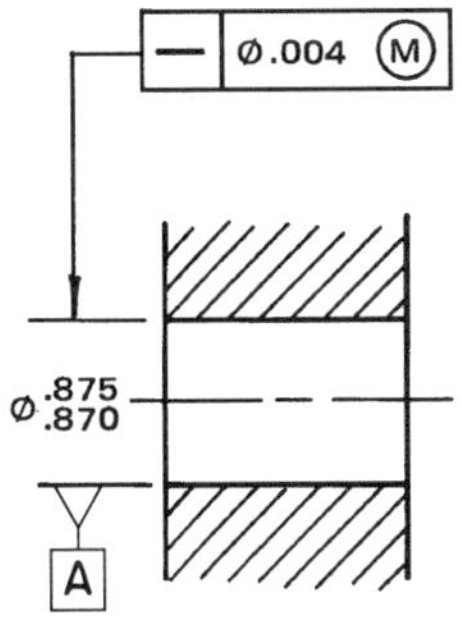

EXAMPLE 9

MEASURED SIZE OF DATUM FEATURE A = .875
GAGE SIZE = Ø.866

EXAMPLE 10

MEASURED SIZE OF DATUM FEATURE A = .872
GAGE SIZE = Ø.866

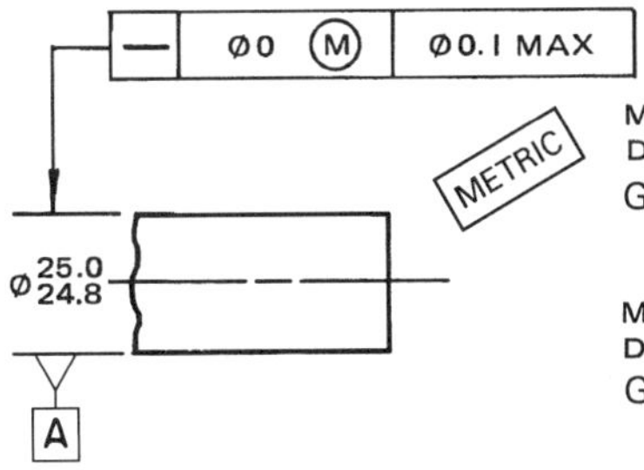

EXAMPLE 5

MEASURED SIZE OF DATUM FEATURE A = 25
GAGE SIZE = Ø25

EXAMPLE 6

MEASURED SIZE OF DATUM FEATURE A = 24.8
GAGE SIZE = Ø24.9

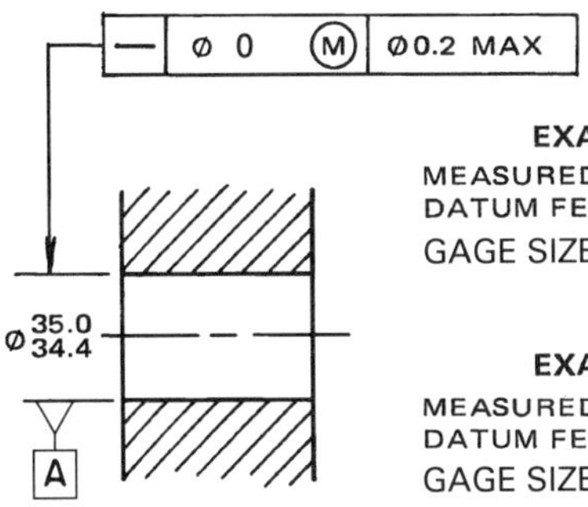

METRIC

EXAMPLE 11

MEASURED SIZE OF DATUM FEATURE A = 34.52
GAGE SIZE = Ø34.40

EXAMPLE 12

MEASURED SIZE OF DATUM FEATURE A = 34.94
GAGE SIZE = Ø34.74

DATUM FEATURES SUBJECT TO SIZE VARIATION	REFERENCE UNIT 16–7	NAME COURSE DATE	DRAWING P–7

Add the following geometric tolerances to the spacer shown.
The scale is 1:1 (decimal inch).

(a) All datums and tolerances to be on an MMC basis unless otherwise specified.
(b) Surfaces marked A, B, and C are datums A, B, and C, respectively.
(c) The .625-.628 diameter slot is perpendicular within .01 to datums B and C, in that order.
(d) Surface D is parallel within .004 of datum B.
(e) The slot is parallel within .002 to datum C and perpendicular within .001 to datum A.
(f) The Ø1.750 hole has an RC7 fit (show the size of the hole as limits) and is perpendicular within .002 to datum A.
(g) Surface E has an angularity tolerance of .010 with datum C.
(h) Surface A should be flat within .002 for any one-inch square surface with a maximum flatness tolerance of .005.

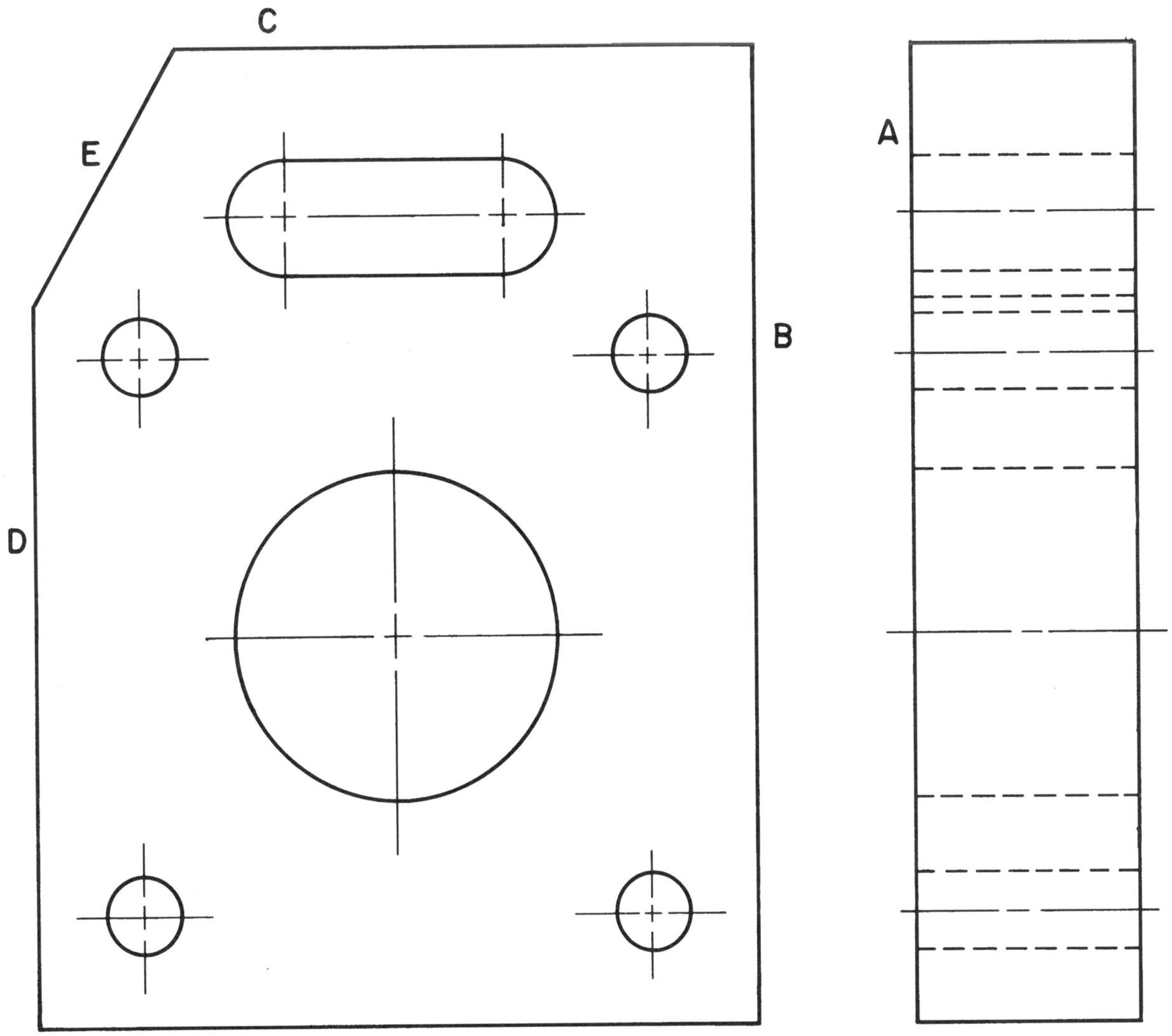

ORIENTATION TOLERANCING FOR FEATURES OF SIZE	REFERENCE UNIT 16–8	NAME COURSE	DATE	DRAWING P–8

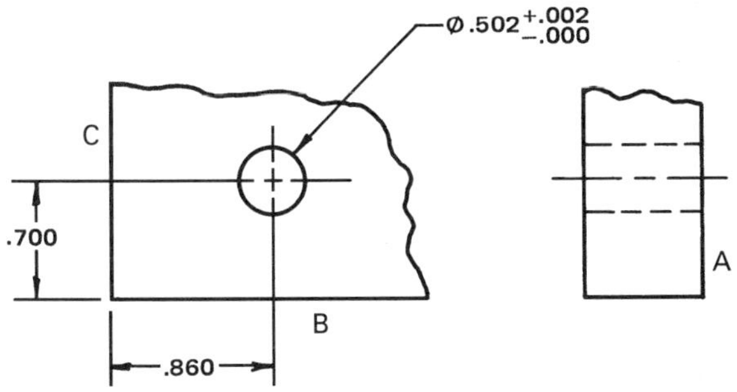

To assemble correctly, the hole shown in the part at left must not vary more than .0014 in. in any direction from its true position when the hole is at its smallest size. The letters A, B, and C shown on the drawing are the primary, secondary, and tertiary datums, respectively. On the drawings below, show suitable tolerances, dimensions, and datums, where required, to achieve this by using:

(a) Coordinate tolerancing
(b) Positional tolerancing—RFS
(c) Positional tolerancing—MMC
(d) Positional tolerancing—LMC

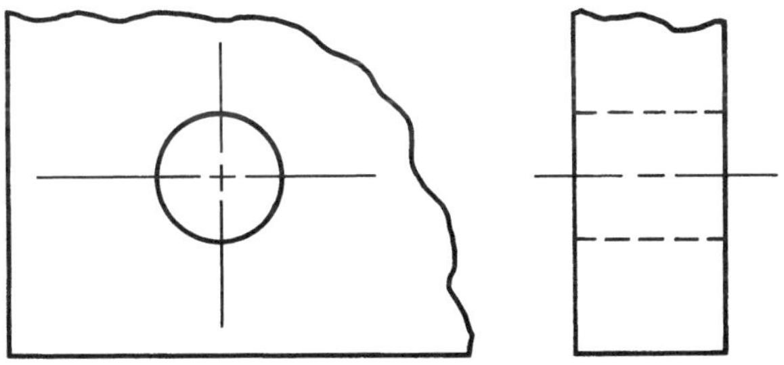

(A) Coordinate Tolerancing

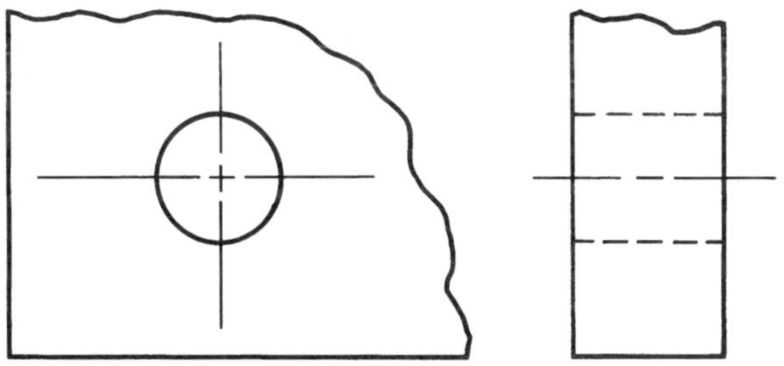

(B) Positional Tolerancing—RFS

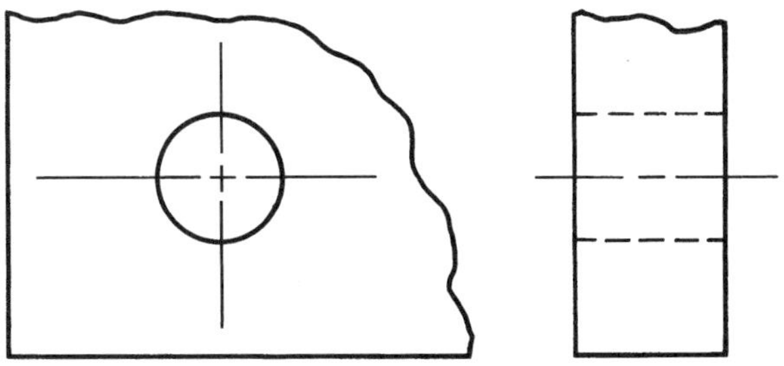

(C) Positional Tolerancing—MMC

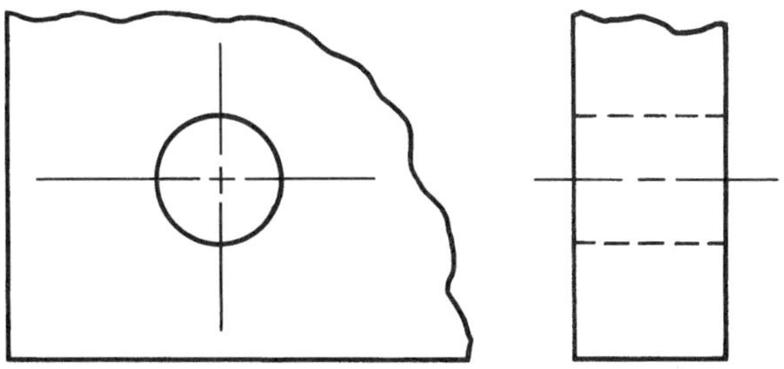

(D) Positional Tolerancing—LMC

POSITIONAL TOLERANCING	REFERENCE UNIT 16–9	NAME COURSE DATE	DRAWING P–9

Add the following geometric tolerances to the cover plate shown. The scale is 1:1 (decimal inch). Add true position dimensions and hole sizes.

(a) All datums and tolerances should be on an MMC basis unless otherwise specified.
(b) Surfaces and features marked A, B, and C are datums A, B, and C, respectively.
(c) The four .502 dia. holes have a +.006 and a –.000 tolerance on their size and must not vary from true position by more than .002 in any direction when the holes are at MMC and are related to datums A, D, and B, in that order.
(d) A flatness tolerance of .010 is required for the underside of the cover plate.
(e) The Ø2.000 hole is datum D and has a +.006 and a –.000 tolerance on its size. It has a positional tolerance of .008 and is referenced to datums A, B, and C, in that order.
(f) A projected tolerance zone of .60 is required for the four holes, the projection being directed away from the top of the cover.

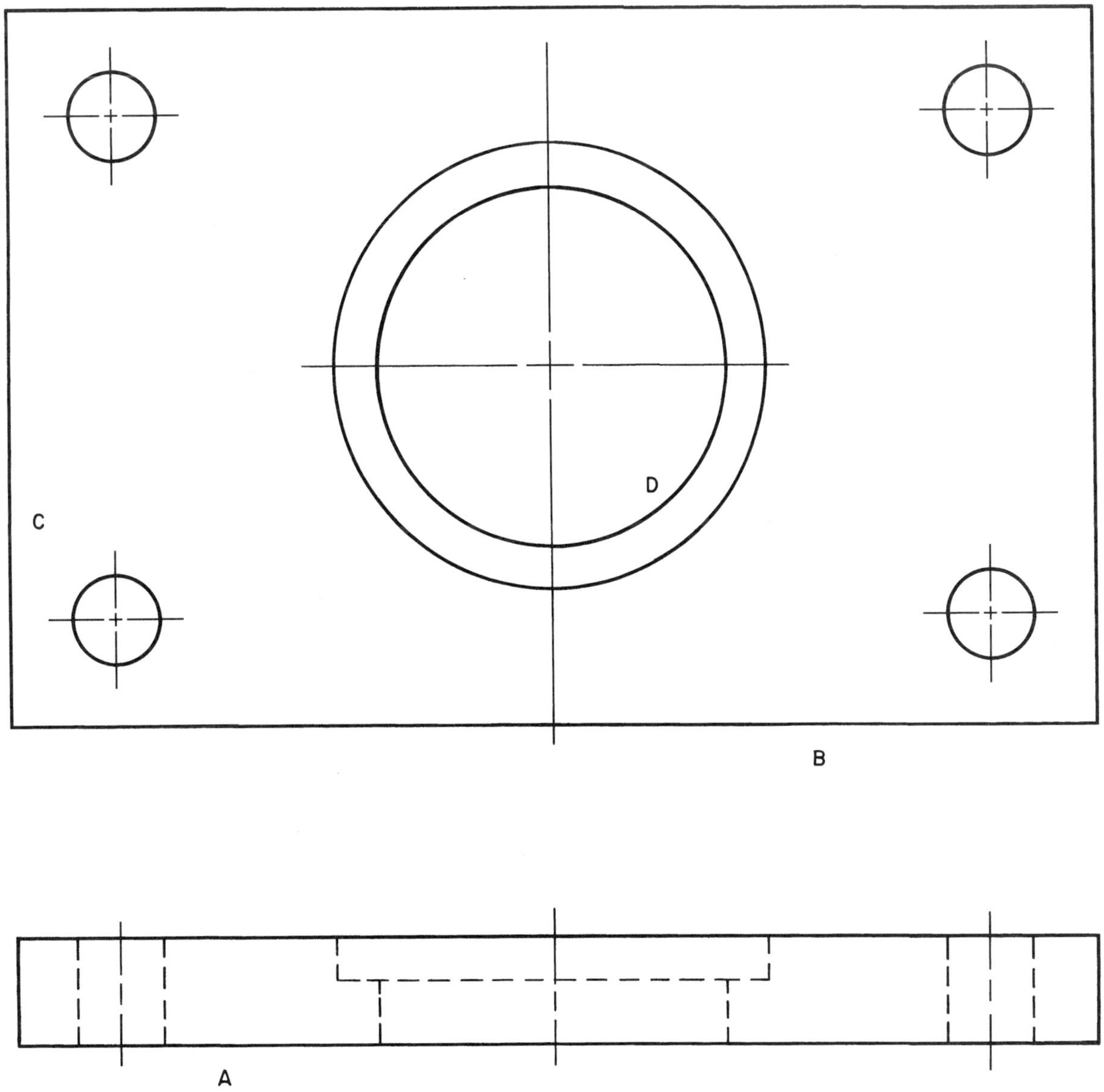

PROJECTED TOLERANCE ZONE— COVER PLATE	REFERENCE UNIT **16–10**	NAME COURSE DATE	DRAWING **P–10**

Add the following geometric tolerances to the bracket guide shown. The scale is 1:1 (metric). Add true position dimensions.

(a) All datums and tolerances should be on an MMC basis unless otherwise specified.

(b) Primary datum A has three Ø3 target areas. A1 and A2 are located on the center of surface M, one-fifth the depth distance from the front and back, respectively. A3 is located on the center of surface N, midway between the center of the hole and the right end.

(c) Secondary datum B is a datum line located at mid-height of surface D.

(d) Tertiary datum C is a datum point located at the center of surface E.

(e) The Ø10 hole has a positional tolerance of 0.2 referenced to datums A, B, and C.

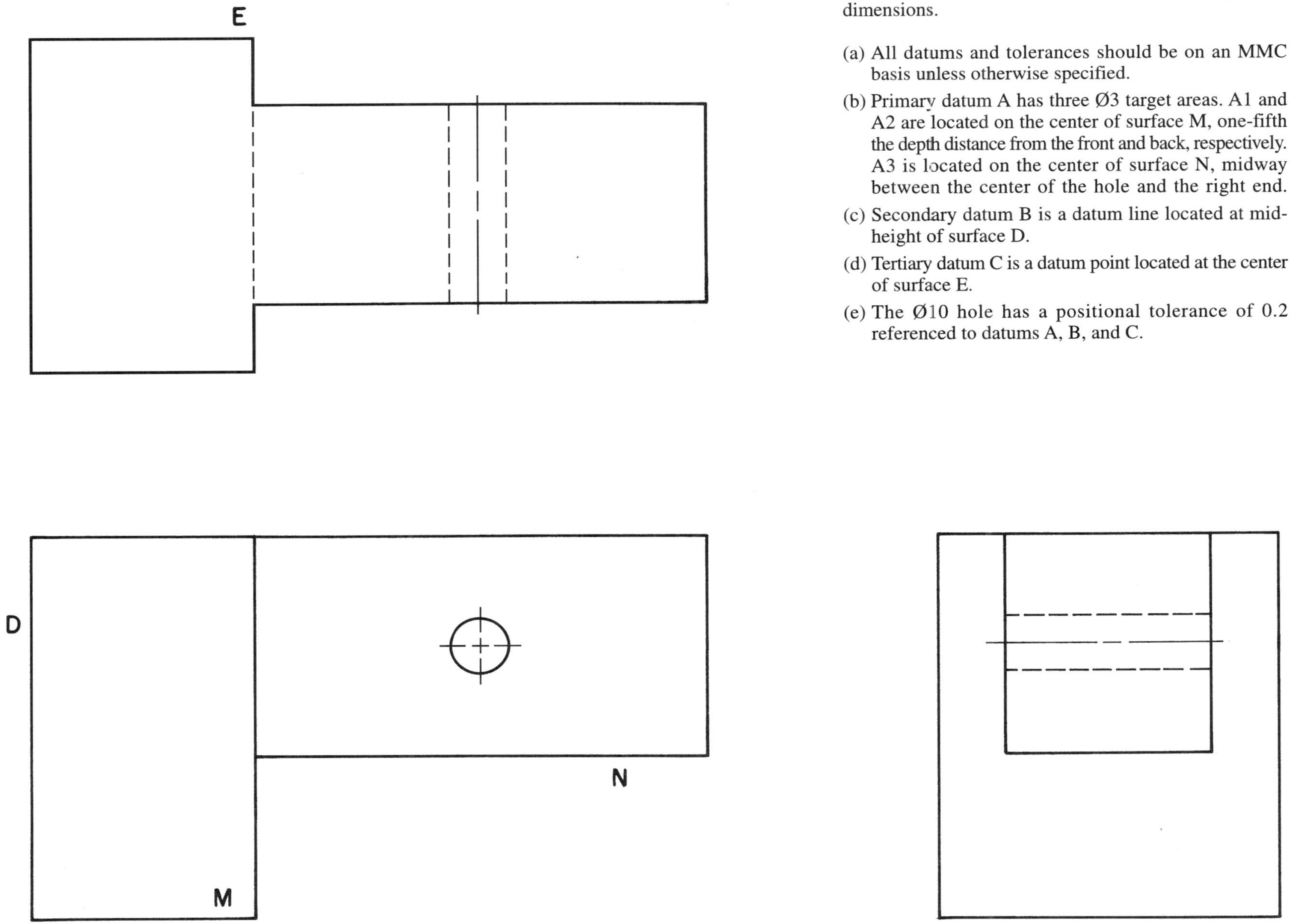

METRIC

DATUM TARGETS— BRACKET GUIDE	REFERENCE UNIT **16–11**	NAME COURSE DATE	DRAWING **P–11**

1. Add circularity tolerances to the diameters shown in Part 1. The circularity tolerances should be one-fifth of the size tolerance for each diameter.
2. Add cylindricity tolerances to Parts 2 and 3. The size of the cylindricity tolerance should equal one-quarter the size tolerance for each diameter.
3. Apply cylindricity tolerances to the three cylindrical features shown on Part 4. The cylindricity tolerance should be 25 percent of the size tolerance.

PART 1

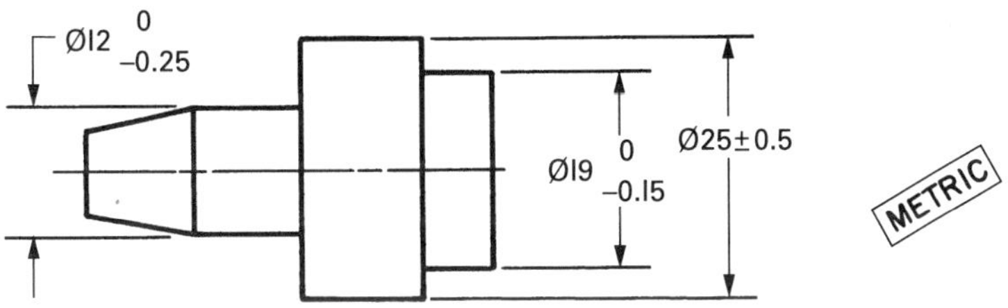

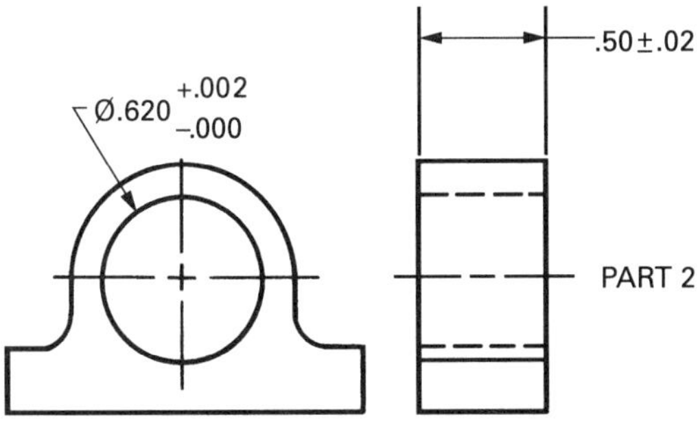

PART 2

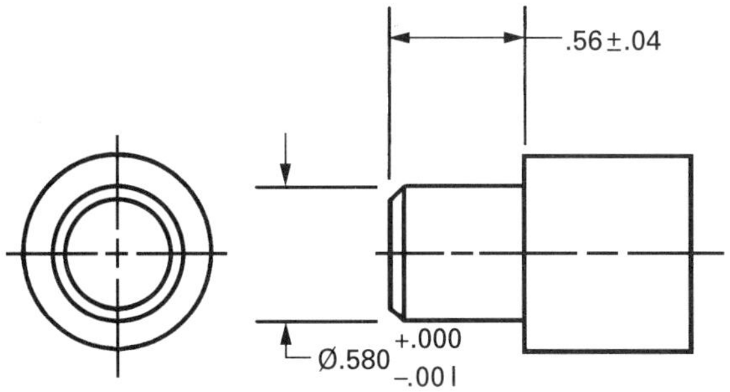

PART 3

PART 4

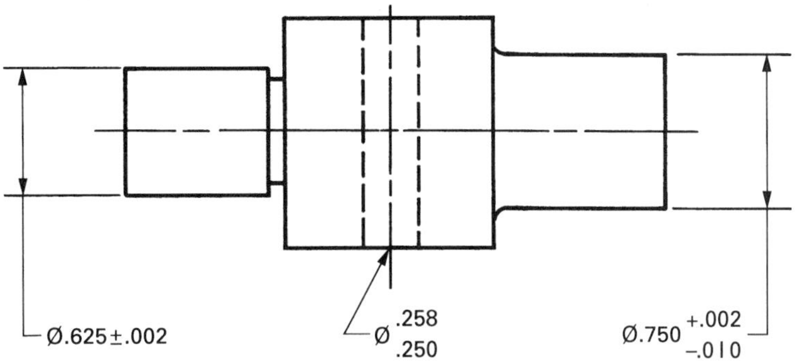

CIRCULARITY AND ROUNDNESS	REFERENCE UNIT 16–12	NAME	DRAWING P–12
		COURSE DATE	

On the drawing below, redimension the slide. Add the following geometric tolerances:

(a) All datums and tolerances should be on an MMC basis unless otherwise specified.
(b) Surfaces A, B, and C are datums A, B, and C, in that order.
(c) All corners on the profile should have a maximum 0.1 radius.
(d) A profile-of-a-surface tolerance of 0.2 should be added all around the profile of the part, and the part cannot exceed the boundary of the dimensions shown.
(e) The profile tolerance should be referenced to datums A and B, in that order.
(f) The hole should have a positional tolerance of 0.12 and be referenced to datums A, B, and C, in that order.

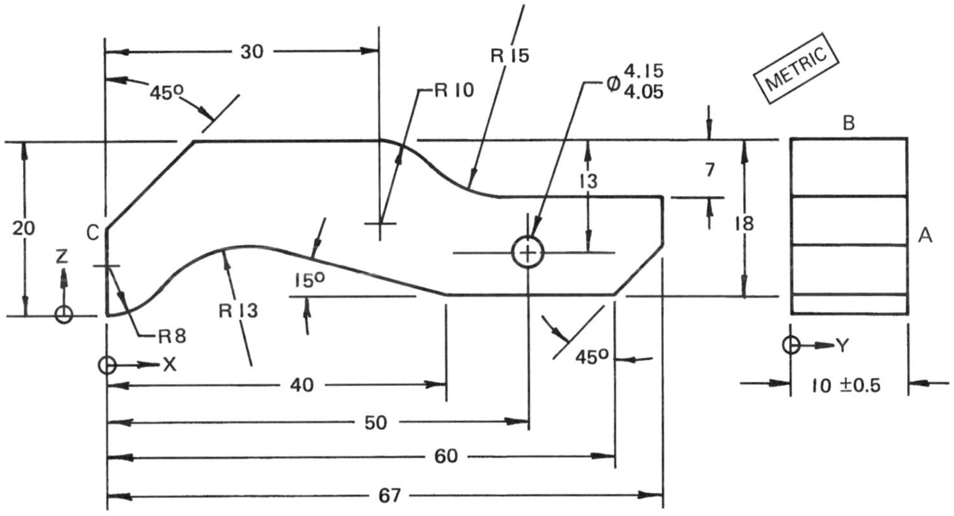

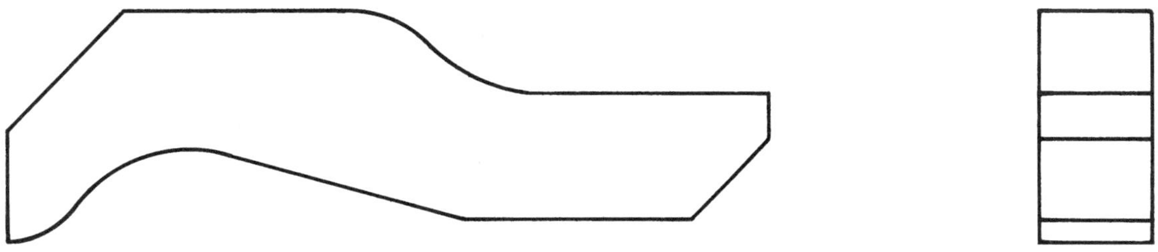

PROFILE TOLERANCING— SLIDE	REFERENCE UNIT 16–13	NAME	DRAWING P–13
		COURSE DATE	

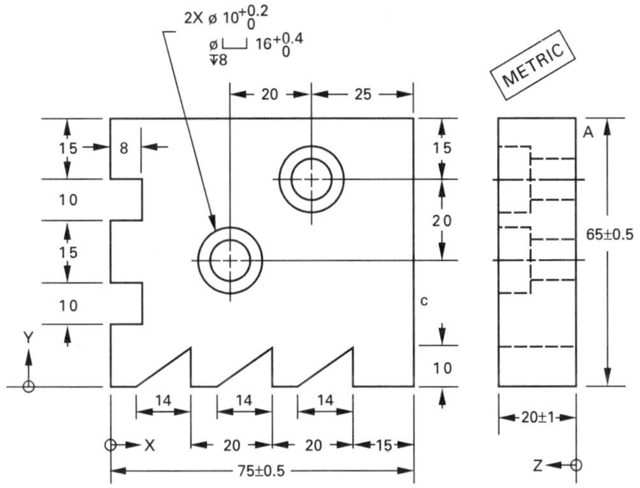

On the drawing below, redimension the adjustable base. Add the following geometeric tolerancing requirements:

(a) All datums and tolerances should be on an MMC basis unless otherwise specified.
(b) Surfaces indicated by letters A, B, and C are datums A, B, and C, respectively.
(c) Surface A should be flat within 0.15.
(d) A profile-of-a-surface tolerance of 0.2 is to be applied to the bottom coplanar surfaces of the two slots shown on the left side of the part.
(e) The bottom surfaces of the part are to be treated as one continuous surface by applying a profile-of-a-surface tolerance of 0.4, with the bottom left and right surfaces to be designated as datums D and E respectively for this tolerance.
(f) The Ø 10 portion of the counterbored holes is datum N and should have a positional tolerance of 0.12 related to datums A, B, and C, in that order.
(g) A coaxial relationship between the Ø10 and Ø16 portions of the counterbored holes is controlled by applying a positional tolerance of zero to the counterbored diameter.

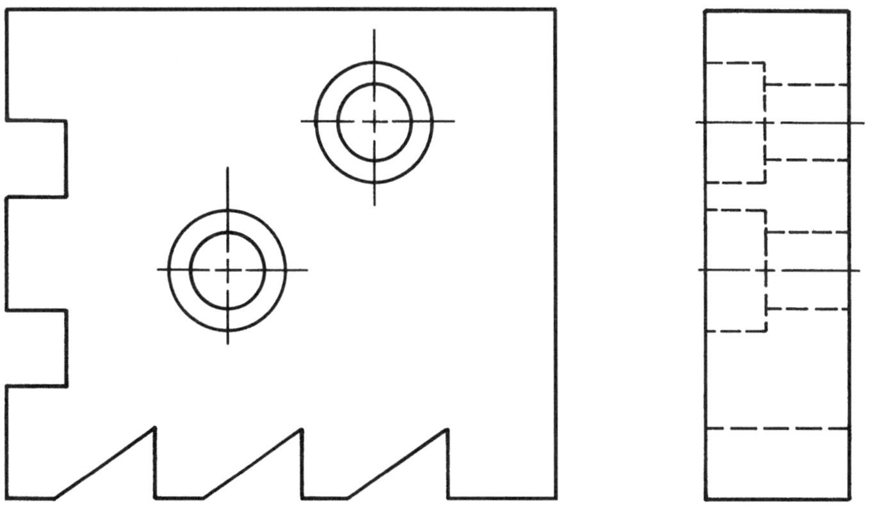

CORRELATIVE TOLERANCES— ADJUSTABLE BASE	REFERENCE UNIT 16–14	NAME COURSE DATE	DRAWING P–14

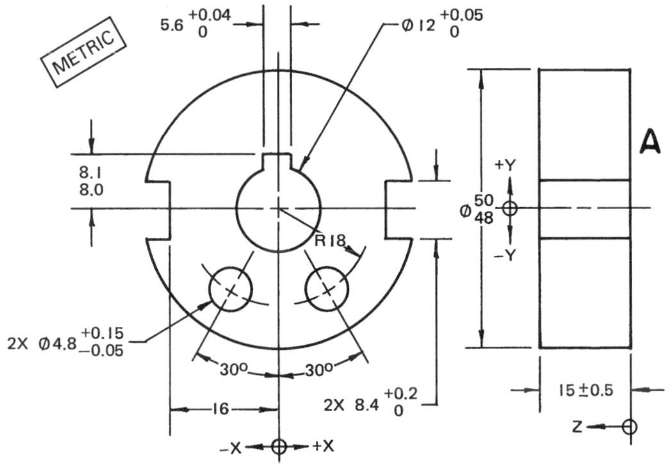

On the drawing below, redimension the locating block. Add the following geometric tolerances:

(a) All datums and tolerances should be on an MMC basis unless otherwise specified.
(b) The primary datum is surface A (datum A).
(c) The secondary datum is the Ø12 hole (datum B).
(d) The tertiary datum is the keyway width (datum C).
(e) The two slots should be located by a positional tolerance of 0.4. They are located on the horizontal center line of the Ø12 hole and referenced to datums A, B, and C.
(f) The axes of the two small holes have a positional tolerance of 0.25 and are referenced to datums A, B, and C.
(g) Datum A has a flatness tolerance of 0.2.
(h) The Ø12 hole has a positional tolerance of 0.08 and is referenced to datum A.
(i) The keyway is to be symmetrically located on the Ø12 hole by a zero positional tolerance.

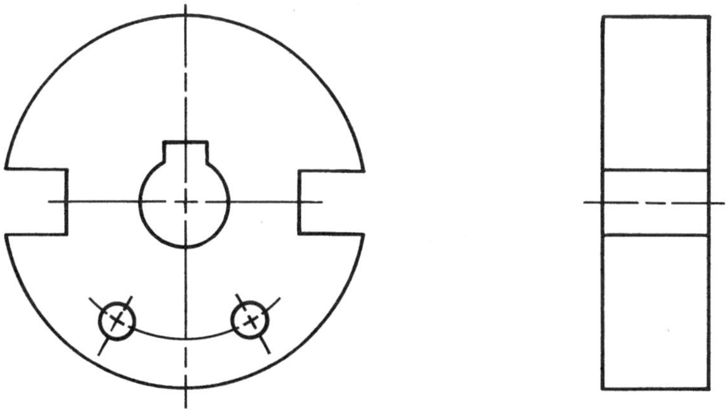

POSITIONAL TOLERANCING— NONCYLINDRICAL FEATURES	REFERENCE UNIT **16–15**	NAME ___ COURSE ___ DATE ___	DRAWING **P–15**

On the drawing below, redimension the locating plate. Add the following geometric tolerances:

(a) All datums and tolerances should be on an MMC basis unless otherwise specified.
(b) Surfaces marked A, B, and C are the primary, secondary, and tertiary datums, respectively.
(c) Composite positional tolerancing is required for the three patterns of holes and is referenced to datums A, B, and C.
(d) The positional tolerances for the location of hole patterns are .010, .008, and .016 for the Ø.250, Ø.188, and Ø.125 holes, respectively.
(e) The positional tolerances for holes within the pattern are .004, .003, and .006 for the Ø.250, Ø.188, and Ø.125 holes, respectively, and are referenced to datum A.
(f) A flatness tolerance of .004 is required for datum B.
(g) The top surface should be parallel within .008 with the bottom surface.
(h) The right side surface should be perpendicular within .005 with the bottom surface.
(i) Surface D should have a perpendicularity tolerance of .008 with the bottom surface.

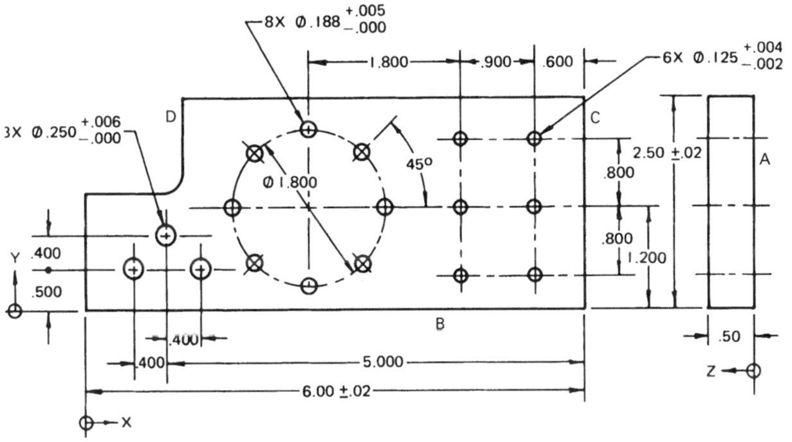

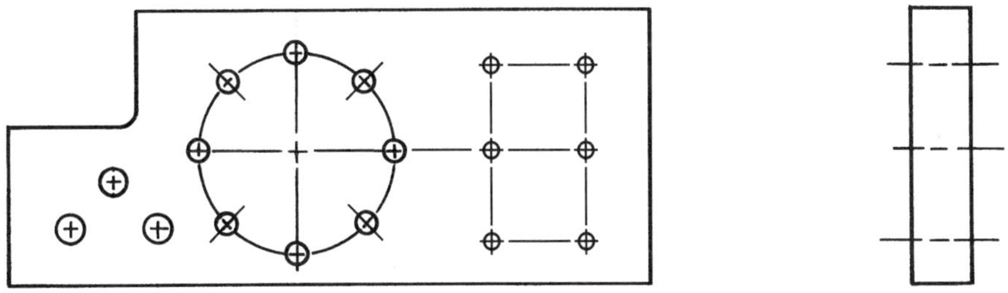

POSITIONAL TOLERANCING— MULTIPLE PATTERNS OF FEATURES	REFERENCE UNIT 16–16	NAME COURSE DATE	DRAWING P–16

1. Refer to Fig. 1 and add the size of the positional tolerance required for the cover and flange to the feature control frames.
2. Complete the feature control frames in Fig. 2 and assume that the positional tolerance for the holes in the flange in assignment 1 is to be twice the size as the positional tolerance of the holes in the cover.
3. Complete the feature control frames in Fig. 3. The positional tolerances for the flange and cover should be equal.
4. Apply equal positional tolerances to the coaxial features in Fig. 4.

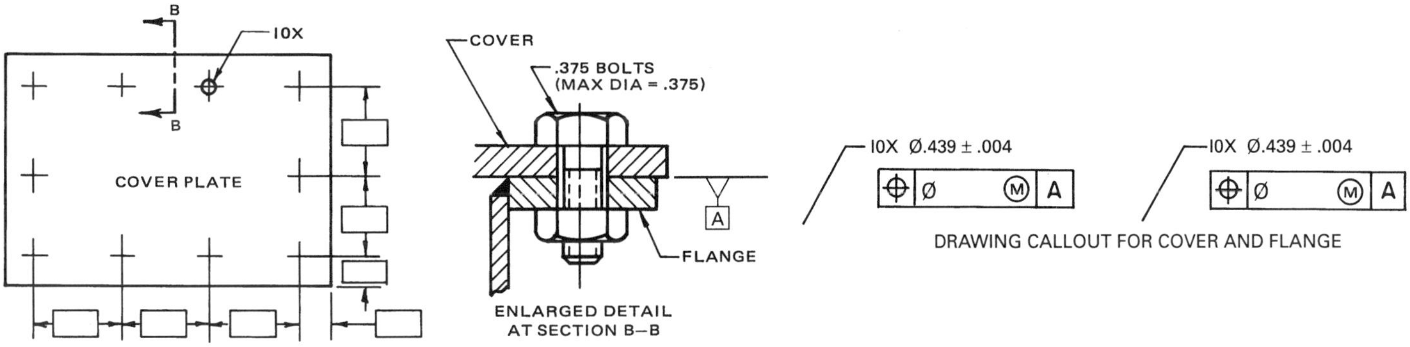

FIGURE 1

FIGURE 2

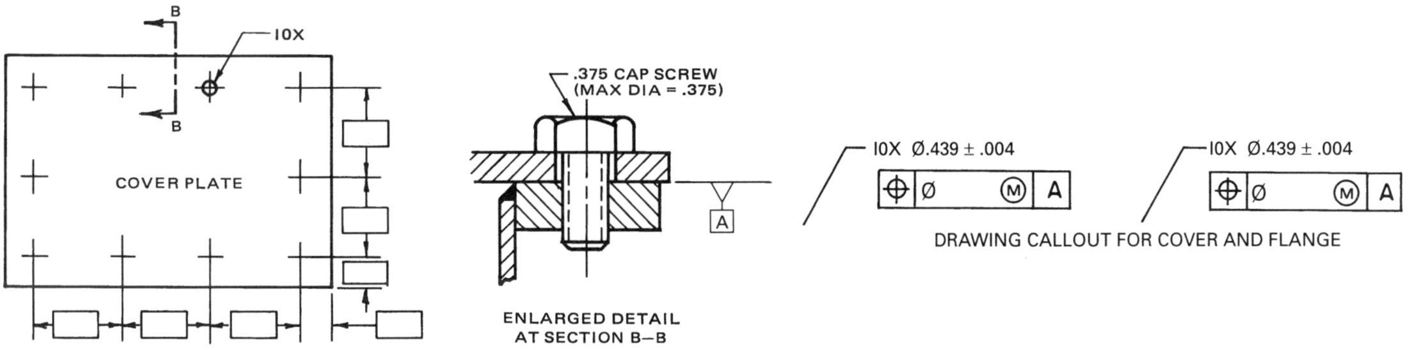

FIGURE 3

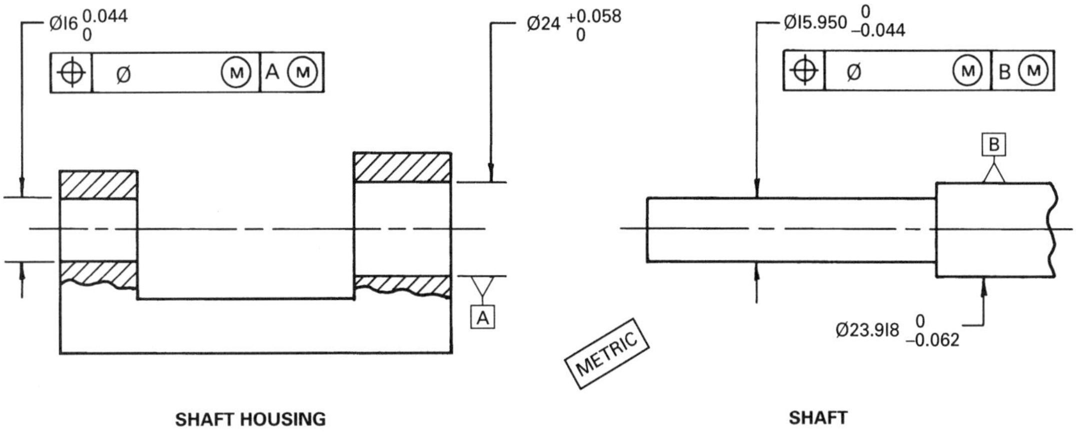

FIGURE 4

FORMULAS FOR POSITIONAL TOLERANCING	REFERENCE UNIT 16–17	NAME COURSE	DATE	DRAWING P–17

On one of the blank sheets from the back of this book, make a two-view working drawing of the housing shown below. The scale is 1:2. Add the following geometric tolerancing requirements.

(a) All datums and tolerances should be on an MMC basis unless otherwise specified.
(b) The bearing housing surface attached by .375 cap screws to the main housing is datum A and has a flatness tolerance of .003.
(c) The bearing seat is datum D and has a parallel tolerance of .004 with datum A.
(d) An LN3 fit is required between the housing and the bearing. This diameter is datum E and has a perpendicularity tolerance of .001 with datum D. It also has a positional tolerance of .002 referenced to datums A and C, in that order.
(e) The surface contacting the cover plate is datum F.
(f) The holes for the spring pins are Ø.375–.376 and are designated as datum C. They are located by a positional tolerance of .001 and referenced to datum A.
(g) The holes for the .375 cap screws are Ø.390–402. They are located by a positional tolerance of .008 and referenced to datums A and C, in that order.
(h) The four tapped holes have a positional tolerance of .010 and are referenced to datums D and E, in that order.

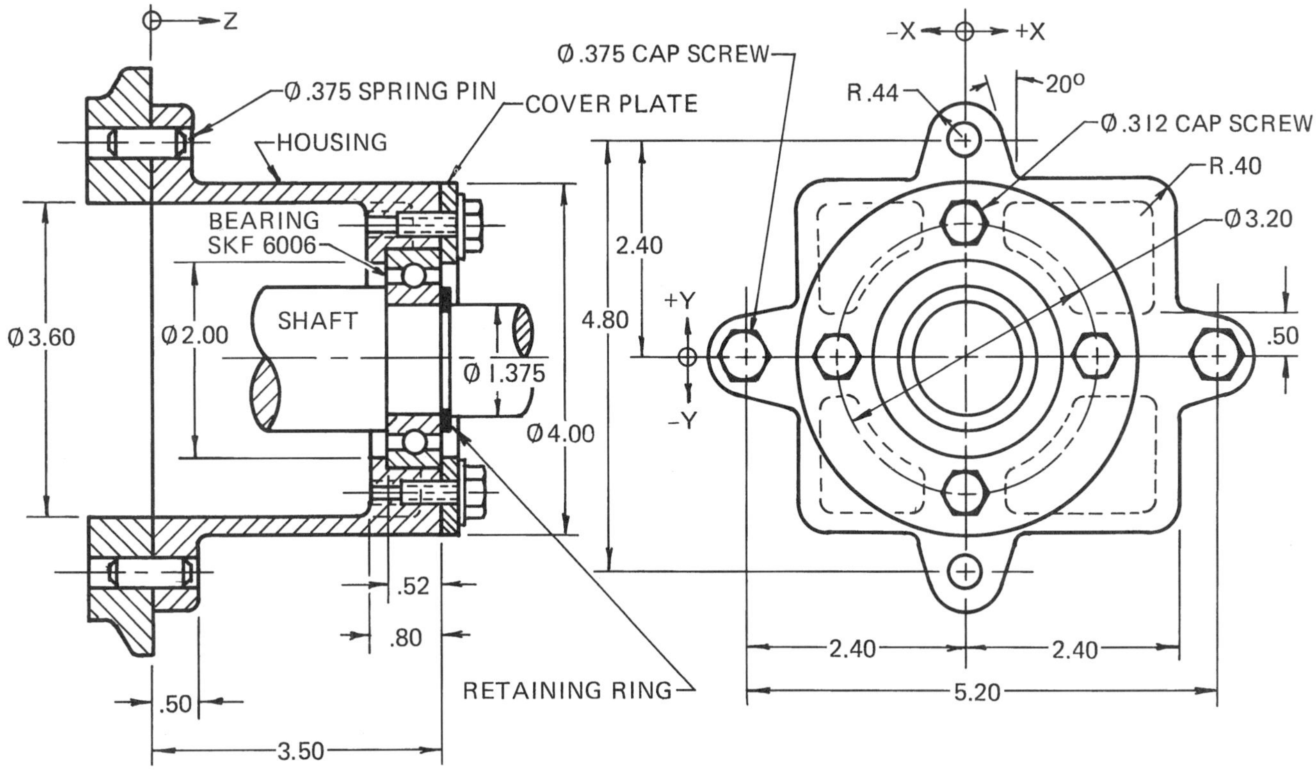

SUMMARY OF RULES GEOMETRIC TOLERANCING—HOUSING	REFERENCE UNIT 16–18	NAME COURSE DATE	DRAWING P–18

On the charts below, insert the X and Y coordinates for the ten holes. The letters beside the holes indicate the sequence in which they should be drilled. The origin is shown on the drawing.

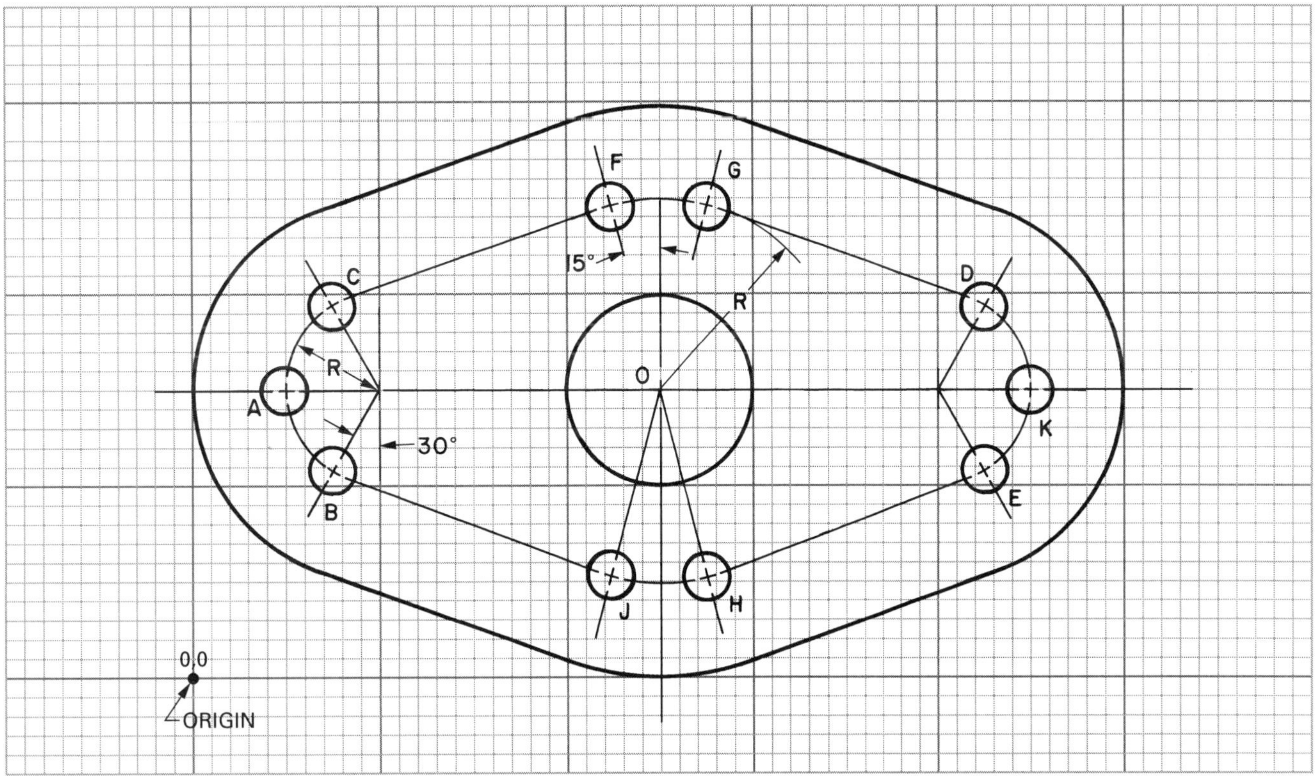

NOTE: GRID 10 X 10 TO THE INCH   NOTE: NO HOLE I (I) SHOWN

COORDINATE DIMENSIONING		
HOLE	X AXIS	Y AXIS
A		
B		
C		
D		
E		
F		
G		
H		
J		
K		

POINT-TO-POINT DIMENSIONING		
HOLE	X AXIS	Y AXIS
A		
B		
C		
D		
E		
F		
G		
H		
J		
K		

DRAWINGS FOR NUMERICAL CONTROL— TWO-AXIS CONTROL SYSTEMS	REFERENCE UNIT 17–1	NAME ___ COURSE ___ DATE ___	DRAWING R–1

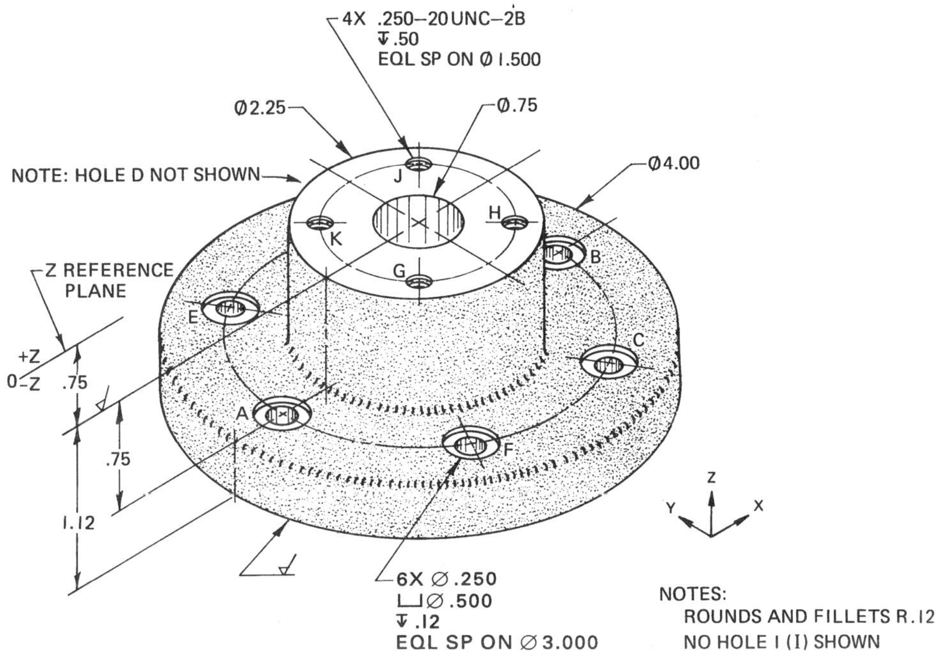

The top-drill holes and the six through holes should be made on a numerical-controlled machine. The depth of the tap drill should be .10 in. below the last complete thread. In the chart below, list the X and Y coordinates for each hole using point-to-point coordinates. The letters at the holes show the sequence in which they are to be drilled. Calculating the Z coordinate should be done in the same manner as for the part shown in Fig. 17–2–4. The origin for the X and Y coordinates is the center of the end plate and are located below. Draw the two views to the scale of 1:1.

POINT-TO-POINT COORDINATES			
HOLE	**X**	**Y**	**Z**
A			
B			
C			
D			
E			
F			
G			
H			
J			
K			

DRAWINGS FOR NUMERICAL CONTROL—
THREE-AXIS CONTROL SYSTEMS

REFERENCE UNIT **17–2**

NAME

COURSE

DATE

DRAWING **R–2**

Redesign the pivot arm shown for fabrication by welding using standard metal sizes and shapes. Make a detail assembly drawing. Welding symbols and sizes are not required. Complete the item list and identify each part on the assembly. The scale is 1:1.

QTY	ITEM	MATL	DESCRIPTION	PT NO.

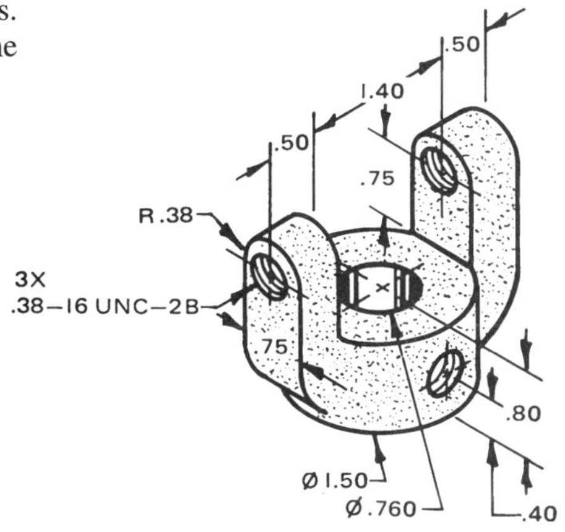

DESIGNING FOR WELDING— PIVOT ARM	REFERENCE UNIT 18–1	NAME COURSE DATE	DRAWING S–1

1. Complete the enlarged views of the welded joints in the drawing callouts shown. Use notes to explain any additional welding requirements.

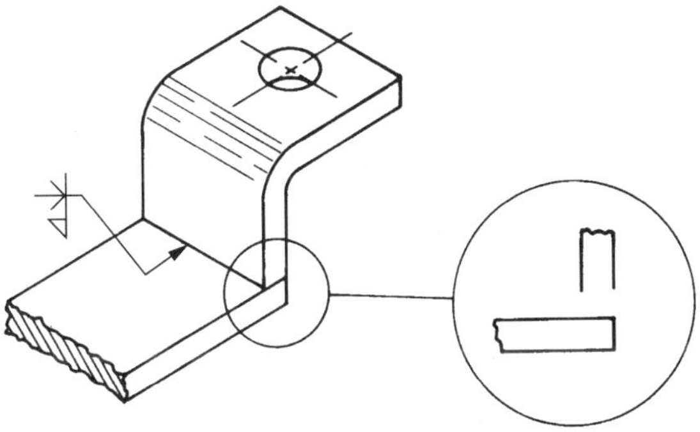

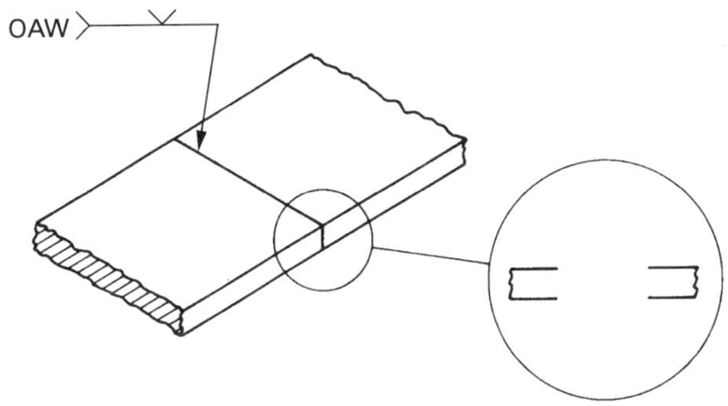

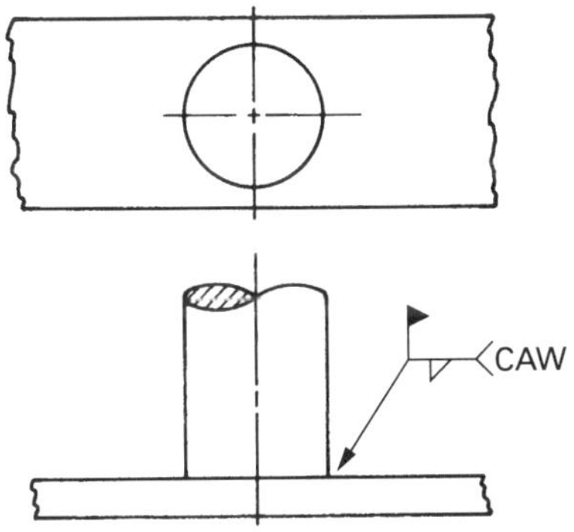

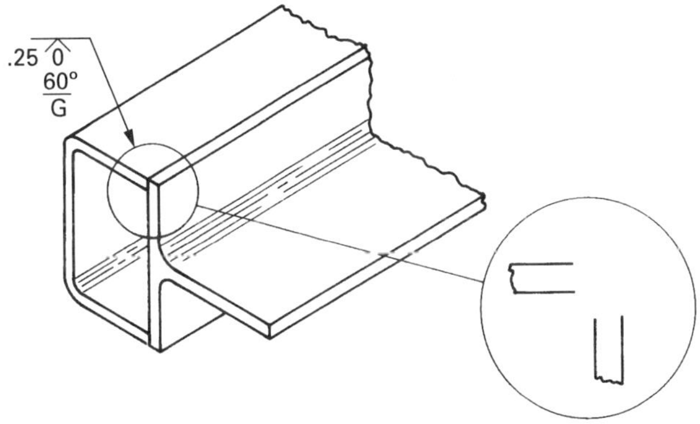

2. Add the information at the right to the seven welding symbols below.

Weld	Welding Process Required	Type of Weld	Additional Requirements
1	**Carbon Arc Welding**	**Bevel**	
2	**Oxyacetylene Welding**	**Durable Fillet**	**Both Sides** **Field Weld**
3	**Oxyacetylene Welding**	**Fillet**	**Both Sides**
4	**No Specifications Required**	**J-Groove**	
5	**Carbon Arc Welding**	**Fillet**	**All Around**
6	**Carbon Arc Welding**	**Fillet**	**All Around** **Field Weld**
7	**Gas Metal Arc Welding**	**Double V-Groove**	

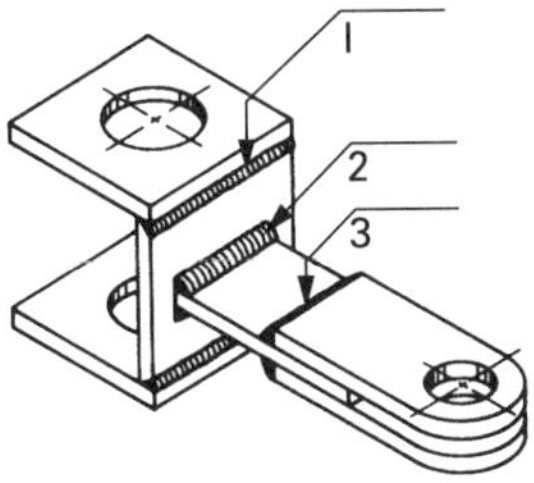

BEVEL AND FILLET WELDS

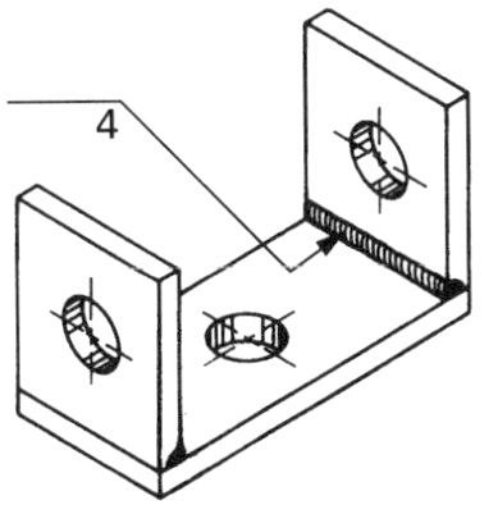

J-GROOVE WELD

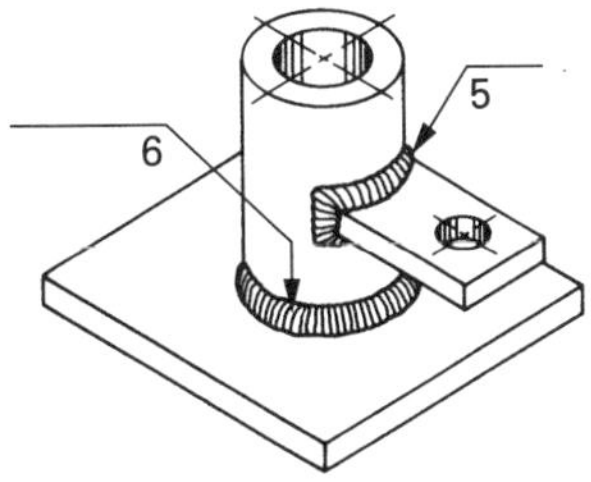

FILLET WELD ALL AROUND

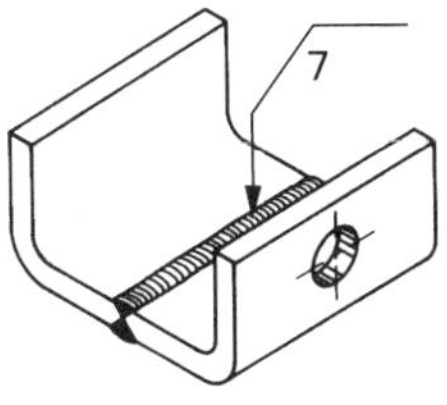

DOUBLE V-GROOVE WELD

WELDING SYMBOLS	REFERENCE UNIT 18–2	NAME COURSE DATE	DRAWING S–2

Make a two-view working drawing, complete with dimensions and welding symbols, of the swing bracket shown. Complete the item list and identify each part on the assembly. Usc full-strength welds. Use a common break to shorten the height. The centers of the two Ø.625 holes on the side view are located on the drawing area below. The scale is 1:1.

QTY	ITEM	MATL	DESCRIPTION	PT NO.

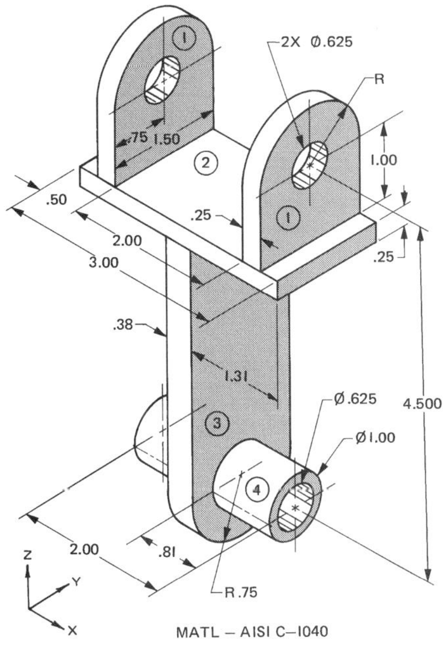

FILLET WELDS— SWING BRACKET	REFERENCE UNIT 18–3	NAME COURSE DATE	DRAWING S–3

On a blank sheet of drawing paper from the back of this workbook, make a two-view welded fabrication drawing of the connecting link shown. Complete the item list below and identify each part on the assembly. Use full-strength welds. The scale is 1:2.

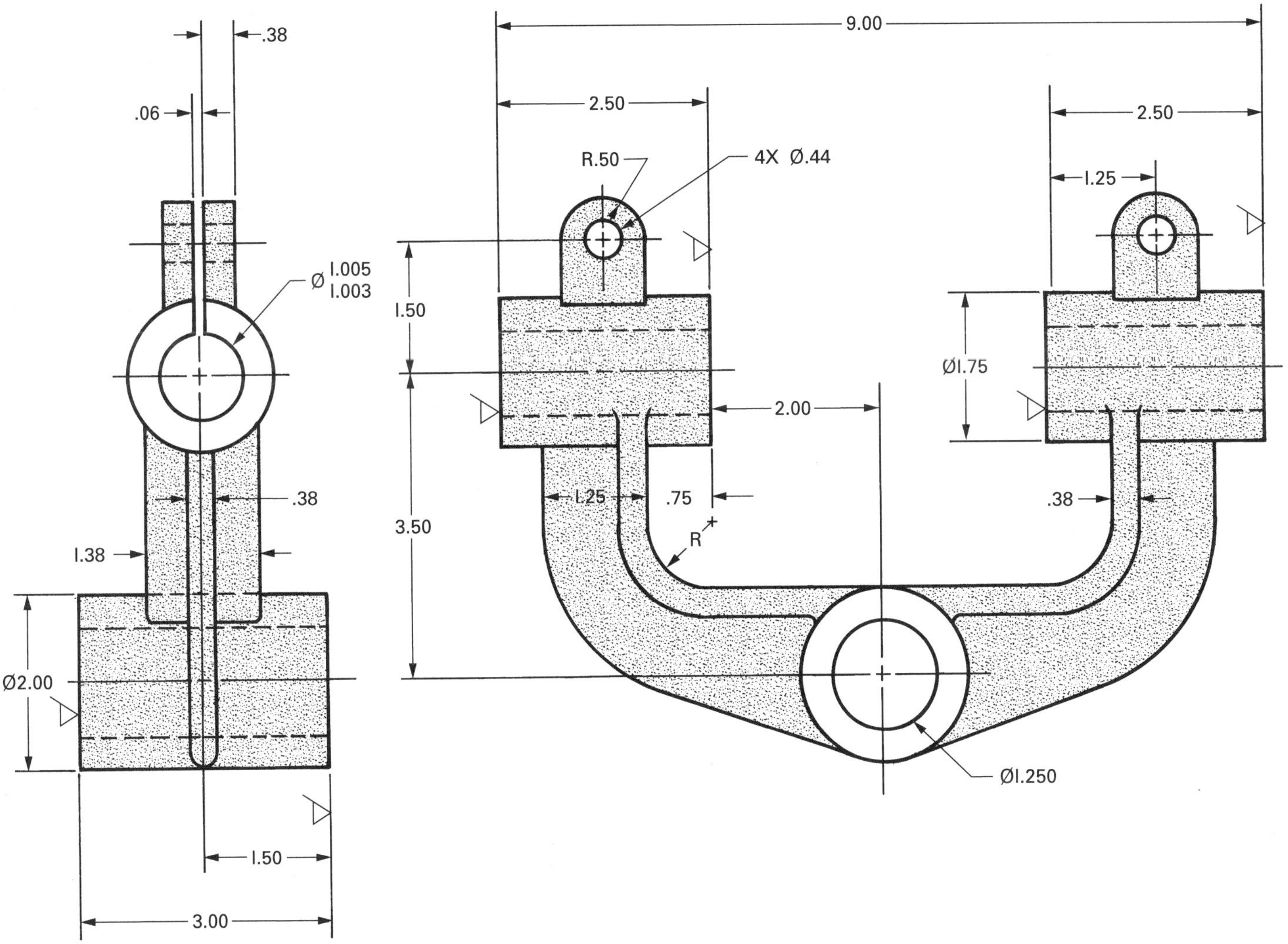

QTY	ITEM	MATL	DESCRIPTION	PT. NO.

GROOVE WELDS—

CONNECTING LINK

REFERENCE UNIT **18–4**

NAME

COURSE

DATE

DRAWING **S–4**

Show the drawing callouts and sketch the weld details on the partial views below.

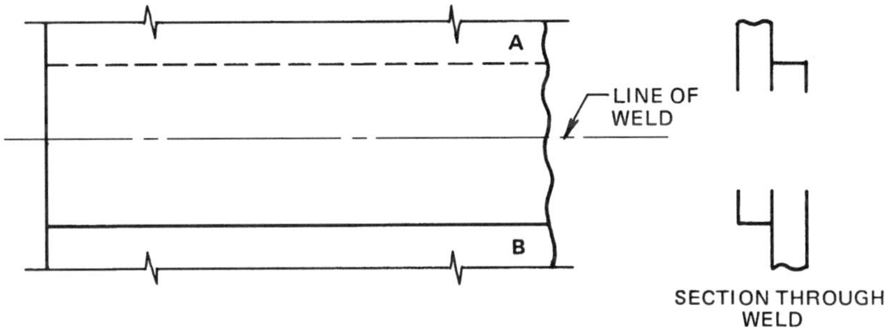

ASSEMBLY I – PLUG WELDS

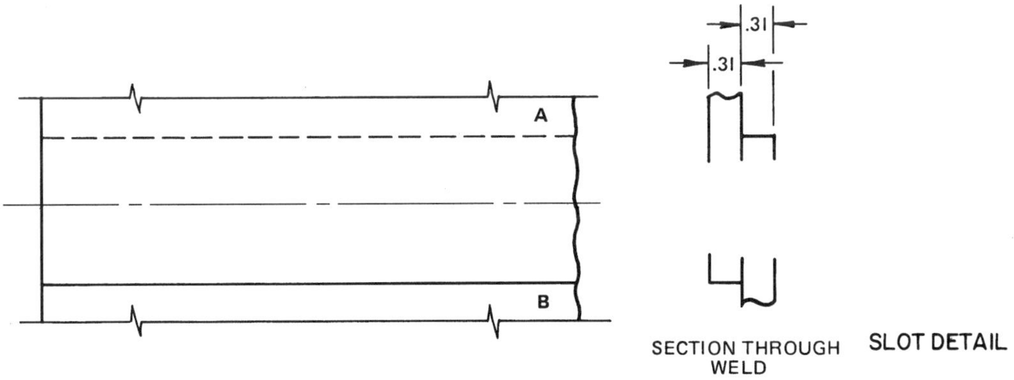

ASSEMBLY 2 – SLOT WELDS

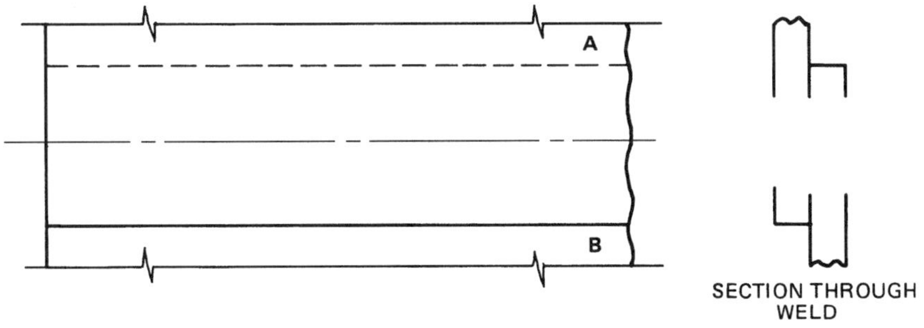

ASSEMBLY 3 – SPOT WELDS

ASSEMBLY I—PLUG WELDS –HOLES IN PART B –HOLES–Ø.62 × 0° CSK –CENTER OF FIRST HOLE –2.50 FROM LEFT SIDE –CENTER SPACING OF WELDS–4.00 –HOLES COMPLETELY FILLED –POSTWELD FINISH–CONVEX BY CHIPPING	ASSEMBLY 2—SLOT WELDS –HOLES IN PART "A" PERPENDICULAR TO LINE OF WELD –SLOT SIZE .75 × 2.00 × 30° CSK –CENTER OF FIRST HOLE 3.00 FROM LEFT SIDE –CENTER SPACING OF WELDS–6.00 –DEPTH OF FILLING–.I9	ASSEMBLY 3—SPOT WELDS –GAS TUNGSTEN ARC WELD –CENTER SPACING OF WELDS–3.00 –CENTER OF FIRST SPOT 2.00 FROM LEFT SIDE –WELD ARROW SIDE –STRENGTH OF SPOT WELDS–300 LBS –TOTAL STRENGTH OF JOINT–2400 LBS

OTHER BASIC WELDS— PLUG, SLOT, AND SPOT WELDS	REFERENCE UNIT **18–5**	NAME COURSE DATE	DRAWING **S–5**

Your drafting supervisor has assigned you the responsibility of designing an attractive single toggle-switch plate for use in kitchens and bathrooms. The toggle plate and clearance hole requirements are shown. The production run will exceed 25,000, and four different color plates are required. Lay out the design of the plate and include on the drawing the production and specification data that you would submit with your design.

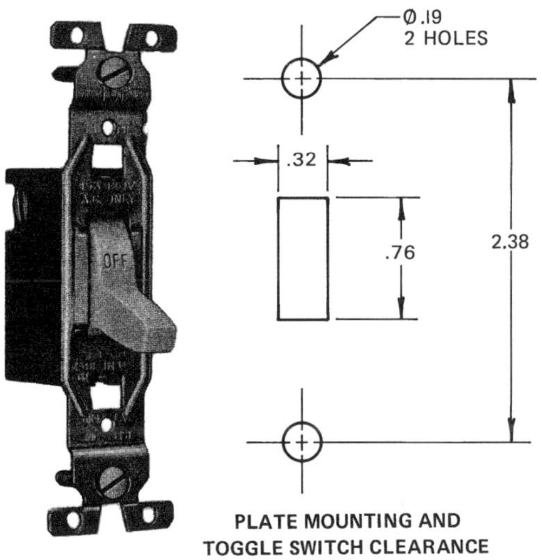

PLATE MOUNTING AND
TOGGLE SWITCH CLEARANCE

THE DESIGN PROCESS— TOGGLE-SWITCH PLATE	REFERENCE UNIT 19–1	NAME COURSE DATE	DRAWING T–1

Two vertical .312 in. diameter electrical conductors should be supported on the inside wall of an oil-filled metal tank. The vertical length of the conductors is 7 ft. The conductors, made of copper, have no insulation on them. The voltage they carry is such that there must be a minimum distance of 2.00 in. between conductors or any other metal when supported by nonconductive material, such as plastic, wood, etc. Although there are no external forces acting on these conductors, they should be supported every 24 in. Prepare an assembly drawing showing the conductors, support, and tank wall. Include an item list. On a second sheet, prepare the details of the parts required. Scale to suit.

ASSEMBLY CONSIDERATIONS— CONDUCTOR SUPPORTS	REFERENCE UNIT **19–2**	NAME COURSE DATE	DRAWING **T–2**

Show your calculations below each of the V-belt drive problems on this page.

A. A .33 hp (0.25 kW), 1750 r/min motor should operate a furnace blower having a shaft speed of approximately 765 r/min. The center distance between the motor and blower shafts is approximately 13.5 in. (340 mm). Select a suitable V-belt.

B. A 1.5 hp (1.1 kW), 1750 r/min motor should operate a band saw whose flywheel turns at approximately 800 r/min. A pulley attached to the flywheel shaft connects, by means of a V-belt, to the pulley on the motor shaft. The center-to-center distance of the shafts is 14.5 in. (368 mm). Calculate the size of the V-belt required.

C. A .5 hp (0.37 kW), 1750 r/min motor drives a power hacksaw. The shaft on the hacksaw should run at approximately 750 r/min, and the center-to-center distance of the shafts is 15.5 in. (400 mm). Calculate the size of the V-belt required.

D. A .75 hp (0.6 kW), 1750 r/min motor is used to drive a punch machine whose flywheel turns at approximately 600 r/min. A pulley is attached to the flywheel shaft and connects to the motor pulley by means of a V-belt. The center-to-center distance is 17 in. (430 mm). Calculate the size of the V-belt required.

BELT DRIVES	REFERENCE UNIT 20–1	NAME COURSE DATE	DRAWING U–1

Show your calculations below each of the chain drive problems on this page.

A. A tumbler barrel is to be driven at approximately 40 r/min by a speed reducer powered by a 5 hp (3.7 kW) electric motor. The reducer output speed is 100 r/min, and the output shaft is 1.75 in. (44 mm) in diameter. The shaft diameter of the tumbling barrel is 2 in. (50 mm). The shaft center distance is approximately 36 in. (900 mm). Select a single chain (heavy shock).

METRIC

B. The head shaft of an apron conveyor, which handles rough castings from a shakeout, operates at 66 r/min and is driven by a gear motor whose output is 7.5 hp (5.6 kW) at 100 r/min. The head shaft has a 2 in. (50 mm) diameter, and the gear motor shaft has a 1.75 in. (44 mm) diameter. The shaft center distance should not exceed 42 in. (1055 mm). Select a multiple chain (moderate shock).

METRIC

C. A gear-type lubrication pump located in the base of a large hydraulic press is to be driven at 860 r/min from a 1.25 in. (32 mm) diameter shaft operating at 1000 r/min. The pump is rated at 3 hp (2.4 kW) and has a 1.375 in. (35 mm) diameter shaft. The shaft center distance must not be less than 10 in. (250 mm).

CHAIN DRIVES	REFERENCE UNIT 20–2	NAME COURSE DATE	DRAWING U–2

Make a working drawing of the spur gear given the following information: tooth form —14.5°, PD = 6.000 in., DP = 5, face width =1.00 in., web—.40 in., shaft = Ø1.10 in., hub = Ø1.90 × 1.50 in. LG, MATL—MI. Select a proper key size, and use your judgment for dimensions not given. Complete the cutting data block. The scale is 1:1.

CUTTING DATA	
NO. OF TEETH	
PITCH DIAMETER	
DIAMETRAL PITCH	
PRESSURE ANGLE	
WHOLE DEPTH	
CHORDAL ADDENDUM	
CHORDAL THICKNESS	

GEAR DRIVES— SPUR GEAR	REFERENCE UNIT 20–3	NAME ___ COURSE ___ DATE ___	DRAWING U–3

Show your calculations below each of the spur gear problems on this page.

A. A 1200 r/min motor drives, by means of a spur gear and pinion, a machine rated at 8 hp and operating under moderate shock 12 hours a day. The reduction in r/min is 4:1. Select a suitable pair of spur gears to transmit the power required.

C. A 1200 r/min motor drives, by means of a spur gear and pinion, a machine rated at 7.5 kW and operating under moderate shock 8 hours a day. The reduction in r/min is 3:1. Select a suitable pair of spur gears to transmit the power required.

METRIC

B. A punch press rated at 22 hp, 900 r/min, is to be driven by a 30 hp, 1200 r/min motor. The punch, which is subjected to moderate shock, will be in operation 16 hours a day. Select a suitable pair of spur gears to transmit the power required.

D. An 1800 r/min motor drives a machine rated at 2 kW. The machine runs at 450 r/min under moderate shock 18 hours a day. Select a suitable pair of spur gears to transmit the power required.

METRIC

POWER TRANSMITTING CAPACITY OF SPUR GEARS	REFERENCE UNIT 20–4	NAME ___ COURSE ___ DATE ___	DRAWING U–4

Make a two-view working assembly drawing of the rack and gear from the data given. Use your judgment for dimensions not given. Show four or five teeth in mesh. The scale is 1:1.

CUTTING DATA	GEAR	RACK
NO. OF TEETH		
PITCH DIAMETER		
MODULE		
DIAMETRAL PITCH		
PRESSURE ANGLE		
WHOLE DEPTH		
CHORDAL ADDENDUM		
CHORDAL THICKNESS		
CIRCULAR PITCH		
LINEAR PITCH		

Gear—N—24
DP=5
Tooth form—14.5°
Web—.50
Shaft—Ø1.25
Hub—Ø2.25 × 1.75 LG
Face width—1.25
Matl—MI
Rack—Matl—Steel

RACK AND PINION

REFERENCE UNIT **20–5**

NAME

COURSE

DATE

DRAWING **U–5**

GEAR DATA
DP–4 Face width–1.10 N–22 Shaft–Ø1.00 Hub–Ø1.90 × 1.50 Lg Web–.75 Tooth form–14.5°

PINION DATA
N–14 Shaft–Ø.75 Hud–1.25 Lg Matl–Steel

CUTTING DATA	GEAR	PINION
NO. OF TEETH		
DIAMETRAL PITCH		
TOOTH FORM		
CUTTING ANGLE		
WHOLE DEPTH		
CHORDAL ADDENDUM		
CHORDAL THICKNESS		

Make a one-view working assembly drawing of the bevel gear and pinion from the data given. Use your judgment for dimensions not shown.

BEVEL GEARS

REFERENCE UNIT **20–6**

NAME

COURSE

DATE

DRAWING **U–6**

WORM CUTTING DATA	
NO. OF THREADS	
PITCH	
PRESSURE ANGLE	
LEAD ANGLE	
LEAD	
WHOLE DEPTH	
ADDENDUM	

GEAR CUTTING DATA	
NO. OF TEETH	
ADDENDUM	
WHOLE DEPTH	
NO. OF THREADS	
PITCH (AXIAL)	
PRESSURE ANGLE	
LEAD ANGLE	
LEAD	

Make a two-view detail assembly drawing of a worm and gear from the following data. Use your judgment for dimensions not given. The scale is 1:2. A worm and worm gear have a pitch of .75 in. The gear, made of phosphor bronze, has 24 teeth, shaft dia = 1.25, hub dia = 2.25, hub length = 2.50, web thickness = .50. The worm, made of steel, has a pitch dia = 2.50, left-hand double thread, shaft dia = 1.00. The location of the worm (bottom left) and the gear (top right) are shown on the drawing area.

WORM AND GEAR	REFERENCE UNIT 20–7	NAME COURSE DATE	DRAWING U–7

On one of the blank sheets from the back of this workbook, make a layout drawing showing a suitable mounting arrangement for the gear and bearing housing mounted on the mounting surface shown. The design is for low-speed moderate use. Draw the gear and one shaft support in full-section, and the pulley and the other shaft support in half-section. A partial top view is required to show the location of the mounting holes. Standard parts should be used wherever possible and an item list should be included on the drawing. The scale is 1:2.

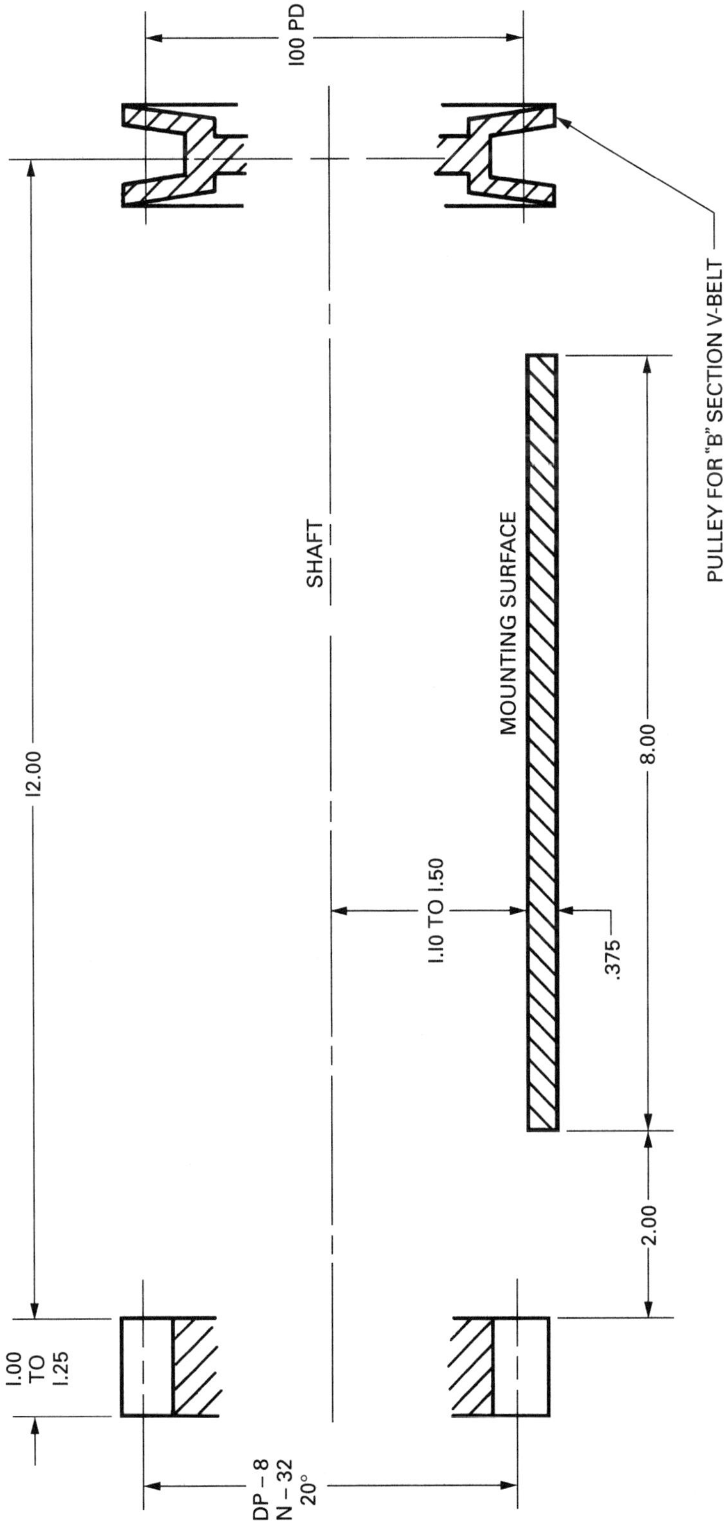

COMPARISON OF CHAIN, GEAR, AND BELT DRIVES	REFERENCE UNIT **20–8**	NAME COURSE DATE	DRAWING **U–8**

Lay out the motor to the gearbox shown. A flexible coupling is required to connect the shaft, and the type of coupling required is shown. Lay out the drive assembly and call for the correct coupling size on the drawing. The scale is 1:4.

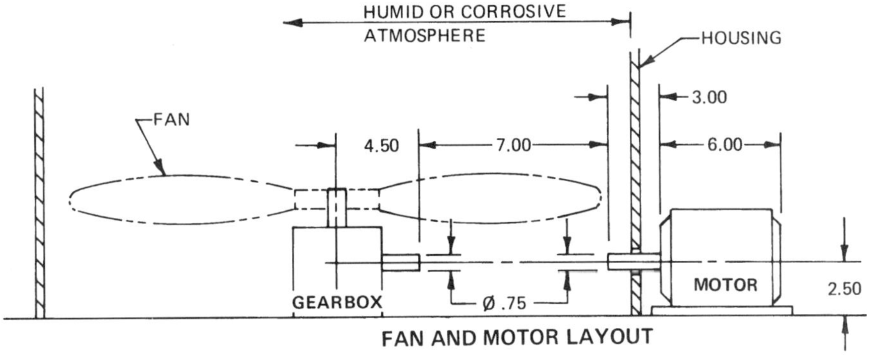

FAN AND MOTOR LAYOUT

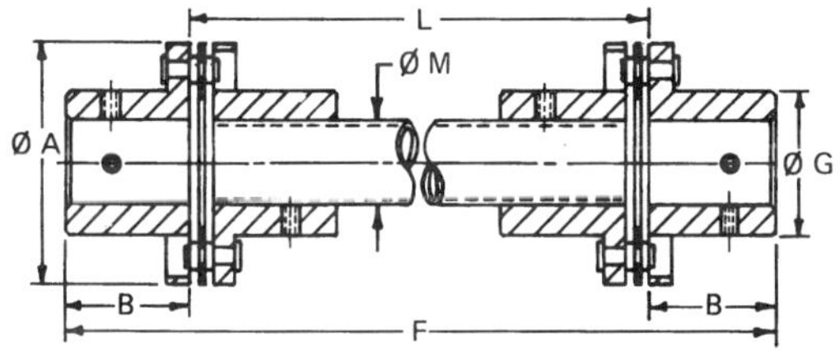

NO.	SHAFT Ø	ØA	B	F	ØG	L	ØM
37 FN	.375	1.50	.70	VARIES WITH DIMENSION L	.88	TO SUIT YOUR REQUIREMENTS	.375
50 FN	.50	1.62	1.00		1.00		.50
62 FN	.62	1.75	1.10		1.38		.62
75 FN	.75	2.00	1.20		1.50		.75

FLEXIBLE COUPLING DATA

COUPLINGS AND FLEXIBLE SHAFTS— FAN AND MOTOR LAYOUT	REFERENCE UNIT 21-1	NAME COURSE DATE	DRAWING V-1

Complete the gearbox assembly below. The shafts for the gearbox are supported by two plain bearings press-fitted into the gearbox. Setscrew collars, as shown in the textbook's appendix, are mounted on the shafts to prevent lateral movement. From manufacturers' catalogs, select suitable 16 DP 20° spur gears that revolve the smaller shaft four times as fast as the larger shaft. Lock the gears to the shafts using square or flat keys. The scale is 1:1.

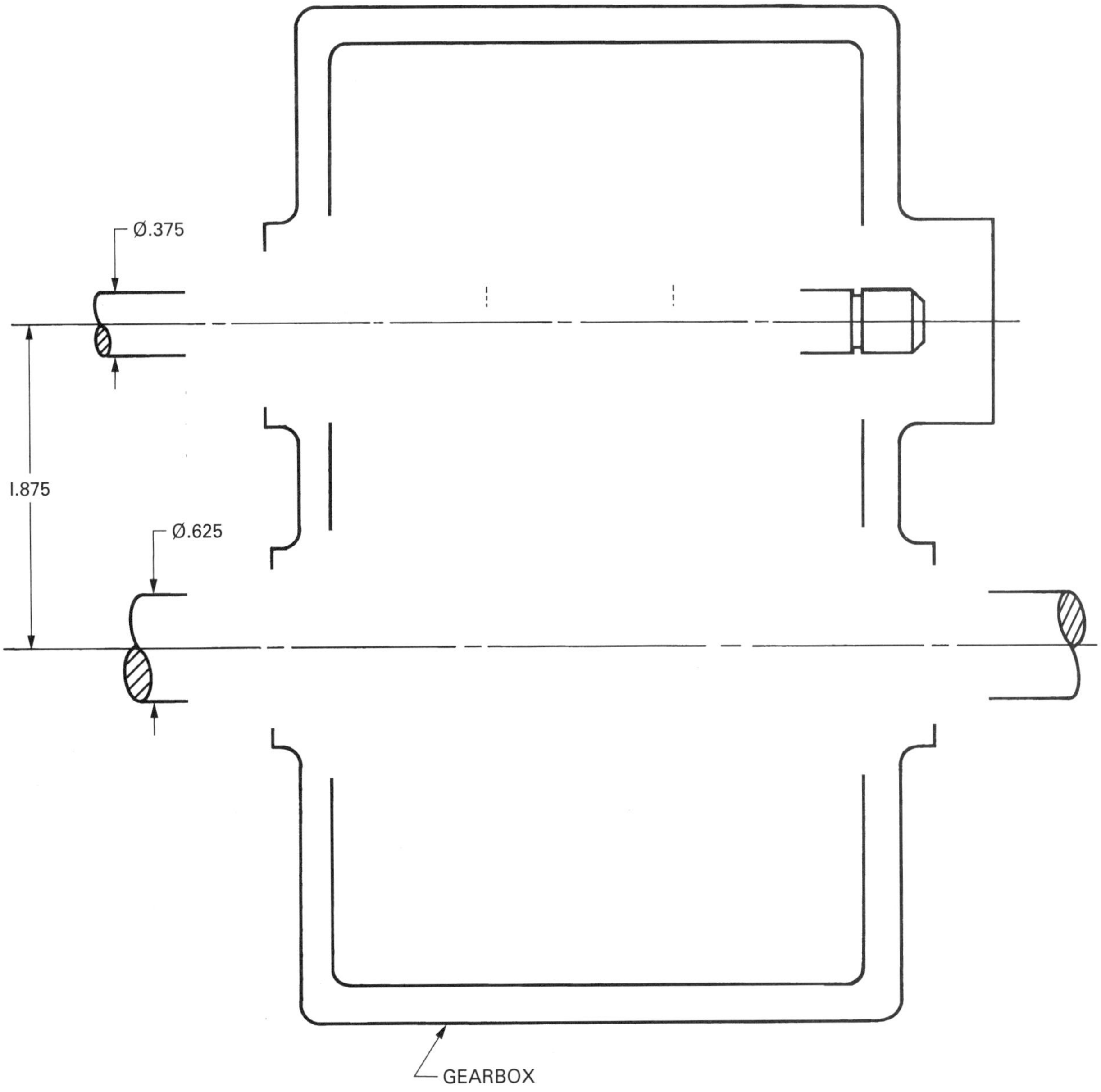

BEARINGS— GEARBOX	REFERENCE UNIT 21–2	NAME COURSE DATE	DRAWING V–2

On a blank sheet, complete this gearbox assembly drawing. The scale is 1:1. Gear 1 and the shaft are cast as a single unit. Gears 2, 3, 6, and 7 are fastened to their respective shafts by keys and are held in location by retaining rings. Gears 4, 5, and 8 are formed as one part that slides along the lay-shaft meshing with gear 3, 6, or 7. Retaining rings located at each end of this sliding-gear assembly locate it in the three positions, and a key locks the assembly to the shaft. Radial ball bearings are positioned at points A and B on each shaft. Each shaft will have one floating and one fixed outer-ring mounting. The gear end of the primary shaft must be designed to house bearing A of the main shaft. Refer to the manufacturers' catalogs or the appendix for standard parts.

SHAFT	A	B	GEAR DATA								
PRIMARY	.394	.591	PITCH=10 20° SPUR GEAR FACE WIDTH=.60								
MAIN SHAFT	.394	.787	GEAR	1	2	3	4	5	6	7	8
LAY SHAFT	.591	.591	N	20	40	25	35	30	30	35	25

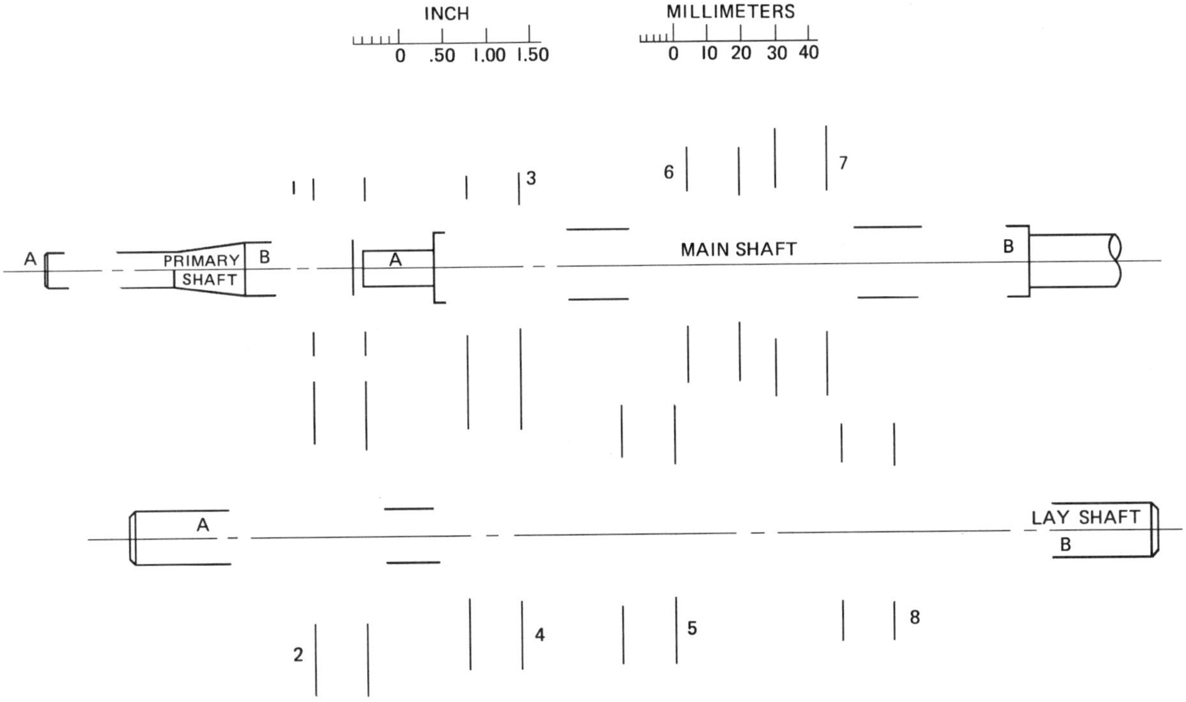

ANTIFRICTION BEARINGS— GEARBOX	REFERENCE UNIT 21-3	NAME COURSE DATE	DRAWING V-3

Make a one-view assembly drawing of the adjustable shaft support from the details shown. Show the bearing housing in its lowest position and a phantom outline of the bearing housing in its highest position. Show only those dimensions that would be used for catalog purchases. The scale is 1:1.

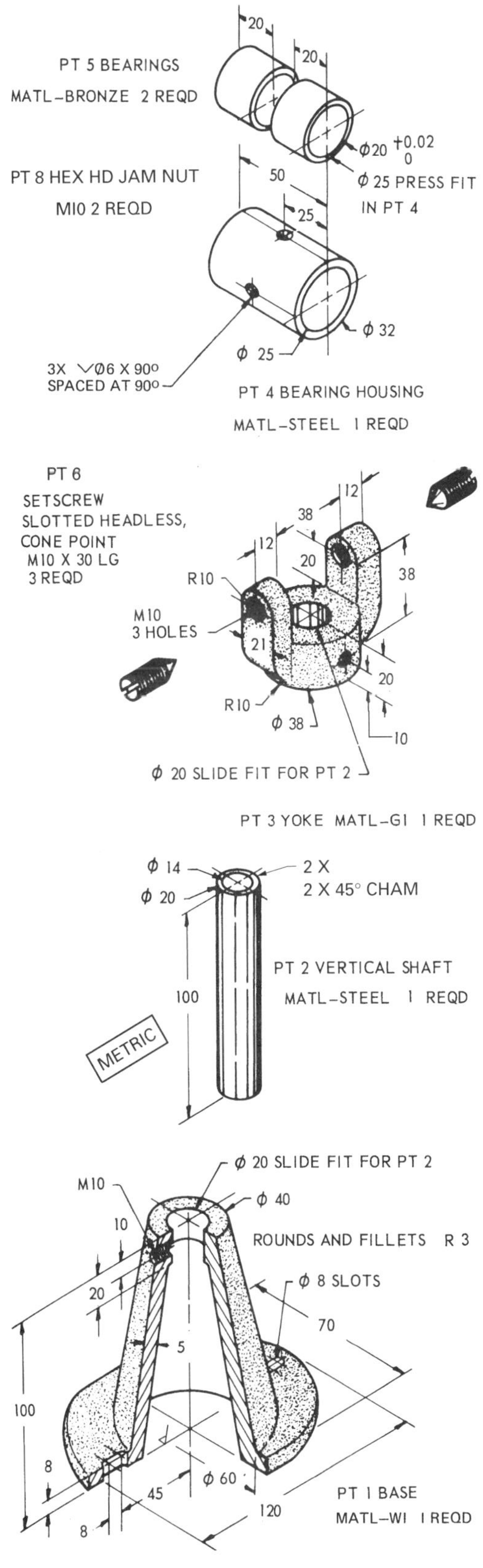

PT 7 SETSCREW
SLOTTED HEADLESS
HEADER POINT M10 X 15 LG
2 REQD

NOTE– DIAMETERS SHOWN FOR SLIDE AND PRESS FITS ARE NOMINAL DIMENSIONS

PREMOUNTED BEARINGS— ADJUSTABLE SHAFT SUPPORT	REFERENCE UNIT 21–4	NAME _ COURSE _ DATE _	DRAWING V–4

Complete the radial oil seal assembly shown below. The inner ring of the tapered roller bearing is held laterally on the shaft by a shaft shoulder. A cover plate bolted to the housing by four cap screws has a stepped shoulder the same diameter as that of the outside diameter of the bearing. This shoulder serves two purposes: it locks the outer ring of the bearing in position, and it locates the cover plate radially on the shaft. The cover plate has a recess to accommodate a metal-cased radial seal. The shaft diameter for the oil seal should be slightly smaller than the diameter of the shaft for the bearing. The outside face of the cover plate and the oil seal should be flush. The scale is 1:1.

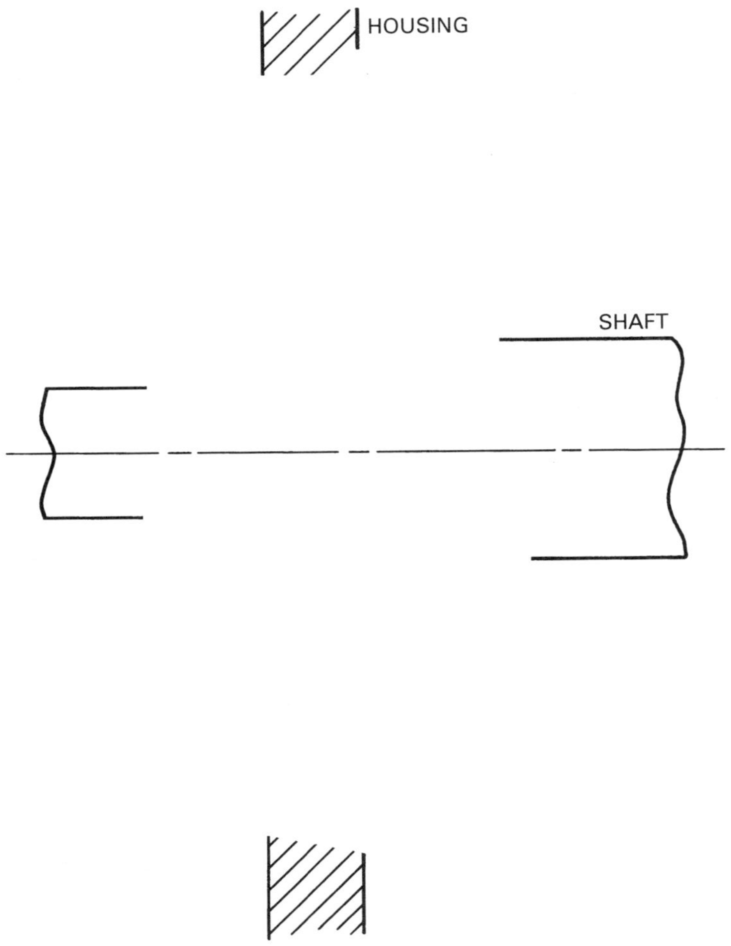

Complete the oil ring seal assembly shown below. A magnetized ring press-fitted into the housing holds the mating ring firmly on the shaft element by magnetic force. The carbon ring in the face of the mating ring, in balanced contact with the lapped surface of the magnet, forms a permanent, self-adjusting face seal. O-rings in both the mating ring and the magnetic shaft element (between the element and housing) prevent leaking of confined fluids. The scale is 1:1.

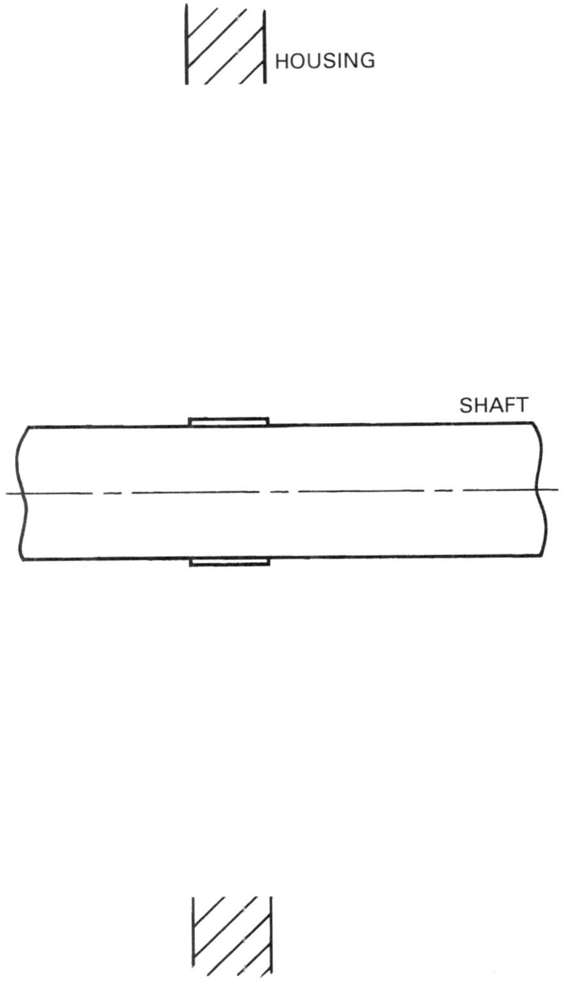

LUBRICATING AND RADIAL SEALS	REFERENCE UNIT 21–5	NAME COURSE	DATE	DRAWING V–5

Complete the cylinder head cap assembly. The cylinder head cap is fastened to the cylinder head by four Ø.375 hex-head cap screws equally spaced. Locking is done with lock-washers. Spotfacing on the cast head cap is required because of the rough finish of the casting. An O-ring provides the seal. The scale is 1:1.

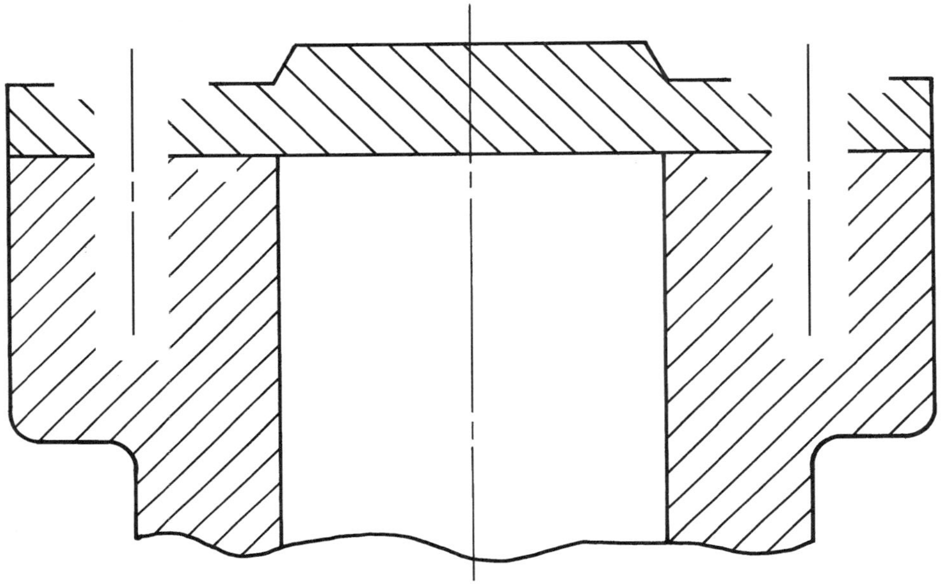

Complete the flanged pipe coupling. The flanges are fastened by hex bolts, nuts, and lock-washers. Alignment is done with a tongue-and-groove joint, similar to that shown in Fig. 21–6–2C, and a gasket positioned in the groove provides the seal. The scale is 1:1.

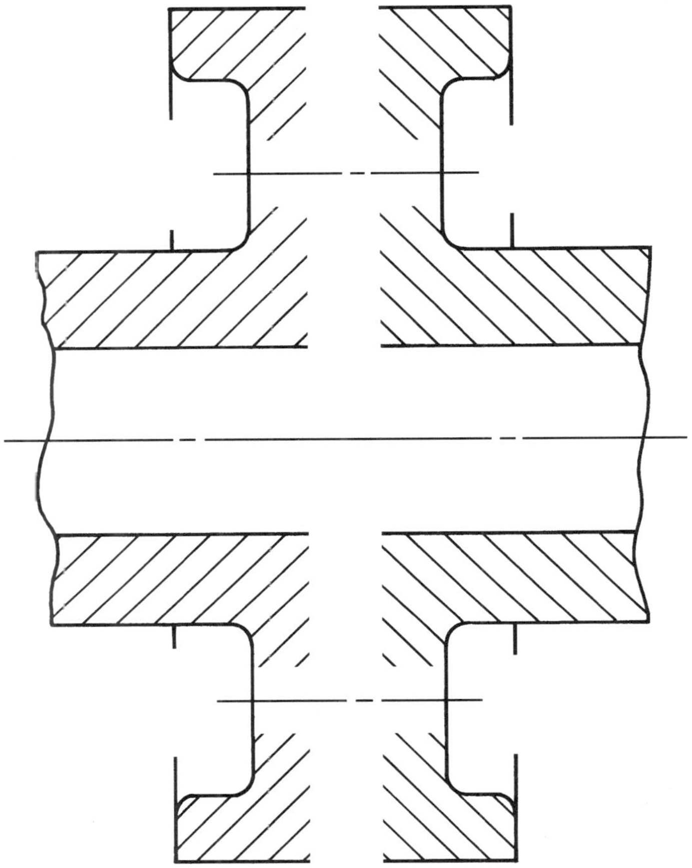

STATIC SEALS	REFERENCE UNIT 21–6	NAME _ COURSE _ DATE _	DRAWING V–6

In the space below, draw the displacement diagram for a cam given the following information: rise 1.60 in. in 150° with harmonic motion, dwell for 45°, drop 1.60 in. in 120° with uniformly accelerated and retarded motion, dwell for the remainder. Displacement diagram size—1.60 in. high × 8.00 in. long.

In the space below, draw the displacement diagram for a cam given the following information: rise 20 mm in 90° with modified uniform motion, dwell for 60°, rise 13 mm in 45° with harmonic motion, drop 33 mm in 120° with parabolic motion, dwell for the remainder. Displacement diagram size—33 mm high × 200 mm long.

CAMS— CAM MOTIONS	REFERENCE UNIT **22–1**	NAME COURSE DATE	DRAWING **W–1**

In the space provided, draw the displacement diagram for a plate cam that will activate a Ø.50 in. roller follower as follows: rise 1.50 in. in 180° with harmonic motion, dwell 30°, drop 1.50 in. in 120° with modified uniform motion, dwell for the remainder.

Draw the front view of this cam given the following additional information: prime circle = Ø3.00 in., shaft = Ø1.00 in., hub = Ø1.75 in., keyway to suit. The scale is 1:1.

When the cam drawing is finished, complete the chart showing the radial displacements for the angular displacements shown. The radial measurements are taken from the prime circle.

ANGULAR DISPLACEMENT (DEGREES)	RADIAL DISPLACEMENT FROM PRIME CIRCLE
0	
15	
30	
45	
60	
75	
90	
105	
120	
135	
150	
165	
180	
195	
210	
225	
240	
255	
270	
285	
300	
315	
330	
345	

0°

CAM DISPLACEMENT DIAGRAM

CAMS— PLATE CAMS	REFERENCE UNIT 22–2	NAME COURSE DATE	DRAWING W–2

In the space provided, draw the displacement diagram for a face cam that will activate a Ø.50 in. roller as follows: rise 1.20 in. in 150° with harmonic motion, dwell 30°, drop 1.20 in. in 120° with parabolic motion, dwell for the remainder. Draw the front and side views of the cam given the following additional information: prime circle = Ø3.00 in., outside diameter of cam = 6.50 in., cam thickness = 1.00 in., groove depth = .38 in., shaft = Ø1.00 in., hub = Ø1.75 × 1.50 in., keyseat to suit. The scale is 1:2.

When the cam drawing is finished, complete the chart on the right showing the radial displacements for the angular displacements shown. The radial displacements are taken from the prime circle.

ANGULAR DISPLACEMENT (DEGREES)	DISTANCE FROM PRIME CIRCLE
0	
15	
30	
45	
60	
75	
90	
105	
120	
135	
150	
165	
180	
195	
210	
225	
240	
255	
270	
285	
300	
315	
330	
345	

0°

DISPLACEMENT DIAGRAM

CAMS— POSITIVE MOTION CAMS	REFERENCE UNIT 22–3	NAME / COURSE / DATE	DRAWING W–3

Draw the top and front views of a drum cam given the following information: rise 2.00 in. in 120° with parabolic motion, dwell for 60°, drop 2.00 in. in 150° with modified sine motion, dwell for remainder. Roller follower = Ø.70 in., cam = Ø4.00 in. × 4.80 in. long, follower groove = .50 in. deep. The scale is 1:2. Use your judgment for dimensions not given. Show the full development of the cam, which will serve as a displacement diagram. When the cam drawing is finished, complete the chart showing the radial displacements taken from the base line.

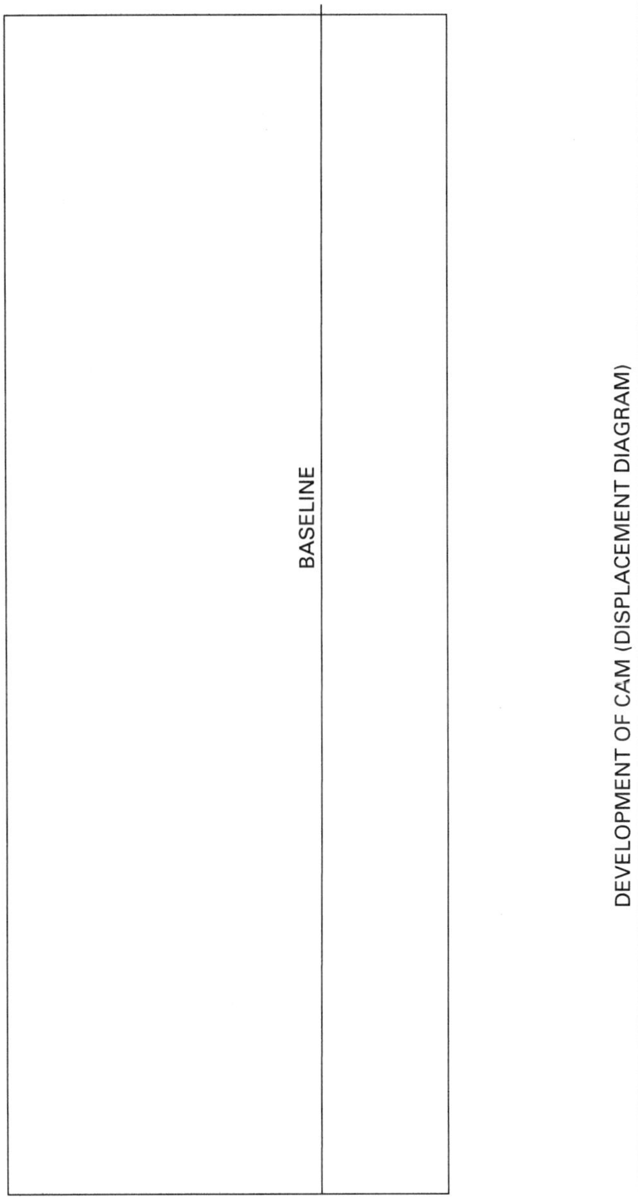

ANGULAR DISPLACEMENT FROM TIMING HOLE	DISPLACEMENT FROM BASELINE
0	
15	
30	
45	
60	
75	
90	
105	
120	
135	
150	
165	
180	
195	
210	
225	
240	
255	
270	
285	
300	
315	
330	
345	

CAMS— DRUM CAM	REFERENCE UNIT 22–4	NAME / COURSE / DATE	DRAWING W–4

DIMENSIONS			MODEL	A	B	C	D	DRIVER					WHEEL		
								E	F	G	H	K	L	M	N
6 STATIONS	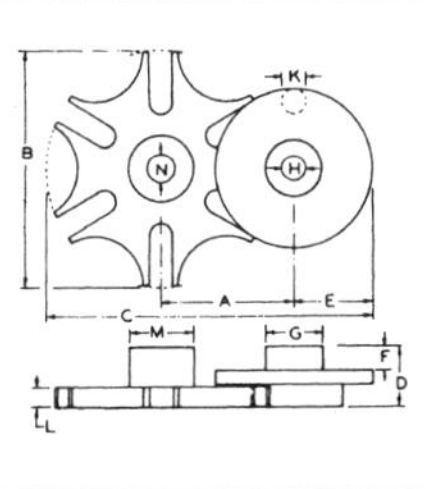	INCH	6S5	5.00	8.75	12.31	1.62	2.94	.50	1.50	.75	.88	.62	2.00	1.00
		INCH	6S5.5	5.50	9.62	13.50	1.62	3.14	.50	1.50	.75	.88	.62	2.00	1.00
STATIONS		INCH	6S6	6.00	10.50	14.75	1.75	3.50	.50	1.50	.75	1.00	.75	2.00	1.00
		METRIC	6S75	75	134	188	35	46	12	40	16	16	12	45	20
TIME RATIO:		METRIC	6S90	90	160	222	35	52	12	40	16	16	12	45	20
120° INDEX		METRIC	6S100	100	180	250	38	60	12	40	20	20	16	48	22
240° DWELL		METRIC	6S115	115	204	274	38	67	12	40	20	20	16	48	22

In the space below, make a two-view drawing of the indexing drive 6S75. Use your judgment for dimensions not given. Add suitable keyseats. The scale is 1:1. Draw the front view in full section. Do not dimension.

INDEXING— INDEXING DRIVE	REFERENCE UNIT 22–5	NAME COURSE	DATE	DRAWING W–5

In the locations below, lay out the two linkages shown and plot the paths at 15° intervals of points A, B, C, and D. The scale is 1:2. The centers of the two Ø2.75 are indicated on the drawing area.

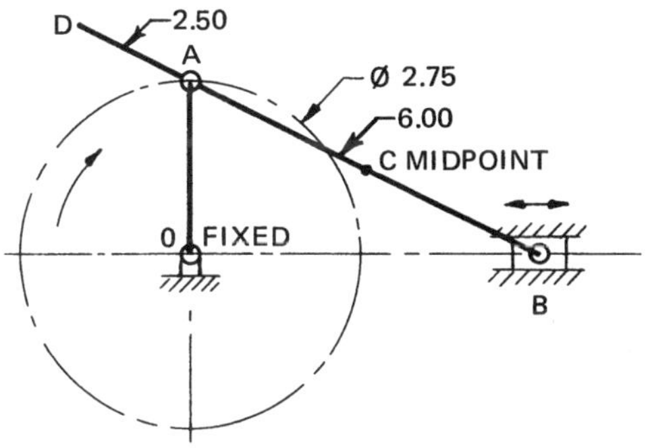

(A) SIMPLE CRANK

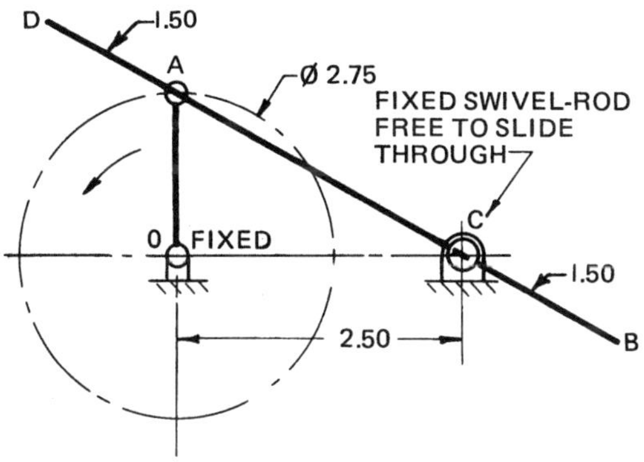

(B) CRANK WITH SLIDING ROD

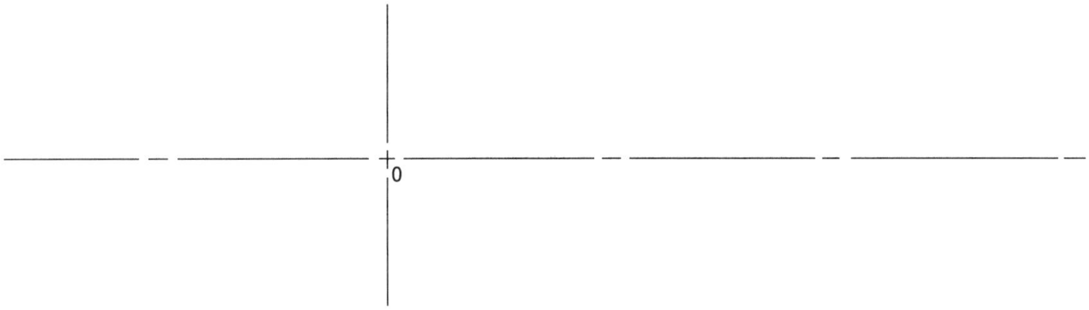

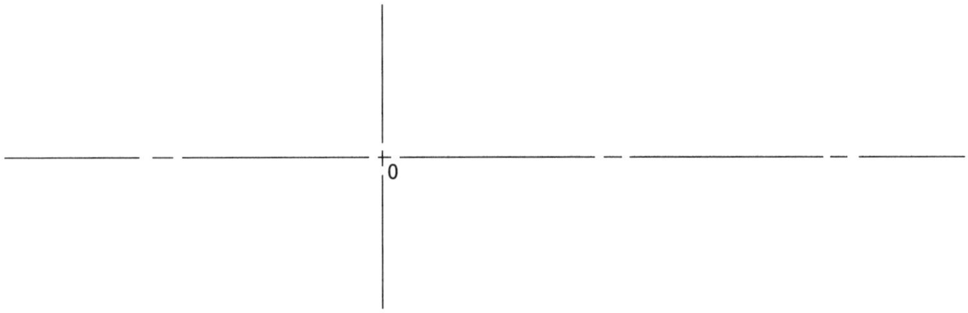

LINKAGES— SIMPLE CRANK MECHANISMS	REFERENCE UNIT 22–6	NAME COURSE DATE	DRAWING W–6

In the space below, lay out a one-view drawing of the ratchet design shown. Two pawls are used, a drive pawl as shown and a holding pawl held in position by a spring. Using crank rotation positions every 22.5°, plot the path of the end of the drive pawl. Use your judgment for dimensions not shown. The scale is 1:2. The center of the ratchet is located on the drawing area.

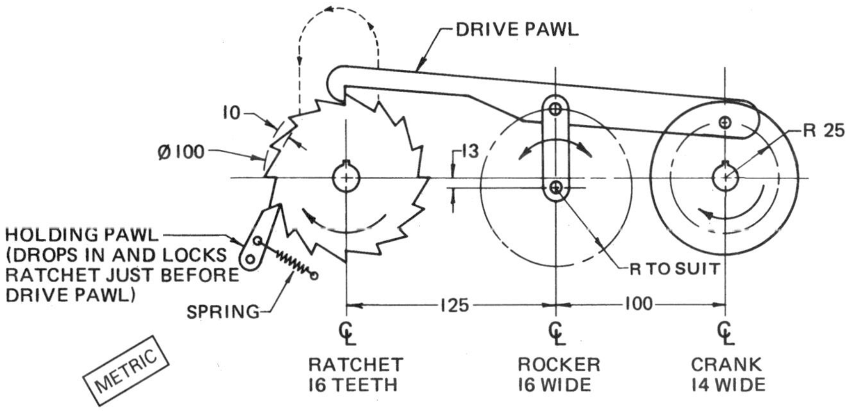

RATCHET WHEELS— RATCHET AND CRANK MECHANISM	REFERENCE UNIT 22–7	NAME _ COURSE _ DATE _	DRAWING W–7

In the space below, make a development drawing, complete with bending instructions and dimensions, of the wall tray. The scale is 1:2. The bottom of the wall tray development is located on the drawing area. Allow 6 mm for safe edges and seams.

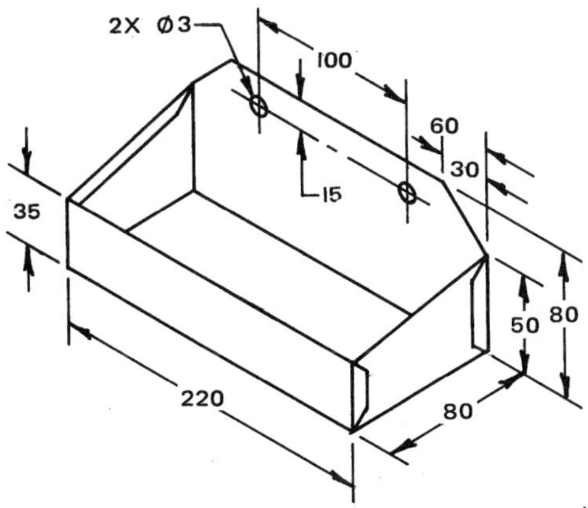

SURFACE DEVELOPMENTS— WALL TRAY	REFERENCE UNIT **23–1**	NAME COURSE DATE	DRAWING **X–1**

Make a development drawing of the hexagon box shown. The scale is 1:1. After the development drawing has been checked by your instructor, add suitable seams and joint allowances. Then cut out the development, score on the bend lines, and form and glue the box together. The bottom of the box is located on the drawing area and is hinged at D–E. The top is hinged at D–E. The seam is at A. All seams are .25 in. wide and placed on the inside.

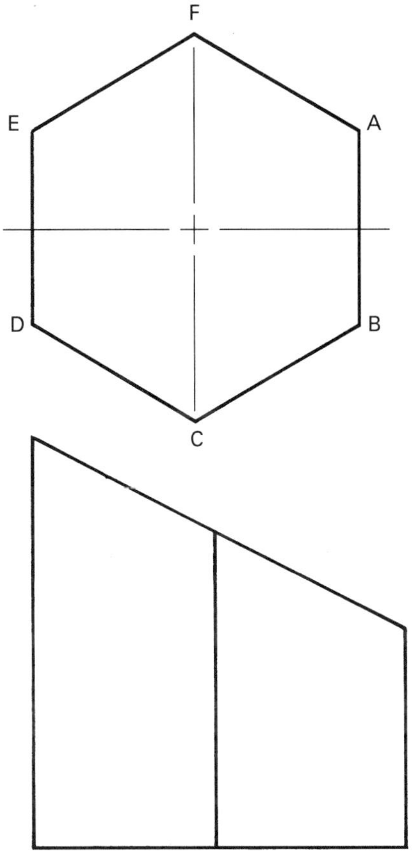

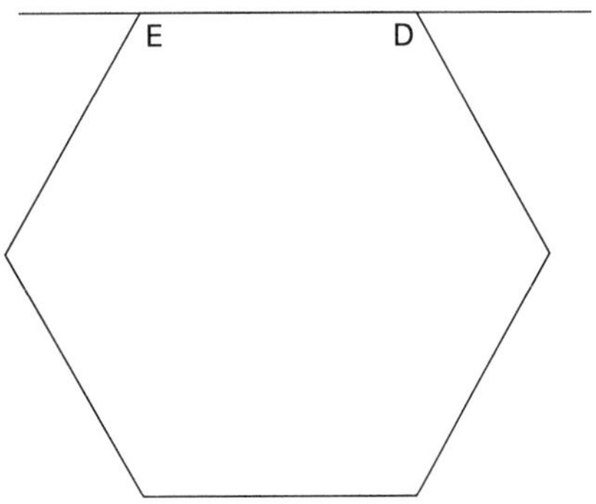

THE PACKAGING INDUSTRY— HEXAGON BOX	REFERENCE UNIT 23–2	NAME COURSE DATE	DRAWING X–2

Make a development drawing of the truncated concentric pyramid shown. The scale is 1:1. Add suitable seams. The bottom of the pyramid is located on the drawing area and is hinged at 1–2. The top is hinged at A–B.

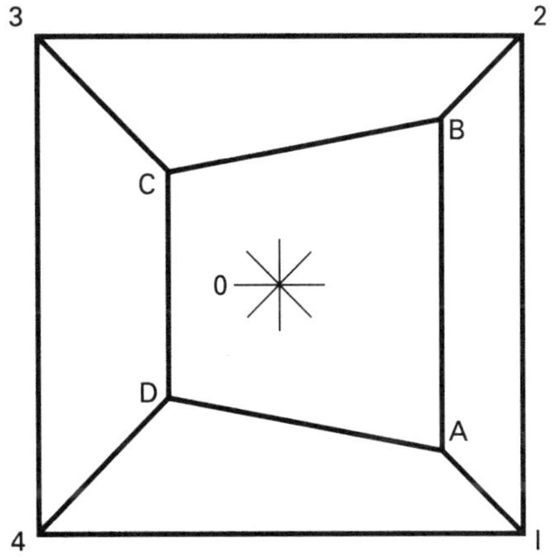

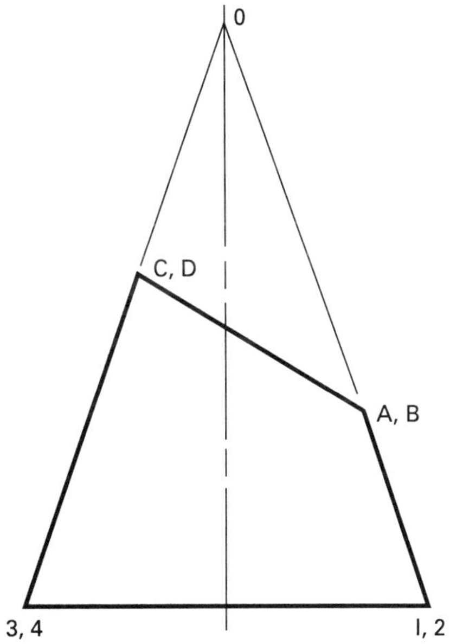

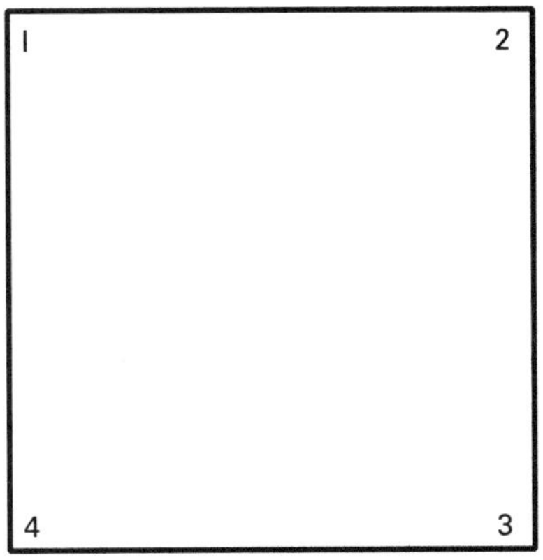

RADIAL LINE DEVELOPMENT— TRUNCATED CONCENTRIC PYRAMID	REFERENCE UNIT **23–3**	NAME COURSE	DATE	DRAWING **X–3**

Make a development drawing of the three-piece elbow in the space below. The scale is 1:1. Add suitable seams. Use construction lines every 30° on the circumference to develop the elbow.

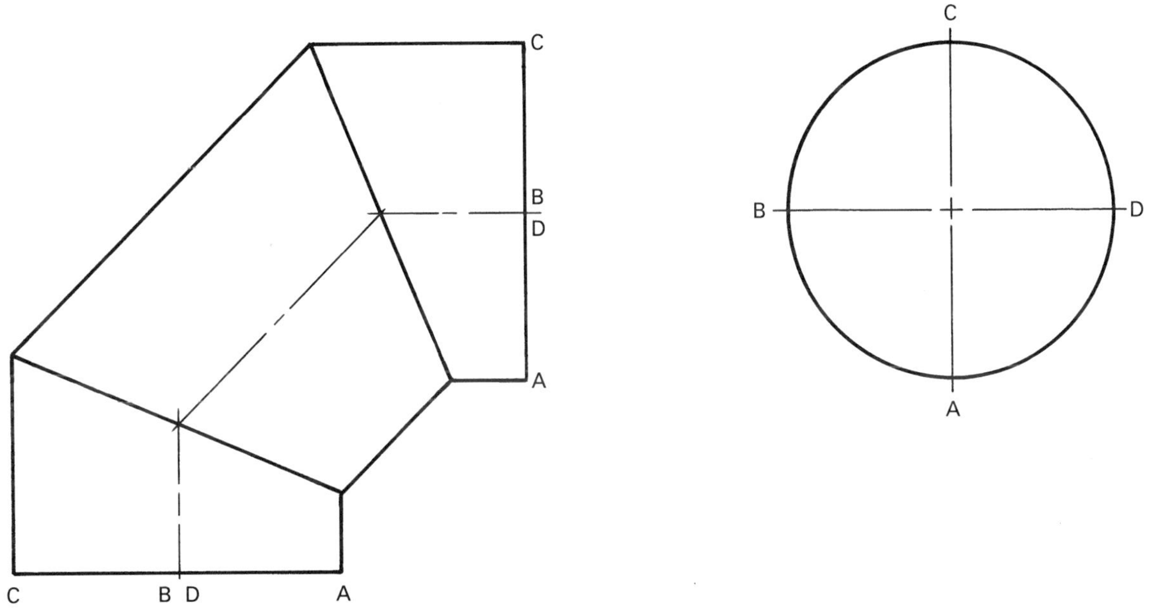

PARALLEL LINE DEVELOPMENT— THREE-PIECE ELBOW	REFERENCE UNIT 23–4	NAME COURSE	DATE	DRAWING X–4

Make development drawings for the two parts of the funnel shown. Add suitable seams. The scale is 1:1. The apexes of the two cones are located on the drawing surface.

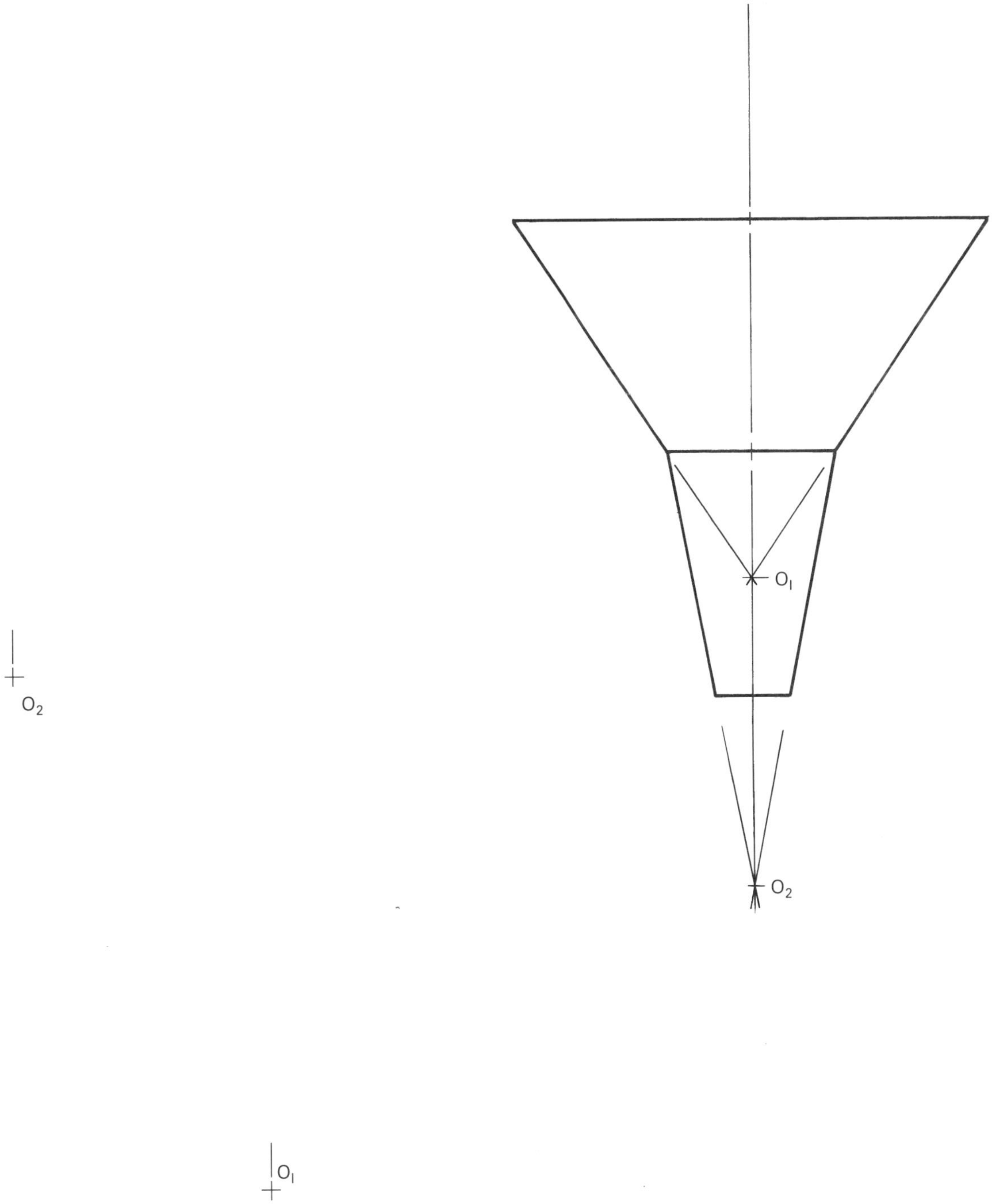

RADIAL LINE DEVELOPMENT—

CONICAL SURFACES

REFERENCE UNIT	NAME	DRAWING
23–5	COURSE	DATE	X–5

Make a development drawing of the transition piece shown. Add suitable seams. The scale is 1:1. Points B and C for flat surface B–C–2 are located on the drawing area. Use development lines every 30° on the diameter.

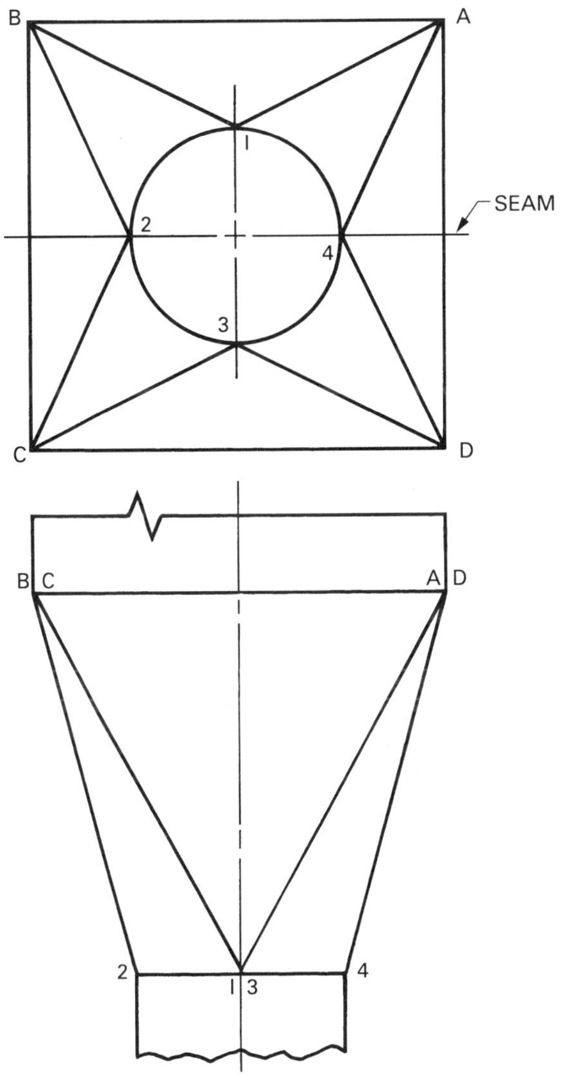

B C

DEVELOPMENT OF TRANSITION PIECES	REFERENCE UNIT 23–6	NAME ___ COURSE ___ DATE ___	DRAWING X–6

In the space below, make a half development of a Ø3.00 in. sphere using the gore method. Show six of twelve sections. The scale is 1:1.

DEVELOPMENT OF A SPHERE— GORE METHOD	REFERENCE UNIT 23–7	NAME _ COURSE _ DATE _	DRAWING X–7

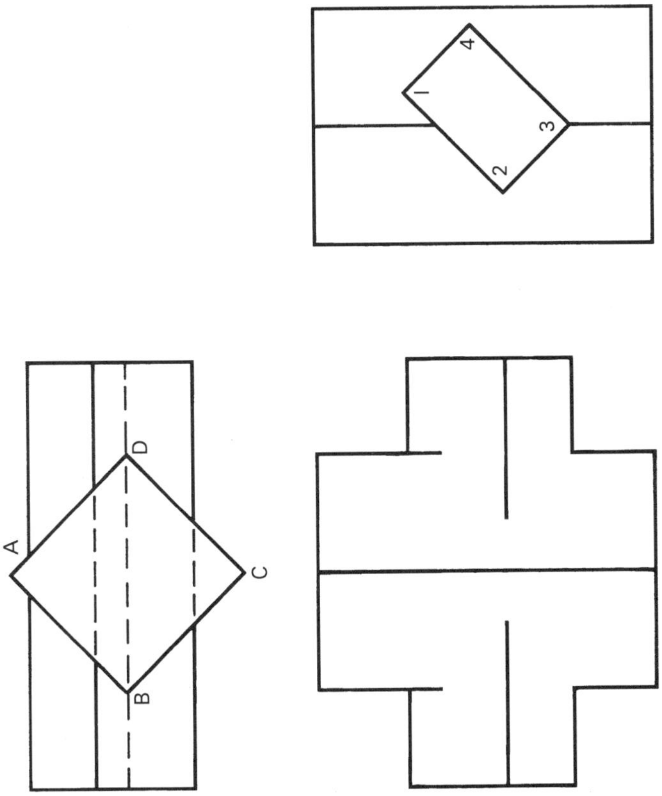

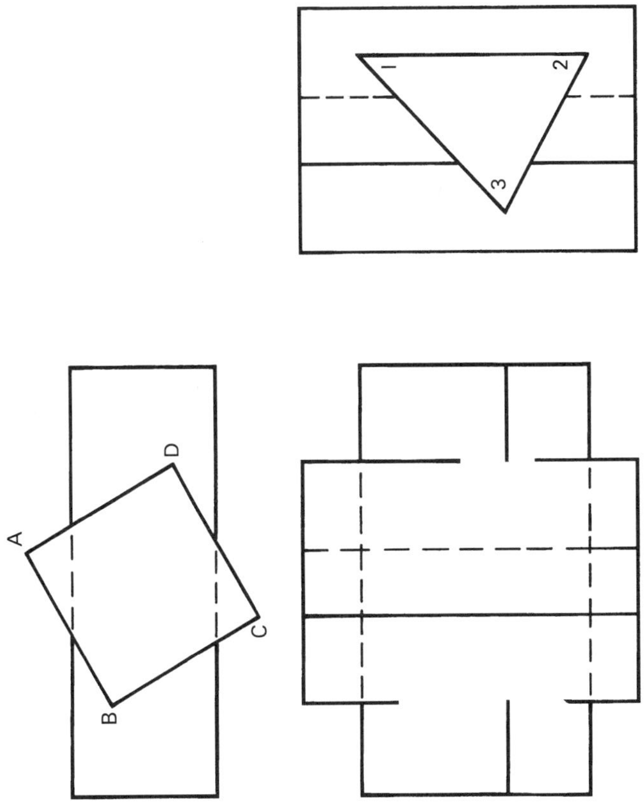

Complete the views and make development drawings of the vertical prisms. The horizontal prisms go through the vertical prisms and the ends are open. Do not show laps or seams. Show the letters and numbers on all views.

INTERSECTIONS OF FLAT SURFACES	REFERENCE UNIT **23–8**	NAME COURSE DATE	DRAWING **X–8**

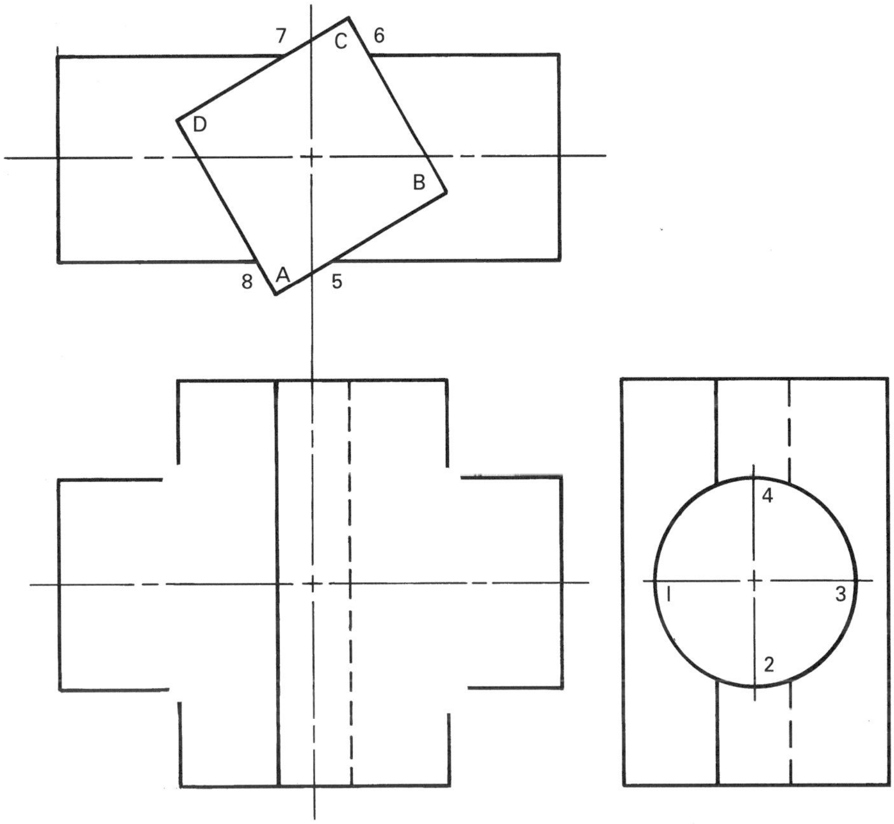

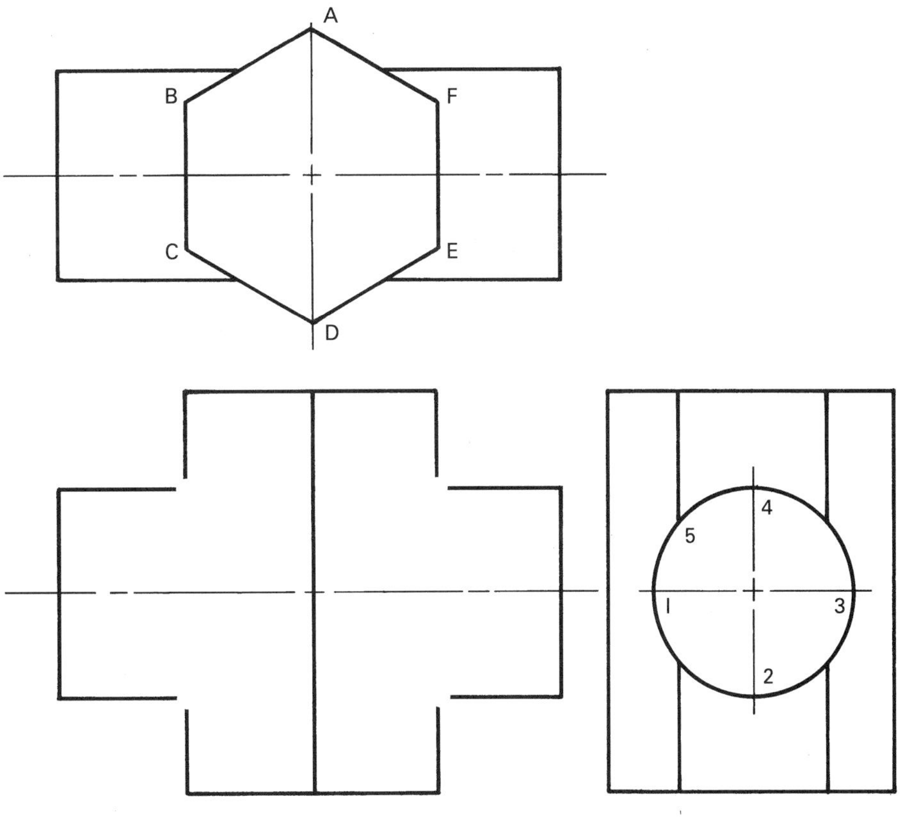

Complete the lines of intersection on all views. The cylinders pass through the vertical prisms. Show the letter and numbers on all views.

INTERSECTIONS OF CYLINDRICAL SURFACES	REFERENCE UNIT 23–9	NAME COURSE DATE	DRAWING X–9

A.

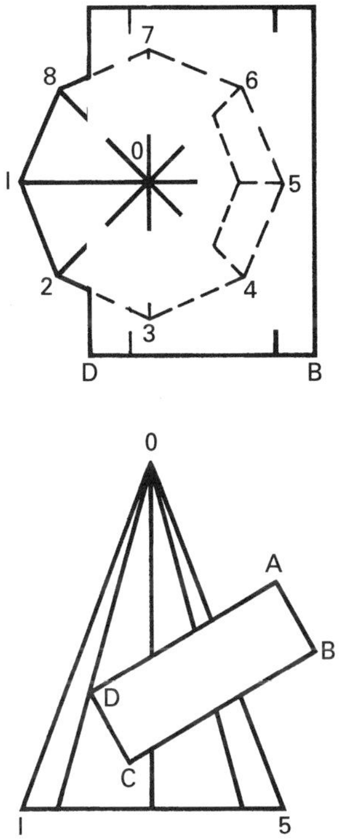

B.

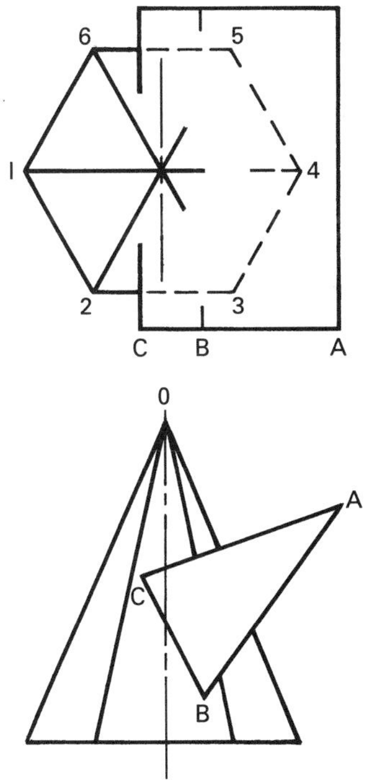

Complete the lines of intersection on the partially completed views, and make a development drawing of the conical prisms, rectangle, and triangle. The seams are at lines 0–1 and A. Do not show seam allowances.

INTERSECTIONS OF PRISMS	REFERENCE UNIT 23–10	NAME COURSE DATE	DRAWING X–10

Using a blank sheet, prepare a working drawing and a developed blank of the bracket and connector shown in the engineer's sketch of the swivel hanger. Place only the overall dimensions on the blank. Scale 1:2. (See Table 23-11-13, Bending allowances for each degree of bend, page 882 in the text.)

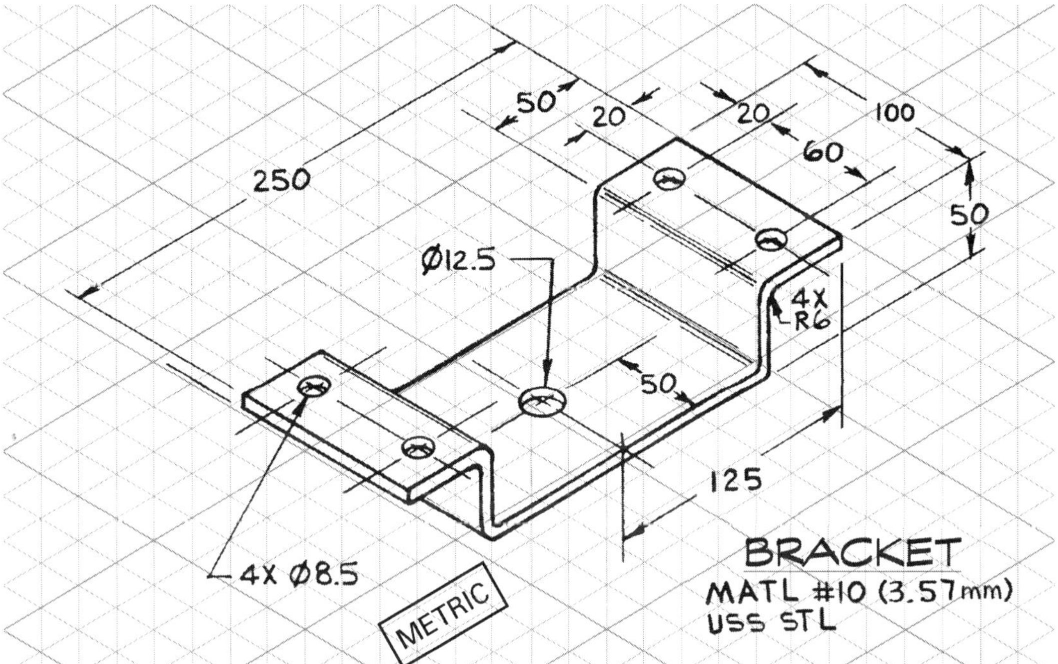

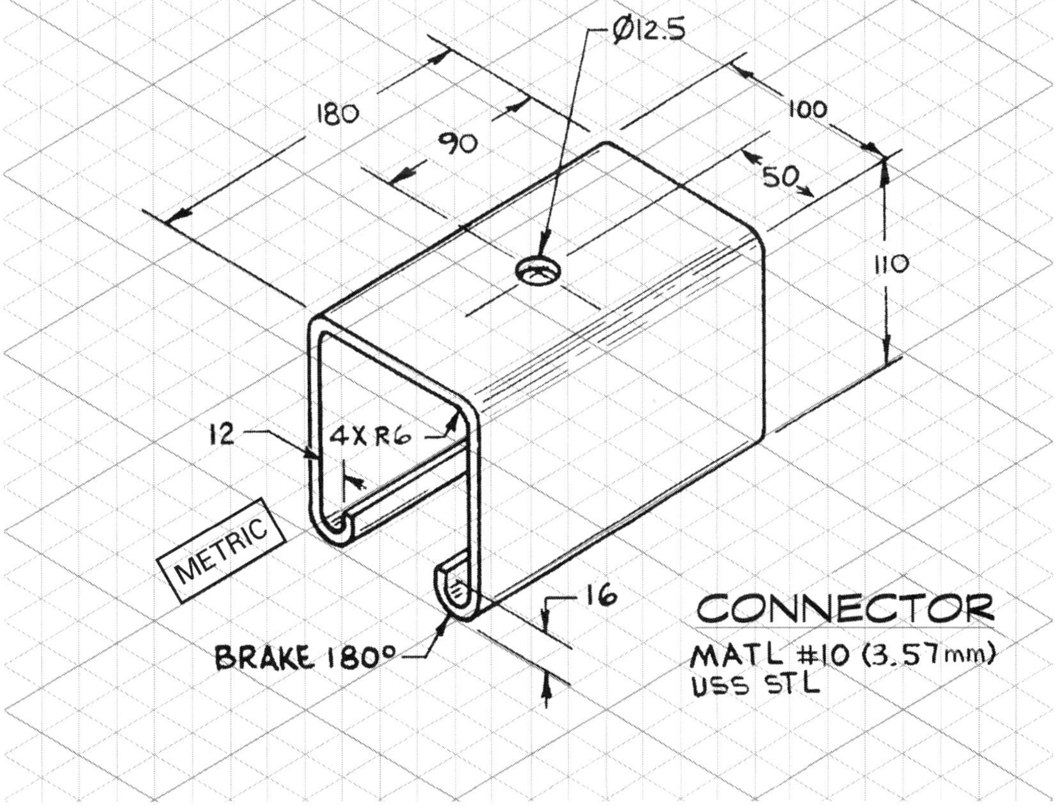

STAMPINGS– SWIVEL HANGER	REFERENCE UNIT 23–11	NAME COURSE DATE	DRAWING X–11

On one of the blank sheets from the back of this book, make a three-view drawing of the fuel oil supply system. Include with the drawing an item list calling for all the pipe fittings and valves. The scale is $\frac{1}{8}$ in. = 1 ft. Scale the pictorial drawing to obtain center-to-center distances and lengths of pipe.

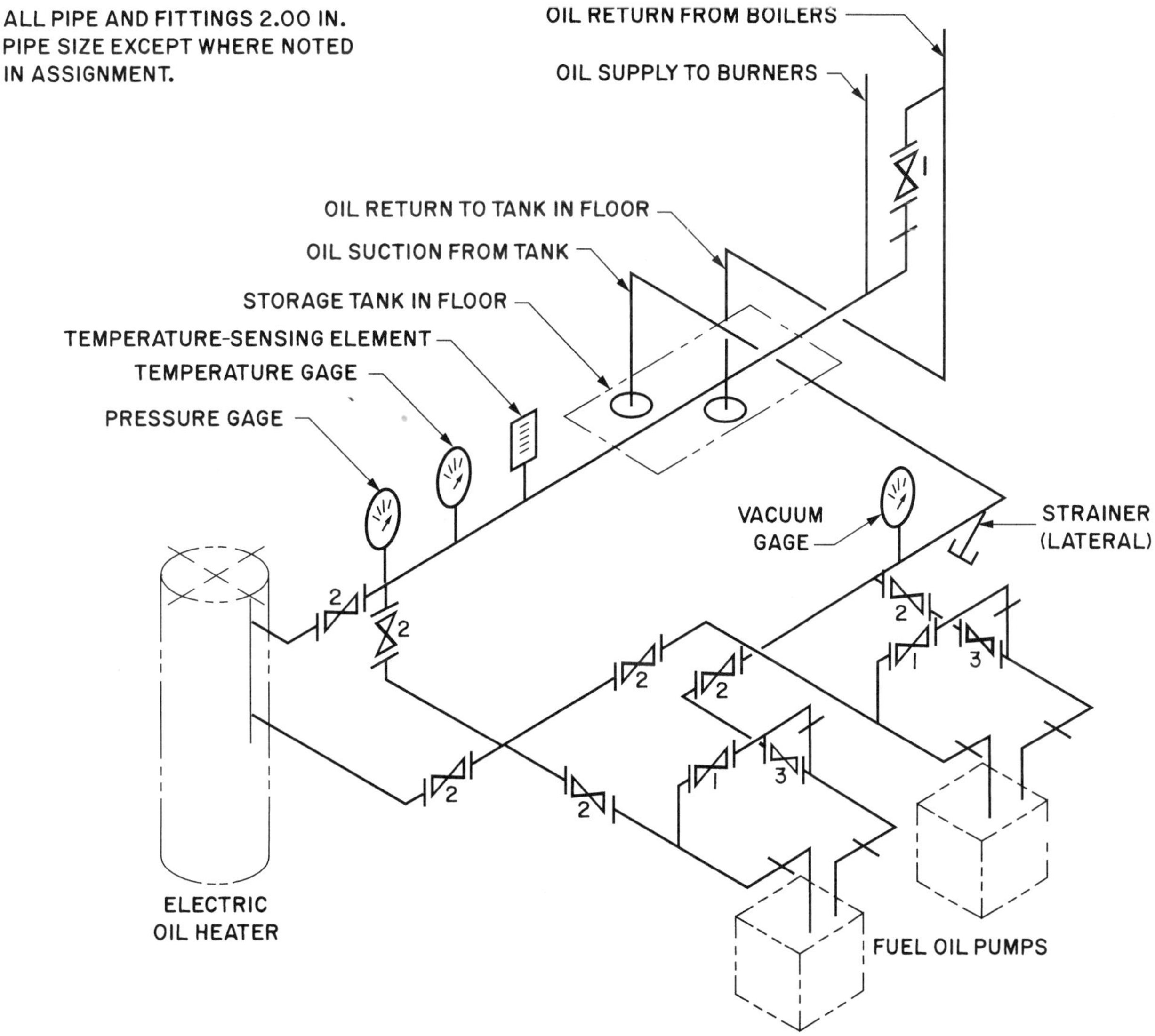

ORTHOGRAPHIC PIPING DRAWINGS— FUEL OIL SUPPLY SYSTEM	REFERENCE UNIT 24–1	NAME COURSE	DATE	DRAWING Y–1

For starting diesel engines, the most dependable and widely used method is an air system of the type illustrated here. With this starting system hooked up to diesel installations generating power and heat for buildings such as factories, hotels, large apartment houses, and stores, interruptions through failure of electric supply or storage cells are avoided.

Safety valves are provided for the compressor and the air storage tanks. Check valves are installed on the air storage tank feed lines and the compressor discharge lines to prevent accidental discharge of the tanks.

Piping is arranged so that the compressor will fill the storage tanks and/or pump directly to the engines. Any of the three storage tanks can be used for starting, and pressure gages indicate their readiness. The engines are fitted with quick-opening valves to admit air speedily at full pressure and shut it off the instant rotation is obtained. A bronze globe valve is installed to permit complete shutdown of the engine for repairs and for regulation of the air flow. Drains are provided at low points to remove condensate from the air storage tanks, lines, and engine feeds.

Globe valves are recommended throughout this hookup except on the main shutoff lines, where gate valves are used because of infrequent operation.

All valves connected to horizontal pipelines 6 ft (1800 mm) or higher above the floor will have their spindles located on the underside for ease of operation. Other horizontally positioned valves will have their spindles in the upright position.

All valves connected to vertical pipes will have their valve spindles oriented to the front of the drawing.

Flanges are to be located on the top pipeline near the three starting air tanks and near the air compressor for assembly and disassembly purposes. Flanges are located on the starting diesel engines.

On one of the blank sheets from the back of this book, make an isometric piping drawing of the diesel-engine starting system. The scale is .50 in. = 1 ft. Use the scale shown on this sheet to measure the center distance and lengths of pipe.

On your drawing, include an item list calling for the pipe fittings and valves. All the fittings are threaded, and 1.50 in. pipe is used throughout.

CODE	VALVE	SERVICE
A	BRONZE GLOBE	AIR STORAGE TANK FEED LINES
B	BRONZE GLOBE	AIR STORAGE TANK DISCHARGE LINES
C	BRONZE GLOBE	DIESEL ENGINE SHUTOFF CONTROL
D	BRONZE GLOBE	AIR COMPRESSOR DISCHARGE
E	BRONZE GLOBE	PRESSURE GAGE SHUTOFF
F	BRONZE GLOBE	DRAIN VALVES
G	SPINDLE GATES	MAIN LINE SHUTOFF
H	BRONZE CHECK	AIR COMPRESSOR CHECK
J	BRONZE CHECK	AIR STORAGE TANK FEED LINES

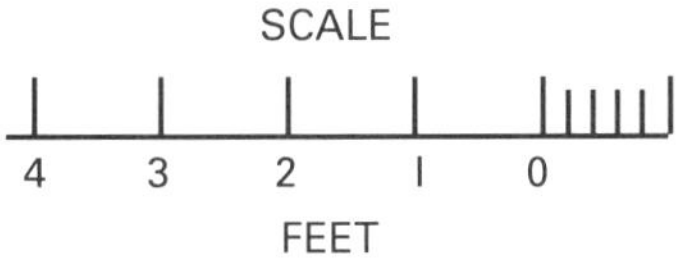

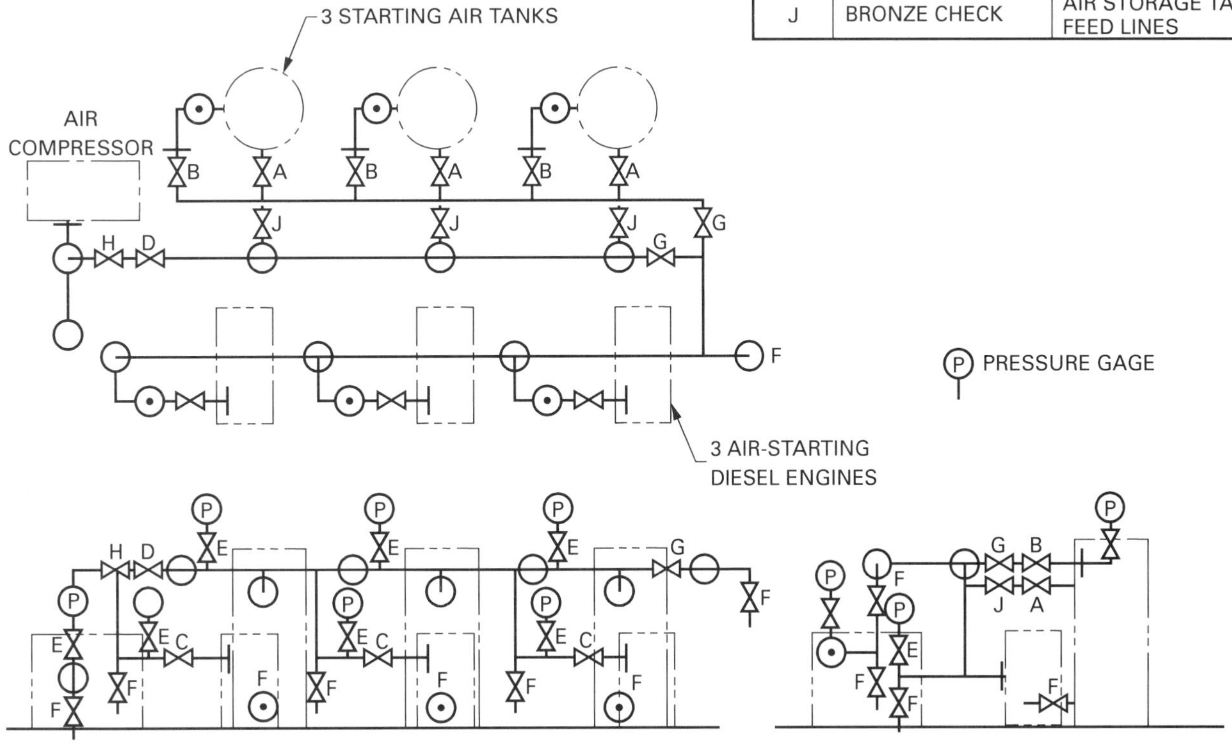

ISOMETRIC PIPING DRAWINGS— DIESEL-ENGINE STARTING SYSTEM	REFERENCE UNIT 24–2	NAME COURSE DATE	DRAWING Y–2

The one-story, taxpayer-type building shown here has been developed and steadily improved as a result of the movement of shopping centers to suburban areas. This type of building, which is multiplying rapidly, is constructed either with or without a basement. It houses retail stores, service establishments, amusement centers, restaurants, and offices.

Heating and plumbing services in such buildings are usually provided by the owner or operator, and for this reason, he or she might give careful consideration to low-cost and trouble-free installations. An oil- or gas-fired steam boiler with automatic control and a separate gas-fired heater for the hot-water supply will generally meet these requirements.

The two-pipe heating system, located in the basement in the installation illustrated, has unit heaters with individual thermostatic controls. Valuable extra floor space is made available for tenants' use because the heaters are hung from the ceiling. Because the heaters in each store or each section of a store are controlled automatically, fuel savings are effected and even heating is ensured.

On one of the blank sheets from the back of this book, make an isometric drawing of the piping layout shown. With the drawing, include an item list calling for all the pipe fittings and valves. Pipe hangers are required for every 8 ft of piping. Direction of flow, level indicators for horizontal piping using the basement floor as zero, indication of pipe runs not in the direction of the coordinate axes, and a drainage slope of 1:20 are to be shown on the drawing. The scale is $\frac{1}{8}$ in. = 1 ft. Use the scale shown on this sheet to measure the center distances and lengths of pipe. Use 1.50 in. pipe.

CODE	VALVE	SERVICE
A	BRONZE GATE	WATER SERVICE SHUTOFF
B	BRONZE GATE	DISTRIBUTION SHUTOFF
C	BRONZE GLOBE	WATER SUPPLY TO BOILER
D	BRONZE SWING CHECK	PREVENT BOILER BACKFLOW
E	BRONZE GLOBE	EMERGENCY BOILER FILL
F	BRONZE GLOBE	DRAINS SHUTOFF
G	BRONZE GLOBE	STEAM SUPPLY TO HEATERS
H	BRONZE GATE	CONDENSATE DRAIN SHUTOFFS
J	BRONZE GLOBE	STEAM MAIN TRAP CONNECTION
K	BRONZE GLOBE	TRAP BYPASS
L	BRONZE GATE	HOT WATER HEATER SHUTOFF
M	BRONZE SWING CHECK	PREVENT WATER HEATER BACKFLOW
N	BRONZE GATE	WATER SUPPLY SHUTOFFS
P	I.B.B.M GATE	STEAM MAIN SHUTOFF
R	BRONZE GATE	RETURN SHUTOFF

SCALE

FEET

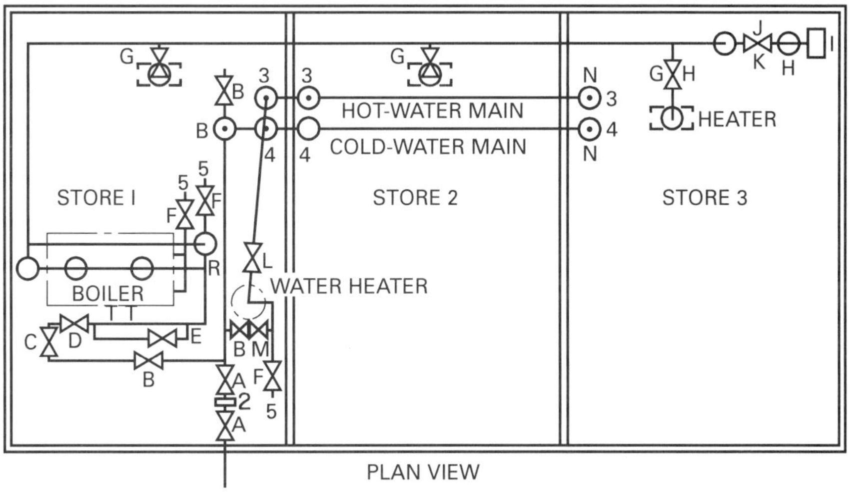

1. FLOAT AND THERMO TRAP
2. WATER METER
3. HOT WATER TO STORE
4. COLD WATER TO STORE
5. DRAIN

PLAN VIEW

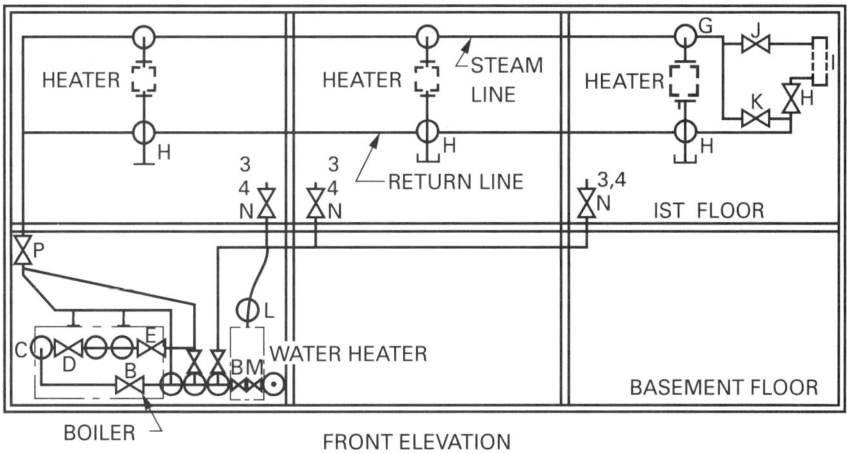

FRONT ELEVATION

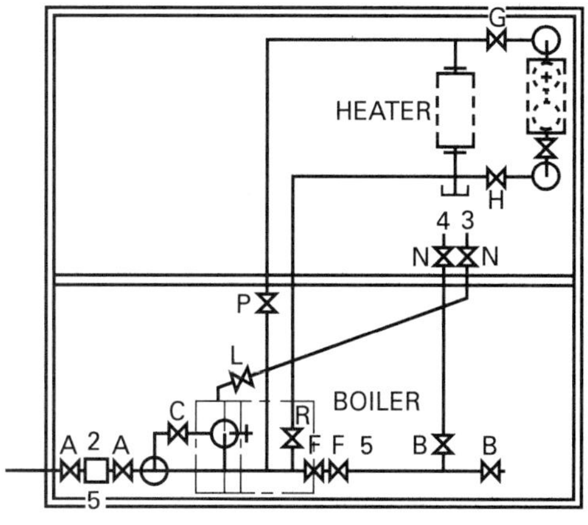

SIDE ELEVATION

SUPPLEMENTARY PIPING INFORMATION —OIL-BURNING PIPING SYSTEM	REFERENCE UNIT 24-3	NAME COURSE DATE	DRAWING Y-3

Calculate the limits, tolerances, and sizes of the beams shown. Refer to Fig. 25–1–8 and structural steel handbooks for the sizes of structural steel shapes.

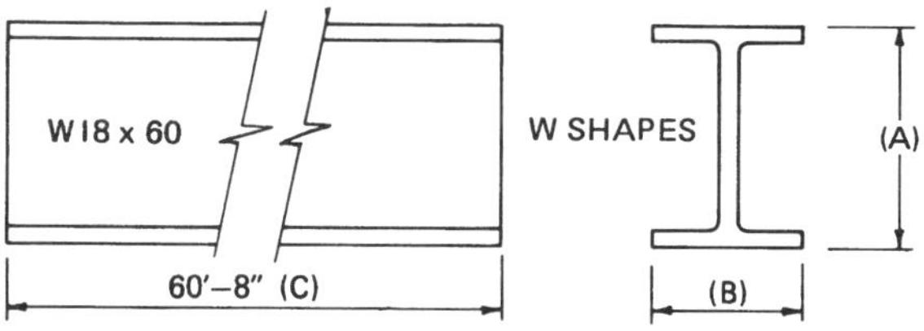

DIM	NOMINAL	TOLERANCE		LIMITS	
				UPPER	LOWER
A		+	−		
B		+	−		
C		+	−		

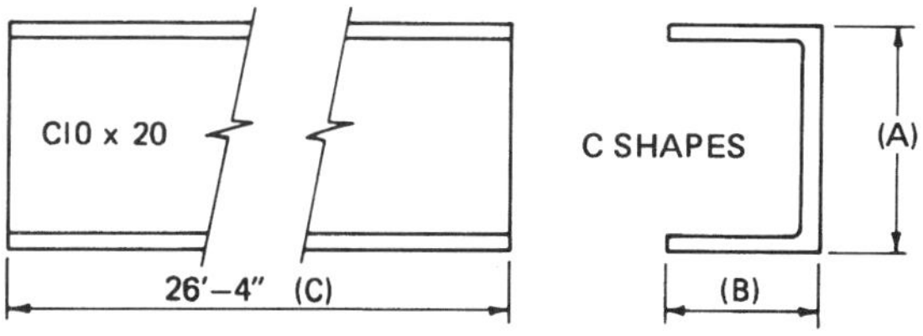

DIM	NOMINAL	TOLERANCE		LIMITS	
				UPPER	LOWER
A		+	−		
B		+	−		
C		+	−		

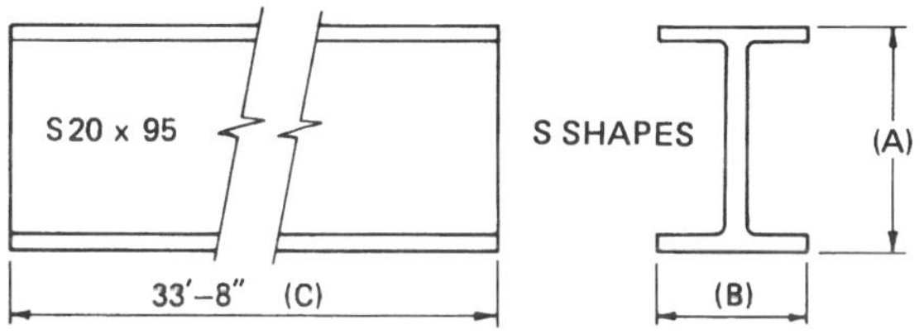

DIM	NOMINAL	TOLERANCE		LIMITS	
				UPPER	LOWER
A		+	−		
B		+	−		
C		+	−		

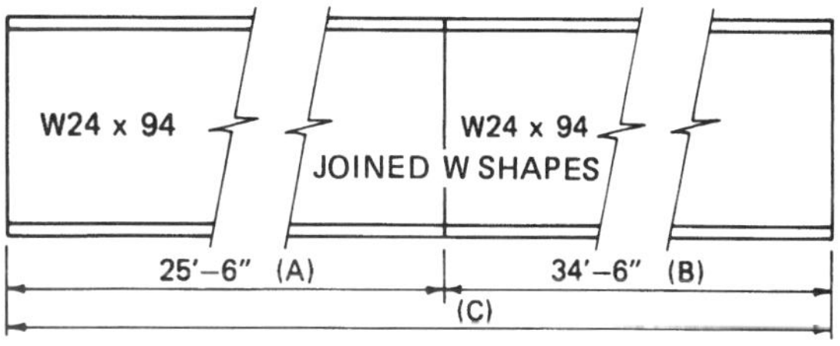

DIM	TOLERANCE		LIMITS	
			UPPER	LOWER
A	+	−		
B	+	−		
C	X	X		

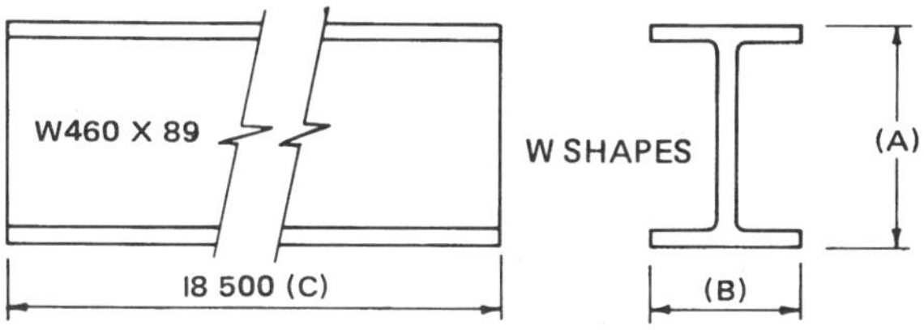

DIM	NOMINAL	TOLERANCE		LIMITS	
				UPPER	LOWER
A		+	−		
B		+	−		
C		+	−		

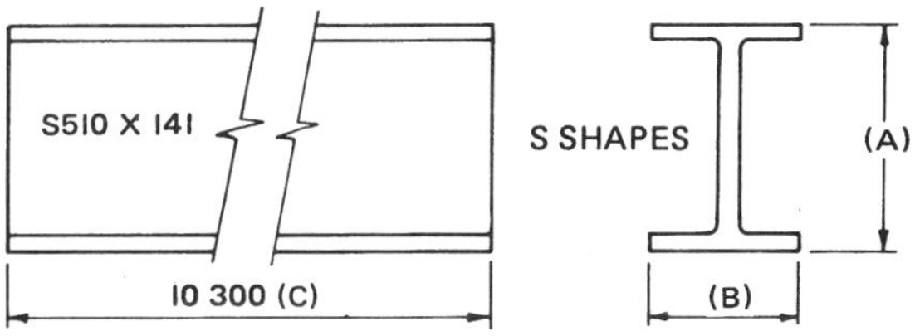

DIM	NOMINAL	TOLERANCE		LIMITS	
				UPPER	LOWER
A		+	−		
B		+	−		
C		+	−		

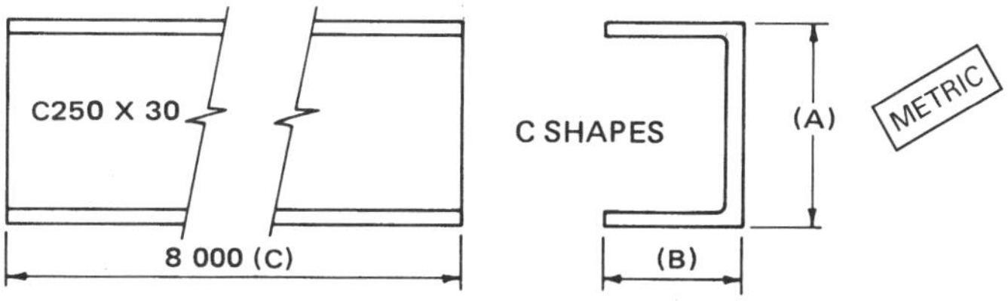

DIM	NOMINAL	TOLERANCE		LIMITS	
				UPPER	LOWER
A		+	−		
B		+	−		
C		+	−		

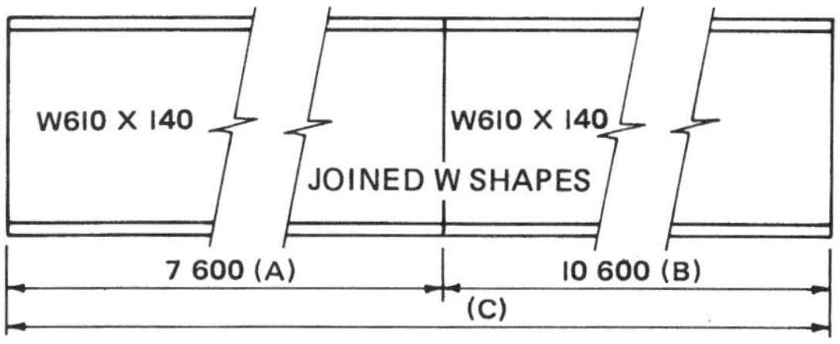

DIM	TOLERANCE		LIMITS	
			UPPER	LOWER
A	+	−		
B	+	−		
C	X	X		

STRUCTURAL DRAFTING— BEAM TOLERANCES	REFERENCE UNIT 25–1	NAME COURSE DATE	DRAWING AA–1

Make detail drawings of the two connections shown. Refer to Fig. 25–1–8 and structural steel manuals. The connection angles are welded to the beam web, and the outstanding angles are bolted to the connecting beam. The bolts and holes need not be shown on these drawings. The scale is 1:8 (U.S. customary). The locations of the W24 × 76 and W24 × 94 are shown on the drawing area.

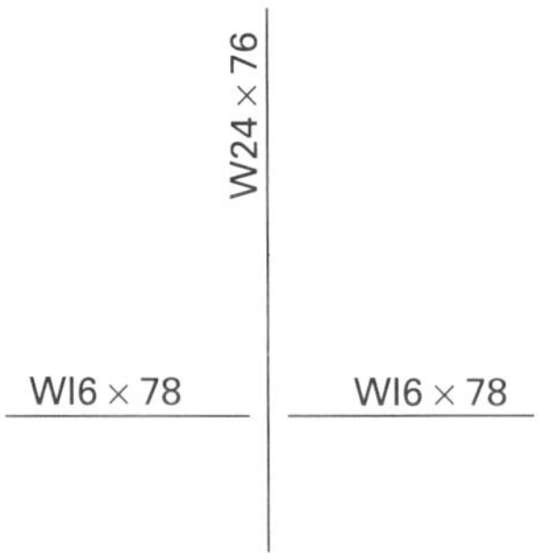

ELEVATION: ALL STEEL FLUSH, TOP AT ELEV. + 33'–6
CONNECTIONS: TWO ANGLES 4 × 3 × $\frac{1}{4}$ × 8 ON BOTH SIDES OF WI6 BEAMS
PARTIAL DESIGN DRAWING

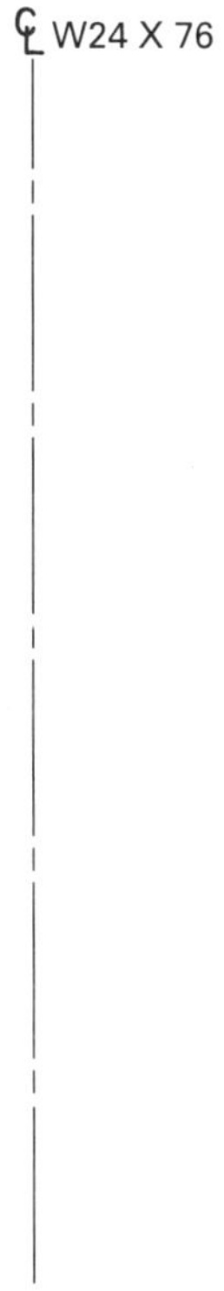

DETAIL OF CONNECTION
(SCALE I:8)

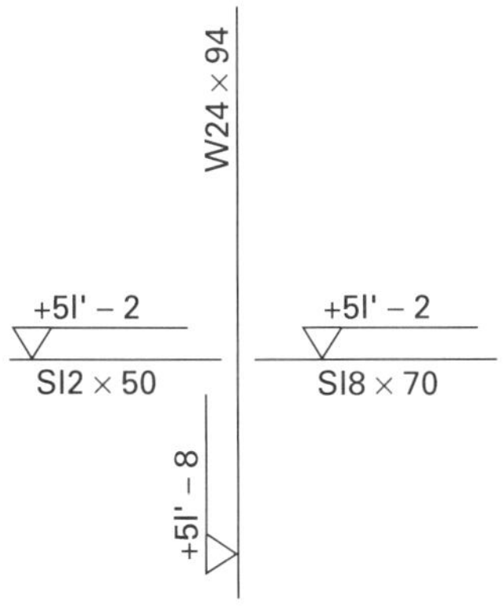

CONNECTIONS: TWO ANGLES 4 × 3 × $\frac{1}{4}$ × 6 ON SI2 × 50
TWO ANGLES 4 × 3 × $\frac{1}{4}$ × I0 ON SI8 × 70
PARTIAL DESIGN DRAWING

DETAIL OF CONNECTION
(SCALE I:8)

STRUCTURAL DRAFTING— BEAMS	REFERENCE UNIT 25–2	NAME ___ COURSE ___ DATE ___	DRAWING AA–2

Sketch the following beam connections from the design sketch shown:

(a) North end connection of beam E3
(b) South end connection of beam K3
The scale is 1:8. The locations of the W18 × 96 beams are shown on the drawing area.

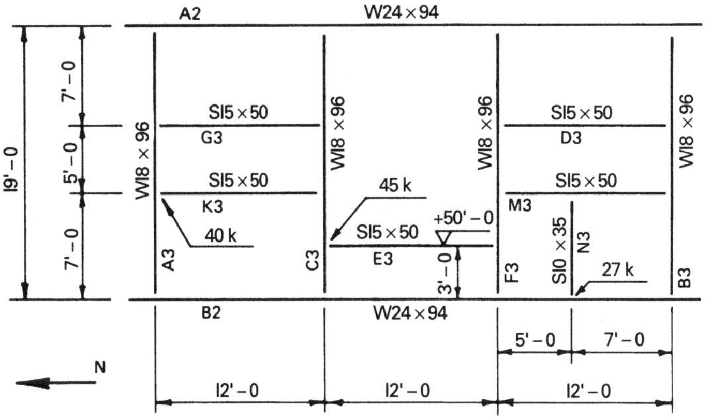

ELEVATION: TOP OF STEEL TO BE (+50'–3") UNLESS OTHERWISE SHOWN
NOTES: ALL HOLES Ø.81 BOLTS Ø.75

℄ W18 X 96

NORTH END CONNECTION OF BEAM E3

℄ W18 X 96

SOUTH END CONNECTION OF BEAM K3

STRUCTURAL DRAFTING— STANDARD CONNECTIONS	REFERENCE UNIT 25–3	NAME COURSE DATE	DRAWING AA–3

Prepare one drawing for the two horizontal W16 × 40 beams shown on the design sketch. Eliminate the top and bottom views. The scale is $1\frac{1}{2}$ in. = 1 ft. The W16 × 40 beam is located on the drawing area.

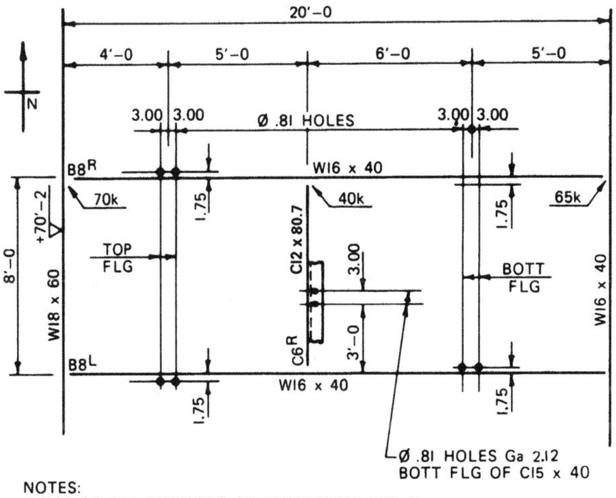

NOTES:
- TOP OF ALL MEMBERS AT ELEVATION +70'–3 EXCEPT WHERE NOTED
- SHOP CONNECTIONS WELDED
- FIELD CONNECTIONS .75–A325 FRICTION TYPE BOLTS
- USE DOUBLE-ANGLE BEAM CONNECTIONS

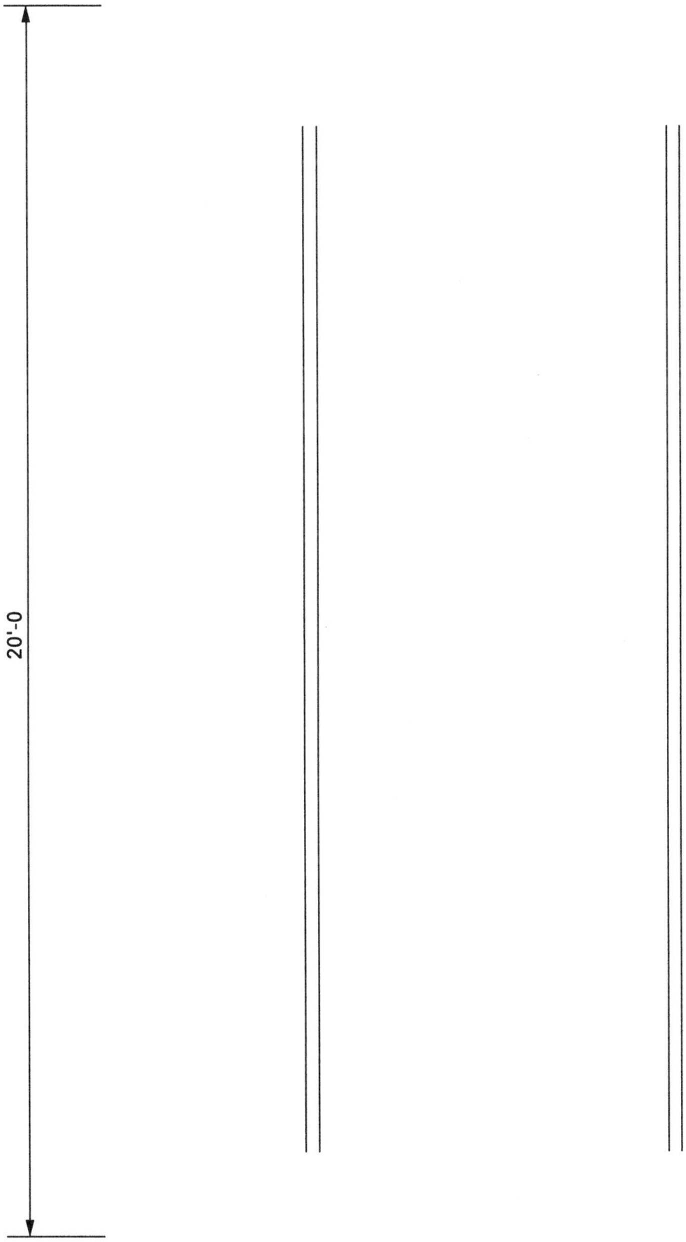

STRUCTURAL DRAFTING— SECTIONING	REFERENCE UNIT 25–4	NAME / COURSE / DATE	DRAWING AA–4

Complete the details of the seated beam connections at both ends of the W18 × 60 beam. The scale is 1:8.

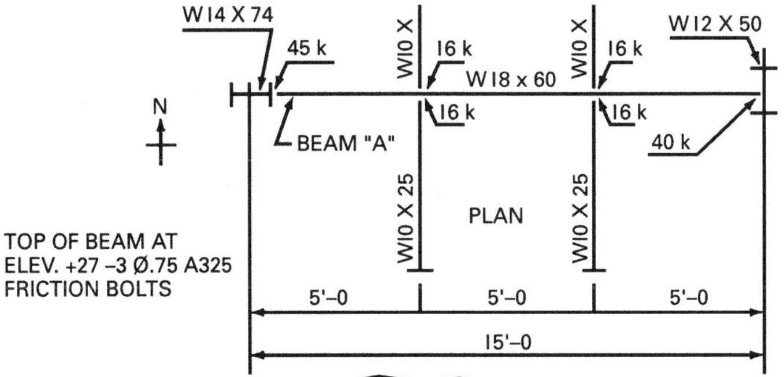

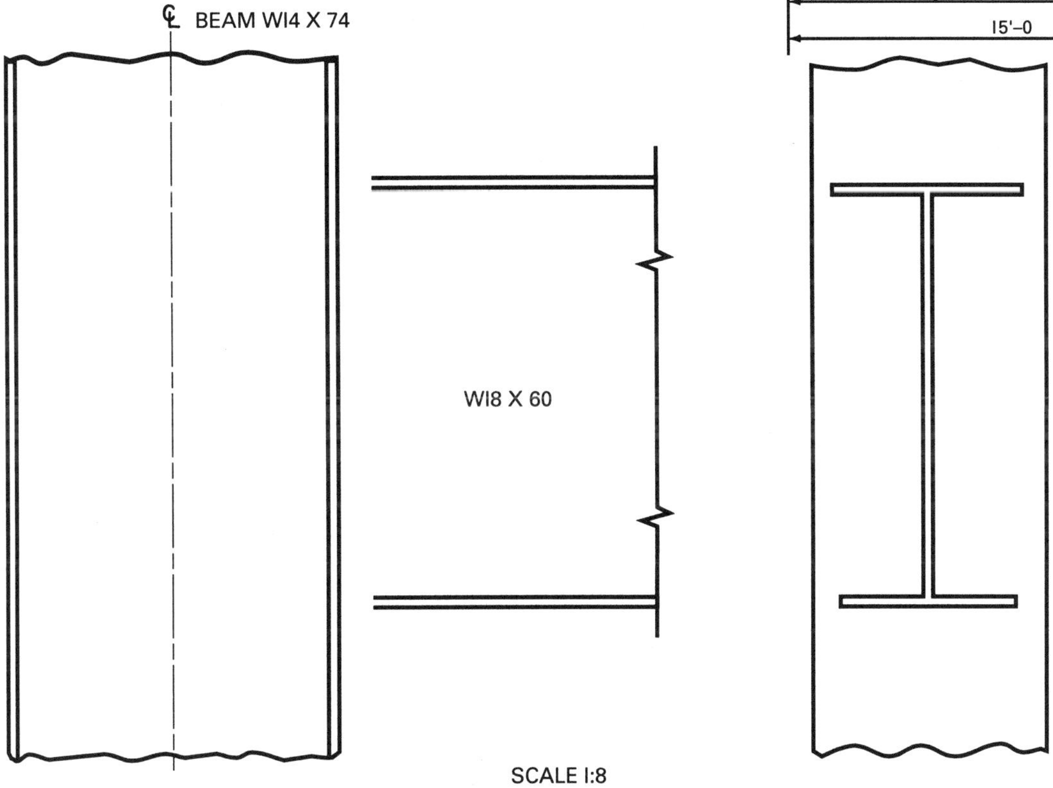

SCALE 1:8
DETAIL OF WEST END OF W18 X 60 BEAM CONNECTION

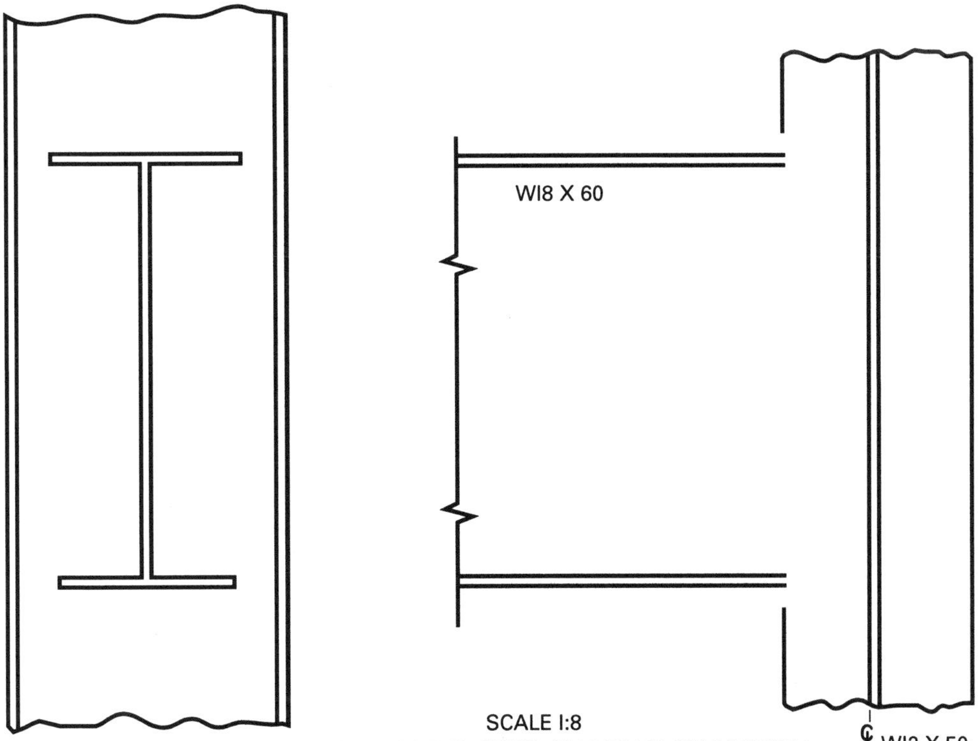

SCALE 1:8
DETAIL OF EAST END OF W18 X 60 BEAM CONNECTION

STRUCTURAL DRAFTING— SEATED BEAM CONNECTIONS	REFERENCE UNIT 25–5	NAME ______ COURSE ______ DATE ______	DRAWING AA–5

On one of the blank sheets from the back of this book, make detail drawings of beams D3, G3, K3, M3, C3, and F3 shown on the layout. *Note:* Only two beam drawings are required. The scale is 1:8.

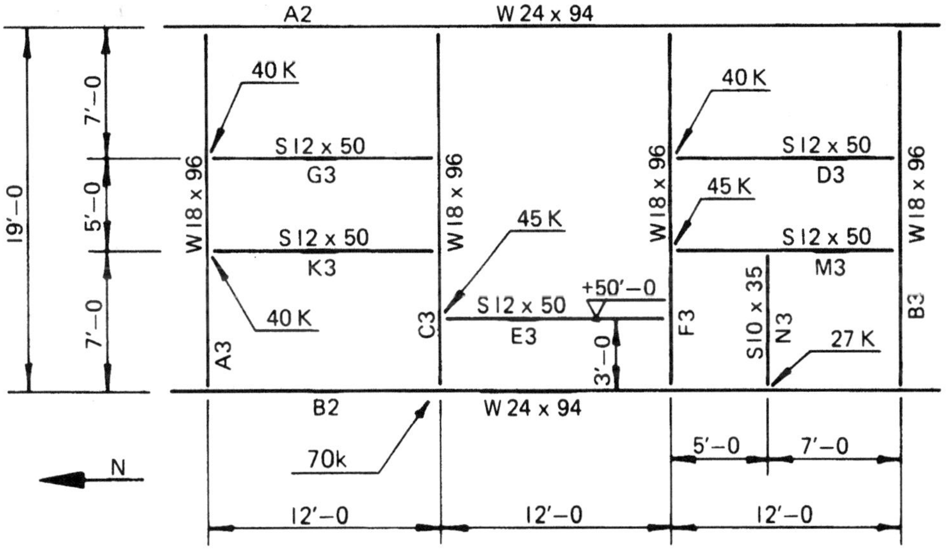

ELEVATION: TOP OF STEEL TO BE (+50'–3) UNLESS OTHERWISE SHOWN
NOTES: ALL HOLES Ø .81   BOLTS Ø .75–A325

STRUCTURAL DRAFTING— DIMENSIONING	REFERENCE UNIT 25-6	NAME _ COURSE _ DATE _	DRAWING AA-6

Design a simple plate jig for drilling the holes in the spacer shown. The scale is 1:1. State the sequence of operations and the time at which locking pins may be required.

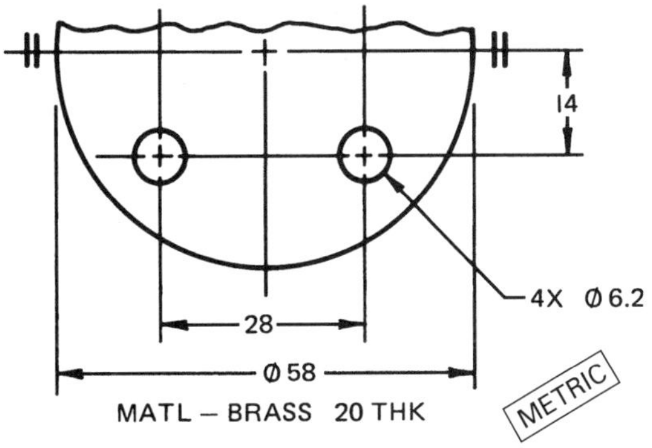

JIGS AND FIXTURES— JIG DESIGN	REFERENCE UNIT 26–1	NAME / COURSE / DATE	DRAWING BB–1

Design a jig for drilling the six small holes shown on the flanged bracket. The large center hole and the finished base should be the features used for locating the part in the jig. A lockpin is recommended for alignment after the first hole is drilled. Standard components should be used wherever possible. The scale is 1:1.

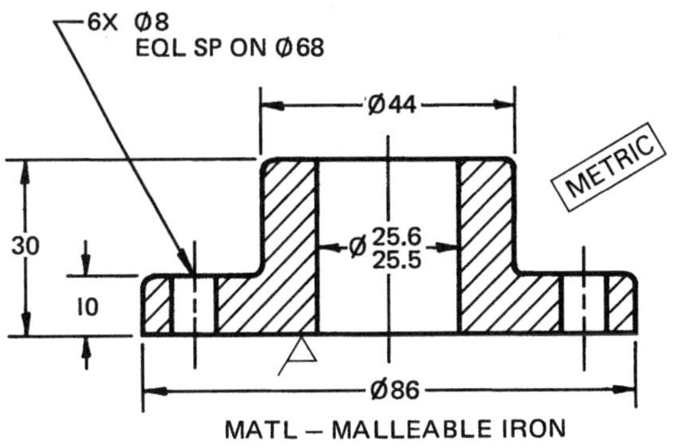

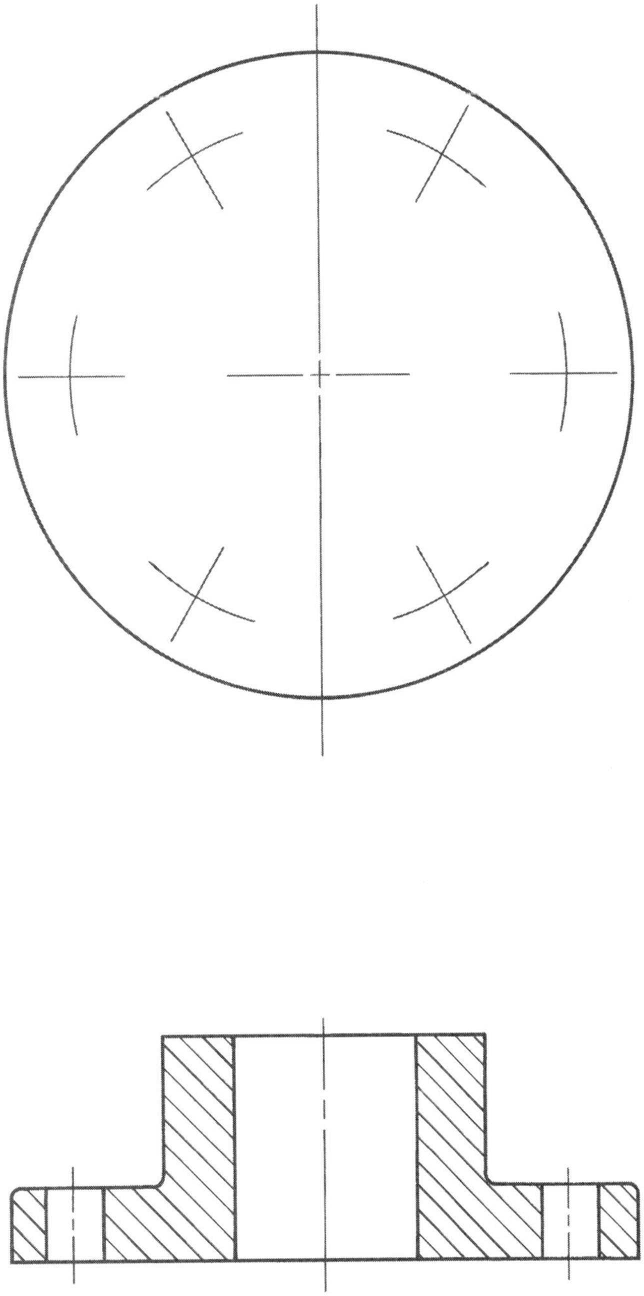

JIGS AND FIXTURES— DRILL JIG COMPONENTS	REFERENCE UNIT 26–2	NAME COURSE DATE	DRAWING BB–2

Design a simple plate jig for drilling the holes in the spacer shown. The size of the dowel pins in the design is Ø6.006 ± 0.003. After your overall design has been approved by your instructor, dimension the jig plate according to the procedures outlined in this unit. The scale is 1:2.

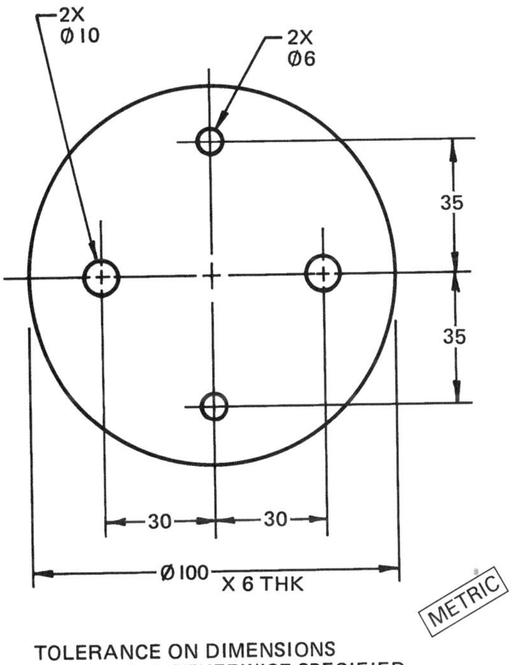

TOLERANCE ON DIMENSIONS
±0.1 UNLESS OTHERWISE SPECIFIED

JIGS AND FIXTURES— DIMENSIONING JIG DRAWINGS	REFERENCE UNIT 26–3	NAME COURSE DATE	DRAWING BB–3

Design a simple milling fixture to mill out the two outside portions of the top of the drive link shown. Use standard components and refer to manufacturers' catalogs wherever possible. The scale is 1:2. The workpiece is partially drawn.

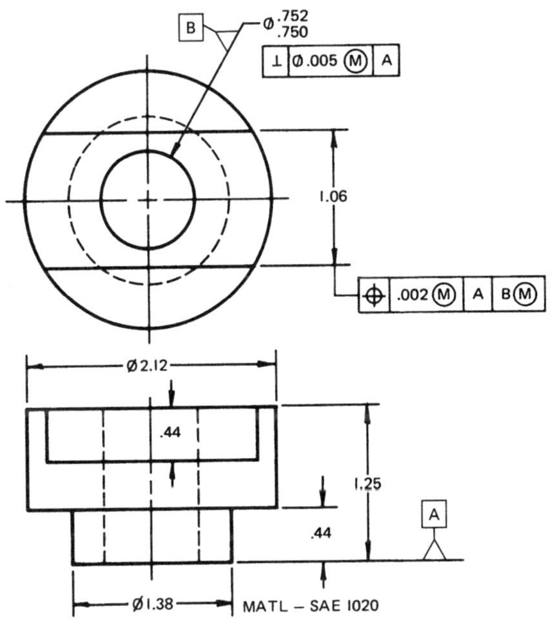

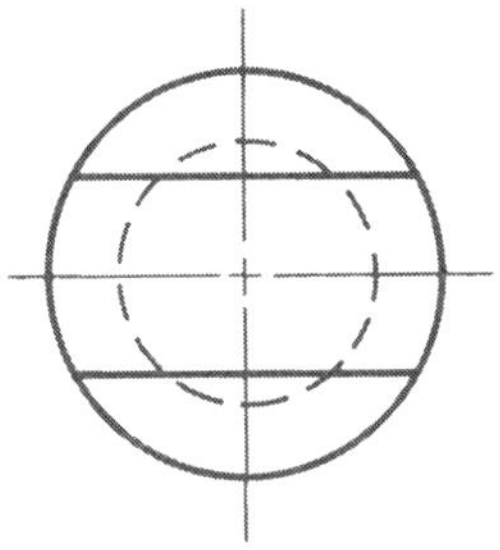

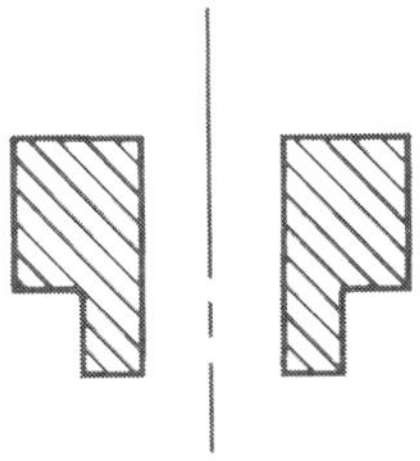

JIGS AND FIXTURES— MILLING FIXTURES	REFERENCE UNIT 26–4	NAME _ COURSE _ DATE _	DRAWING BB–4

On a blank sheet of drawing paper from the back of this workbook, make a schematic diagram of the sound-effects generator layout shown. For electrical symbols not shown in Table 49 of the appendix, use the symbols shown on the layout.

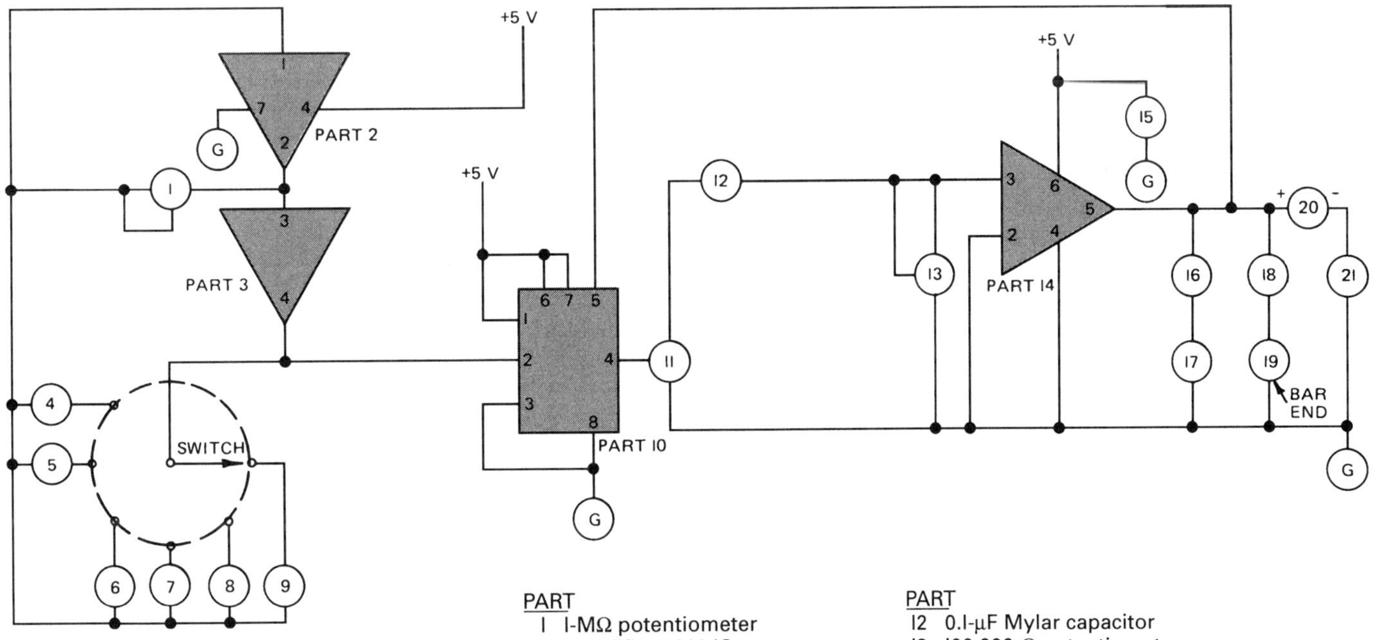

PART
1 1-MΩ potentiometer
2 Amplifier 4069 IC
3 Amplifier 4069 IC
4 1-μF capacitor
5 0.47-μF capacitor
6 0.1-μF capacitor
7 0.05-μF capacitor
8 500-μF capacitor
9 100-μF capacitor
10 TL507C IC (integrated circuit)
11 100 000-Ω potentiometer

PART
12 0.1-μF Mylar capacitor
13 100 000-Ω potentiometer
14 amplifier LM386 IC4
15 0.1-μF Mylar capacitor
16 0.047-μF Mylar capacitor
17 10-Ω resistor
18 1000-Ω resistor
19 Light-emitting diode
20 250-μF capacitor
21 8-Ω speaker
Ⓖ Ground

SCHEMATIC DIAGRAMS— SOUND-EFFECTS GENERATOR	REFERENCE UNIT 27–2	NAME ______ COURSE ______ DATE ______	DRAWING CC–1

On a blank sheet of drawing paper from the back of this workbook, make a highway-type connection diagram of the boat's electrical system shown here.

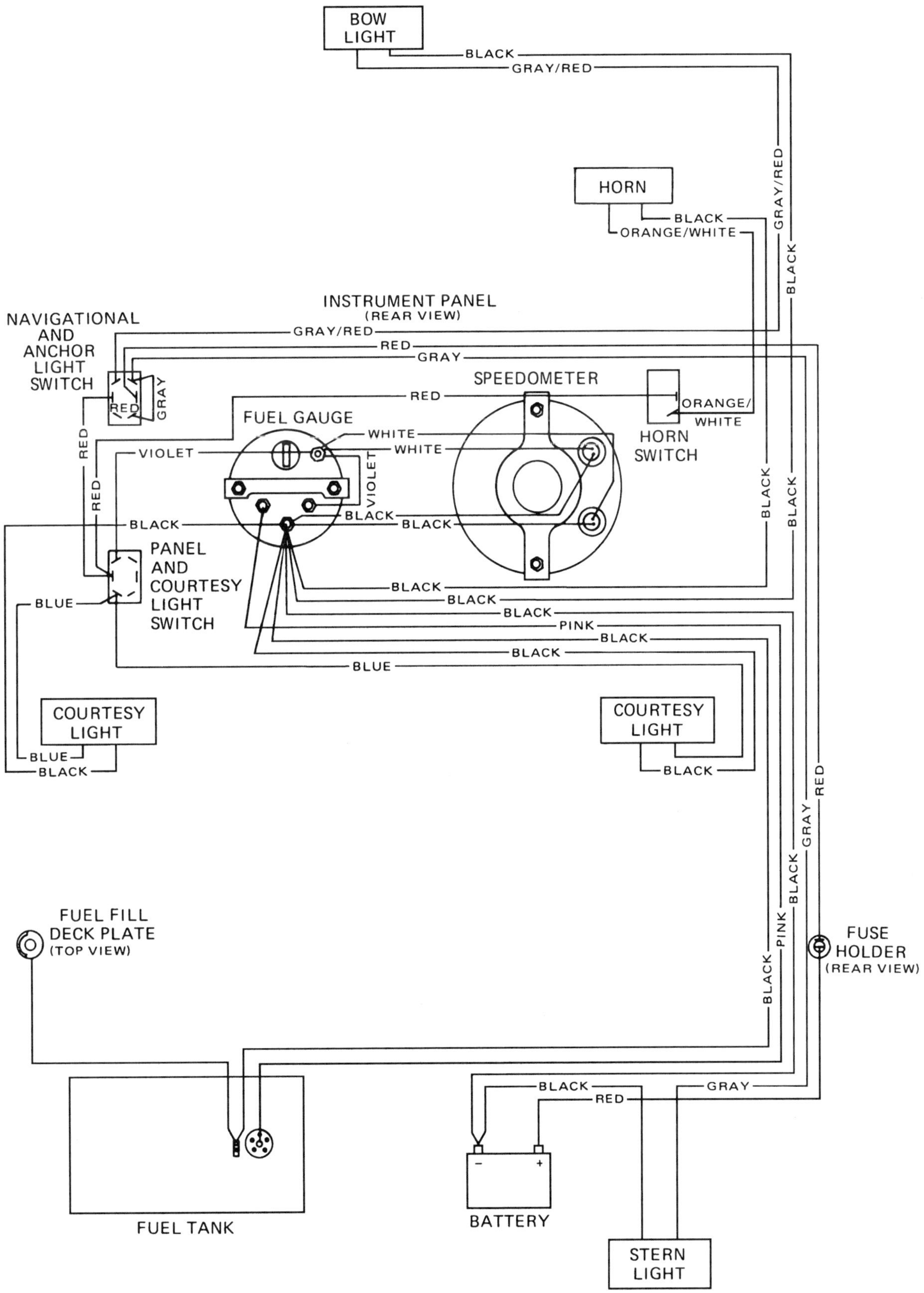

WIRING DIAGRAMS— WIRING DIAGRAM FOR BOAT	REFERENCE UNIT **27–3**	NAME ____ COURSE ____ DATE ____	DRAWING **CC–2**

On a blank sheet of drawing paper from the back of this workbook, prepare a logic diagram from the information shown. Convert to distinctive shaped logic symbols and add the missing connections listed below.

LOGIC SYMBOL	ICs
AMPLIFIER	IC2, IC7, IC9, ICI0
OR	IC3, IC4, IC5, IC6, ICII, ICI3
AND	IC8, ICI2
NEGATION INDICATOR	IC2 PORT 8 ICI0 PORT I0 IC7 PORT 2 ICI2 PORT 8 IC8 PORT 6

<u>CONNECTIONS TO BE MADE</u>:
ICI TERM. 9 TO IC3, TERM. I2
IC3 TERM. II TO IC4 TERM. 4 AND ICII TERM. I2
IC4 TERM. 5 TO IC8 TERM. 6
IC5 TERM. I0 TO IC7 TERM. I. ICI0 TERM. I0, AND ICI2 TERM. 9
IC6 TERM. I TO IC7 TERM. 2
IC9 TERM. I2 TO ICI0 TERM. II
ICII TERM. II TO ICI3 TERM. I0
ICI2 TERM. 8 TO ICI3 TERM. 9
GROUND TERM. 7 ON ICI, IC3, IC9, AND ICI3
GROUND TERM. 8 ON ICI

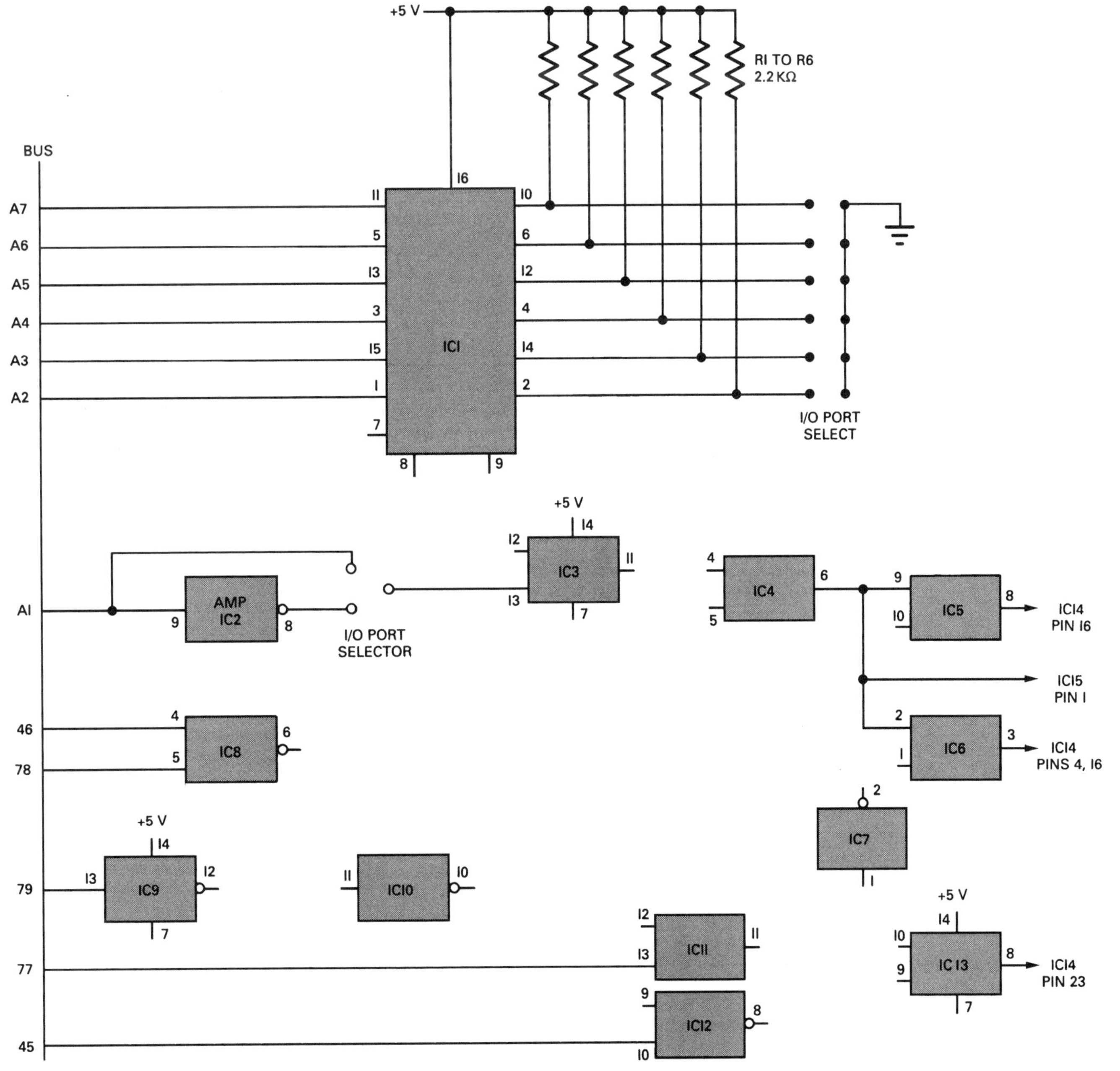

BLOCK AND LOGIC DIAGRAMS— REMOTE CONTROL HOUSING WIRING	REFERENCE UNIT 27–5	NAME COURSE DATE	DRAWING CC–3

Answer the questions using the package drawing given below.

DIP24: plastic dual in-line package; 24 leads (300 mil) **SOT222-1**

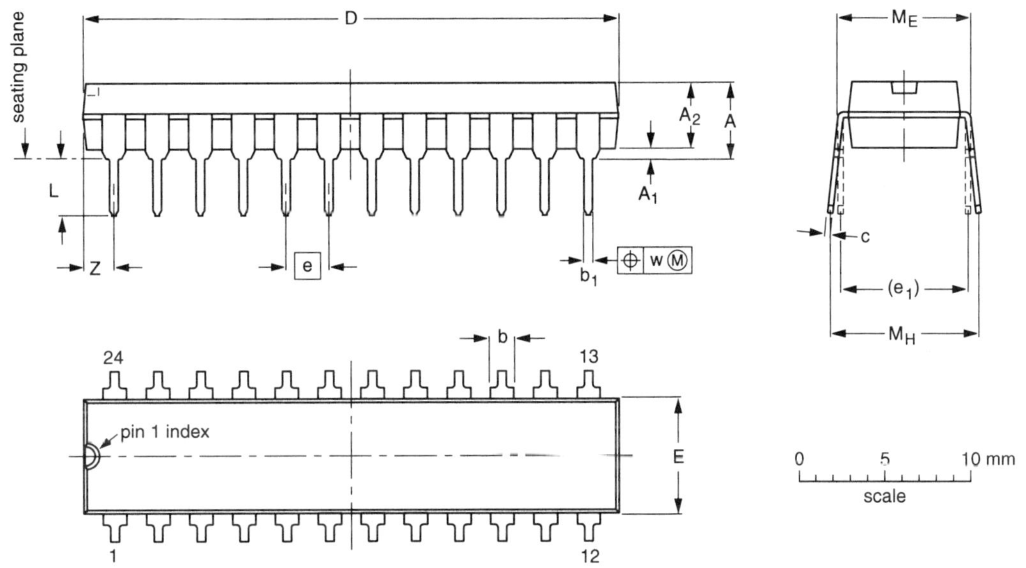

DIMENSIONS (millimeter dimensions are derived from the original inch dimensions)

UNIT	A max.	A_1 min.	A_2 max.	b	b_1	c	D[(1)]	E[(1)]	e	e_1	L	M_E	M_H	w	Z[(1)] max.
mm	4.70	0.38	3.94	1.63 1.14	.56 .43	.36 .25	31.9 31.5	6.73 6.25	2.54	7.62	3.51 3.05	8.13 7.62	10.03 7.62	0.25	2.05
inches	.185	.015	.155	.064 .045	.022 .017	.014 .010	1.256 1.240	.265 .246	.100	.300	.138 .120	.32 .30	.395 .300	.01	.081

Note

1. Plastic or metal protrusions of .01 inches maximum per side are not included.

OUTLINE VERSION	REFERENCES				EUROPEAN PROJECTION	ISSUE DATE
	IEC	JEDEC	EIAJ			
SOT222-1		MS-001AF				99-04-28

1. What is the maximum length of the package? ______

2. What is the minimum length of the package? ______

3. What is the nominal pin to pin distance? ______

4. What is the nominal pin row to pin row distance? ______

5. How is the location of pin 1 indicated? ______

6. How large is the air gap under the component when seated? ______

7. How far do the pins extend beyond the PCB board when the component is seated in .51 in. (1.30 mm) thick PCB board?

______